Follow-Up for NICU Graduates

Howard Needelman · Barbara J. Jackson
Editors

Follow-Up for NICU Graduates

Promoting Positive Developmental and Behavioral Outcomes for At-Risk Infants

Springer

Editors
Howard Needelman
Munroe-Meyer Institute
University of Nebraska Medical Center
Omaha, NE, USA

Barbara J. Jackson
Munroe-Meyer Institute
University of Nebraska Medical Center
Omaha, NE, USA

ISBN 978-3-030-10345-3 ISBN 978-3-319-73275-6 (eBook)
https://doi.org/10.1007/978-3-319-73275-6

Printed on acid-free paper

This Springer imprint is published by the registered company Springer International Publishing AG, part of Springer Nature
The registered company address is: Gewerbestrasse 11, 6330 Cham, Switzerland

Contents

About the Editors

Howard Needelman is associate professor of pediatrics at the University of Nebraska Medical Center and co-director of Nebraska's statewide NICU Follow-Up Program, TIPS (Tracking Infant Progress Statewide). He received his doctorate at Rush Medical College in Chicago. Afterward, he completed a pediatric residency at the University of Michigan, a Neonatal/Perinatal Fellowship at the University of Cincinnati, and a Developmental Behavioral Fellowship at the University of Nebraska. He is board certified in pediatrics, neonatal perinatal medicine, and developmental behavioral pediatrics.

Barbara J. Jackson is professor at the Munroe Meyer Institute (MMI) of the University of Nebraska and director of the Department of Education and Child Development. She received her doctorate in psychology and cultural studies, with an emphasis in developmental psychology, from the University of Nebraska at Lincoln. Dr. Jackson is the primary investigator and program evaluator for multiple regional and state-funded projects, including mental health consultation family support projects; a Newborn Screening and Genetics Regional Collaborative; and statewide early childhood programs. With Dr. Howard Needelman, she is the co-director of the Nebraska TIPS NICU Follow-Up Program.

Contributors

Robin H. Adair Division, Developmental and Behavioral Pediatrics,University of Rochester School of Medicine and Dentistry, Rochester, NY, USA

Elizabeth Asztalos Department of Paediatrics, University of Toronto, Toronto, ON, Canada

Department of Newborn and Developmental Paediatrics, Sunnybrook Health Sciences Centre, Toronto, ON, Canada

Glen P. Aylward Pediatrics and Psychiatry, Southern Illinois University School of Medicine, Springfield, IL, USA

Rudaina Banihani Department of Paediatrics, University of Toronto, Sunnybrook Health Sciences Centre, Department of Newborn & Developmental Paediatrics, Toronto, ON, Canada

Lindsay Bartram Nationwide Children's Hospital, Columbus, OH, USA

Lauren A. Boyd Department of Pediatrics, Developmental and Behavioral Pediatrics, Loyola University Chicago Stritch School of Medicine, Loyola University Health System, Maywood, IL, USA

John Cairney Department of Family Medicine, McMaster University, Hamilton, ON, Canada

Faculty of Kinesiology of Physical Education, University of Toronto, Toronto, ON, Canada

Paige Terrien Church Department of Paediatrics, University of Toronto, Toronto, ON, Canada

Department of Newborn and Developmental Paediatrics, Sunnybrook Health Sciences Centre, Toronto, ON, Canada

Marie A. Clark UH Rainbow Babies and Children's Hospital, Cleveland, OH, USA

Case Western Reserve University, Cleveland, OH, USA

Linda Copeland UCSF Fresno Center for Medical Education and Research, San Francisco, CA, USA

Anne DeBattista Developmental-Behavioral Pediatrics, Stanford Children's Health, Palo Alto, CA, USA

Frances Page Glascoe Department of Pediatrics, Vanderbilt University, Nashville, TN, USA

Benjamin Kennert University of Nebraska Medical Center, Munroe-Meyer Institute, Omaha, NE, USA

Maureen Luther Sunnybrook Health Sciences Centre, Neonatal Follow Up Clinic, Toronto, ON, Canada

Patricia Maddalena Sunnybrook Health Sciences Centre, Neonatal Follow Up Clinic, Toronto, ON, Canada

Cheryl A. Milford Cheryl Milford Consulting, Orange County, CA, USA

National Perinatal Association, Lonedell, MO, USA

Kerry Miller University of Nebraska Medical Center, Munroe-Meyer Institute, Omaha, NE, USA

Christine Mirzaian Children's Hospital of Los Angeles, University of Southern California, Los Angeles, CA, USA

Keck School of Medicine of the University of Southern California, Los Angeles, CA, USA

Anita Narayanan Nationwide Children's Hospital, Columbus, OH, USA

Howard Needelman University of Nebraska Medical Center, Munroe-Meyer Institute, Omaha, NE, USA

Shalin Parekh UCSF Fresno Center for Medical Education and Research, San Francisco, CA, USA

Eric S. Peeples University of Nebraska Medical Center, Omaha, NE, USA

Holly Roberts University of Nebraska Medical Center, Munroe-Meyer Institute, Omaha, NE, USA

Nina Sand-Loud Dartmouth Geisel School of Medicine, Hanover, NH, USA

Children's Hospital at Dartmouth-Hitchcock, Lebanon, NH, USA

G. Bradley Schaefer Department of Pediatrics, Section of Genetics and Metabolism, University of Arkansas for Medical Sciences, Rock, AR, USA

Elizabeth A. Sellars Department of Pediatrics, Section of Genetics and Metabolism, University of Arkansas for Medical Sciences, Rock, AR, USA

Kelly Schifsky Children's Hospital of Los Angeles, University of Southern California, Los Angeles, CA, USA

Katherine Steingass Nationwide Children's Hospital, Columbus, OH, USA

Brian Tang Stanford University, Stanford, CA, USA

Douglas Vanderbilt Children's Hospital of Los Angeles, University of Southern California, Los Angeles, CA, USA

Keck School of Medicine of the University of Southern California, Los Angeles, CA, USA

Sandra Willett Munroe-Meyer Institute of the University of Nebraska Medical Center, Omaha, NE, USA

Part I
Introduction

Introduction

Howard Needelman

Abstract No longer is the success of care in the neonatal intensive care unit measured in terms of survival. As mortality statistics have continued to become more reassuring, success must now be measured in terms of morbidity. This effort should then be used to help the individual survivor and also be used to evaluate current NICU practices. The NICU follow-up clinic has been developed as a means to accomplish these dual goals. In fact, such a program is a requirement for a graduate medical education program in neonatal-perinatal medicine. There are, however, no clear guidelines to describe the optimal follow-up program and, for example, how to structure it, staff it, and pay for it. Other issues not delineated are which graduates are truly at risk and how long they should be followed. Finally, the relationship between the follow-up clinic and the medical home must be clarified in order to not fragment care.

Published in 1985, Jeff Lyon wrote *Playing God in the Nursery* [1]. It is a book discussing the issues faced by parents and professionals who must deal with premature infants and infants with a high risk for developmental disabilities. He quotes the mother of premature twins as reporting that "modern medicine has given me a beautiful son who has great potential to fully enjoy a good life, and it has also given me a beautiful daughter whose life will never bloom. I feel like saying 'Thank you, modern medicine, and damn you, modern medicine.'" One of the challenges in dealing with the high-risk population who have been in the newborn intensive care unit (NICU) is to be certain that their developmental outcome is optimal. Minimally, it is the role of the NICU follow-up clinic to help evaluate these infants and their performance and to help parents or their surrogates navigate an often confusing health-care and educational system. Additionally, the observed outcomes can be used to evaluate the quality of therapies used in neonatal/perinatal medicine.

H. Needelman (✉)
University of Nebraska Medical Center, Munroe-Meyer Institute, Omaha, NE, USA
e-mail: hneedelm@unmc.edu

H. Needelman, B. J. Jackson (eds.), *Follow-Up for NICU Graduates*,
https://doi.org/10.1007/978-3-319-73275-6_1

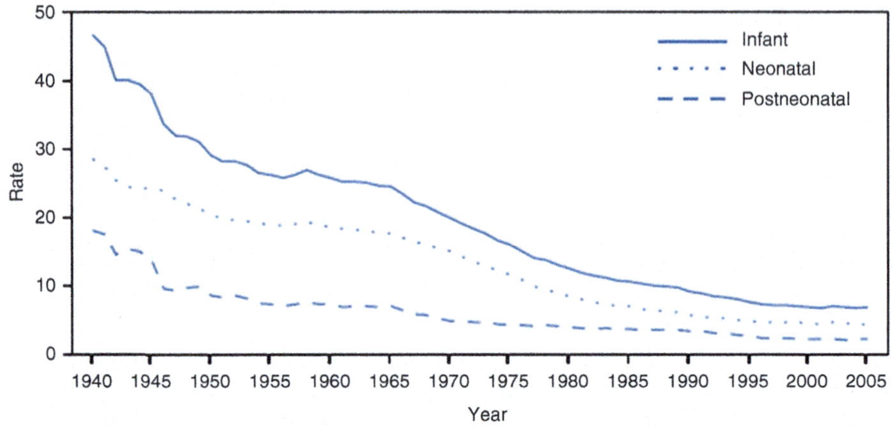

Fig. 1 Infant/neonatal/postnatal mortality

Neonatology has, in its relatively short history, helped make tremendous strides in improving the outcome of its patients. The neonatal death rate, i.e., deaths of infants <28 days, a most basic measure of outcome, has declined from approximately 30 per 1000 in 1940 to approximately 5 per 1000 in 2000 (see Fig. 1). This dramatic decrease has been aided by the advent of interventions in the care of the at-risk neonate. A list of these interventions would include, to name only a few, antibiotics, incubators, ventilators, and hyperalimentation. Subsequent investigations have always found, however, that these interventions come with unforeseen and often tragic cost. While these interventions may be thought to be available to the general population, significant discrepancies in mortality are noted among different ethnic and socioeconomic groups (Fig. 2). This raises the important question of health-care disparities and its effect on neonatal outcome.

While there remains an interest in reporting mortality, there has been a growing movement to report outcomes on the basis of sequelae, both major including cerebral palsy, intellectual disability, blindness, and deafness and minor including speech and language delays, learning disorders, and behavioral sequelae such as ADHD. Concern over neurodevelopmental sequelae is ultimately at the heart of the frequently repeated debates over "how small is too small?" NICU follow-up helps answer this type of question.

Two different agendas must be coordinated for follow-up clinics to be effective. The clinics must work minimally to augment the education system and act as a part of the "Child Find" system described in the Individuals with Disabilities Education Act, thus satisfying the need to help families navigate for services and service providers to recognize those infants at risk or with proven delays. The clinics ideally should also be the repository of information to allow perinatologists, neonatologists, and any other professionals who deal with the at-risk NICU graduate population to determine directions of appropriate perinatal, neonatal, and postnatal care. In fact, studies now appear to describe life course interventions as much as NICU interventions.

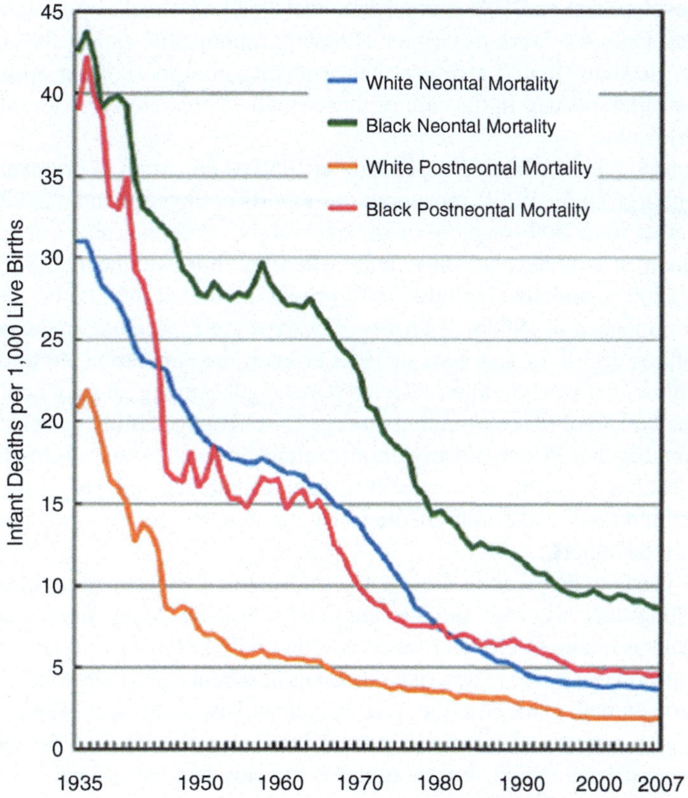

Fig. 2 Mortality by ethnicity

The Need to Evaluate Neonatal and Perinatal Interventions

In *Pediatrics* in 1953, Hess [2] described a 30-year experience in the outcome of premature infants at Michael Reese Hospital in Chicago. He commented regarding worse outcome in those infants not receiving immediate care. He also noted the need to use breast milk. United States Public Health Service recommendations for square feet required per bassinet were also discussed. He stated that "One frequently reads in the public press of the birth of one of these small infants and occasionally its early progress is reported but usually its ultimate development is left to the imagination. This, together with the fact that we must all answer the question of whether these very small premature infants are worth saving, encouraged us to believe that a long range study of our smallest babies would be of instructive interest." He described the outcome of a large cohort of infants with birthweight <1251 g whose outcome from approximately 1 to 28 years was known. NICU discharge mortality was noted to be 10%. Approximately 70% were felt to have normal to superior mental development and 55% to have normal or superior physical development. Only 15% had mental development and 10% had physical development rated as poor to extremely subnormal. Only 6% were described as having retinopathy, and of these, half had spasticity. He stated that "I give you these statistics in support of our original belief that any effort expended in the care of these smallest prematurely born infants is of inestimable value."

Drillien [3] reported later in the decade of 1950 on outcomes of premature infants in an Edinburgh study. While most previous reports of physical and mental development showed some 50% or more of survivors to be of normal physical and mental development, it is noted that these were often low birthweight infants, i.e., those closer to 2500 g, and also included term growth-restricted infants. For Drillien, in the paper published in 1958 in *Archives of Diseases of Childhood*, in the population of prematures of 3 lb or less born in 1953 or later, the population showed frequent growth failure, with weight more effected than height. Among 38 preschool children, only 14 had a DQ of 90 or over, and among 21 school age children, only 3 were of average intelligence. Most, however, could attend "normal primary schooling." Two thirds had behavior problems. One half had a visual defect, and 5 of 31 had cerebral palsy. Drillien does make note of the lower socioeconomic status of many of the families in the cohort.

Credit needs to be given in these early studies to an understanding of the importance of long-term sequelae and the length of follow-up. Many subsequent studies have looked at infant and toddler outcomes, and only in the past decade or so has a renewed effort been made to describe outcomes at school age and beyond. Silverman, in an interview with Gartner in the AAP American History Project [4] in 1997 states that "the total number of infants who were blinded in the 1942–1954 epidemic of RLF has never been tallied. Ten thousand is the closest number I have been able to find as a rough guess; 7000 of these were in the US, more than in any other country." As he spoke about the RLF epidemic, however, he noted that the perception of the visually impaired survivor has changed from "'disabled' to "self-sufficient persons participating fully in everyday life." Yet, "these people, many of them in their 40s, face an unemployment rate of almost 80%. Their parents are my age, in their late 70s, and are saying to me now, 'What is going to happen to my children when I die?'"

While many of the errors in neonatal care were documented from cases while the children were in the nursery, there are a few notable exceptions. The CDC in 1979 was informed of several cases of a Bartter-like syndrome with metabolic alkalosis with failure to thrive, anorexia, or constipation. When other cases were found, the cause was felt to be Neo-Mull-Soy, a soy-based formula, and improvement was noted when the patients received chloride supplementation. The infants ranged in age from 2 to 9 months.

The search for a way to treat and prevent chronic lung disease (CLD) or bronchopulmonary dysplasia (BPD) in the premature has been a frustration of the neonatal community for decades. Because of the presumed inflammatory component to the condition, among the therapies used over the years has been the use of postnatal steroids. A 42-day course of dexamethasone was among the protocols used in the 1980s. A 2002 statement from the American Academy of Pediatrics and the Canadian Paediatric Society [5], partially based on infant outcomes at 1 year of age or older, stated that the routine use of dexamethasone for prevention or treatment of BPD was not recommended. The follow-up studies generally showed an increased risk of neurodevelopmental impairment including cerebral palsy, blindness, or a developmental score more than 2 standard deviations below the mean in dexamethasone-treated patients. The strength of this recommendation was modified some in a revised AAP statement [6] in 2010, but the fact remained that "high daily doses of dexamethasone (approximately 0.5 mg/kg per day) … have been associated with numerous short- and long-term adverse outcomes, including neurodevelopmental impairment…." These recommendations led to discussions of modification in steroid use in the NICU.

There are, of course, success stories in the NICU that follow-up has helped identify. The use of therapeutic hypothermia for hypoxic-ischemic encephalopathy has generally become a standard of care, and when evaluating for the outcome of death or disability at 18–24 months of age, a significant reduction in poor outcomes is found when the infant at risk is cooled. A conclusion of an AAP Clinical Report is that "data from large randomized clinical trials indicate that therapeutic hypothermia, using either selective head cooling or systemic cooling, is an effective therapy for neonatal encephalopathy" [7].

While attempts were being made to improve medical care in the NICU to ideally improve developmental outcomes, others were investigating the environment in which that care was provided. White [8] and others have written extensively regarding the appropriate design of the NICU, discussing issues such as ambient noise and light. Als [9] has written about the potential of long-term sequelae in exchanging the intrauterine environment for the NICU environment. In an attempt to ameliorate some of the negative effects of this transition, she established the Newborn Individualized Developmental Care and Assessment Program (NIDCAP), perhaps the best known of many efforts of developmental care.

While neonatal intensive care services were being provided to premature infants at earlier and earlier gestational ages and neonatology progressed through phases that Robertson in *Journal of Perinatology* [10–12] has described as the "hands off years" from 1920 to 1950 to the "heroic years" of Silverman's "therapeutic exuberance" from 1950 to 1970 to the "experienced years" of 1970–2000, the public was alternatively enamored and disgusted with the events taking place in the NICU. Newspapers would describe miracle babies delivered and surviving at extremely low birthweights and later describe the monumental costs involved in saving these infants and their subsequent apparent poor quality of life.

Therefore, while strides were being made to improve neonatal care, the twentieth century also saw tremendous changes in the perception of the rights and potential of those children felt at risk for a poor neurodevelopmental outcome. Progress occurred primarily through educational and legislative channels, and the history has been well documented in Shonkoff and Meisels [13] and summarized here. While the philosophical groundwork came from Europe, the first public kindergarten came to the United States in St Louis in 1872, and early support for kindergarten came from advocates for poor children. Maria Montessori, a physician in Italy with experience with children with intellectual disability, developed a preschool educational program for the urban poor in the early twentieth century. Rather than using Montessori's self-teaching methods, when the nursery school concept began to grow in the United Sates in the 1930s, the schools were generally designed after the parent involvement models of the Macmillans of London. The recognition of the value of early education has helped underlay the concept of the value of early intervention for all children, including those with disabilities, and explore therapies aimed at ameliorating delays.

Occurring in concert with the realization of the educational needs of the disadvantaged were legislative mandates to deal with the plight of the poor. In 1912, the Children's Bureau was established by Congress to serve children including those with disabilities, and data from this program recognized deficiencies in the care of this latter group. Subsequently, in 1935 the Social Security Act's Title V included assistance to states to promote the health of mothers and children, federal matching funds to states for services for "crippled children," and funds to state welfare agencies to develop programs for at-risk children. Medicaid as part of Social Security became law in the 1960s and with it the Early and Periodic Screening Diagnosis and Treatment Program (EPSDT), mandating, among other things, developmental screening. Also in the 1960s, Head Start began based on the belief that early childhood intervention would be useful in optimizing later development. Certainly, credit for some of the advances of the decade of the 1960s goes to the commitment of President Kennedy to the advocacy of those children at risk including those with intellectual disability and the continuation of that commitment through the Johnson administration.

With this historical backdrop, in 1975, Congress passed P.L. 94-142, the Education for All Handicapped Children Act. This law guarantees a free and appropriate public education (FAPE) to all children, regardless of disability. It included efforts to improve the identification of children with disabilities. It also addressed special education and related services and provision of due process for parents and children to protect their rights. The subsequent years have seen modifications of this groundbreaking law. While P.L. 94-142 guaranteed the right for an Individualized Education Plan (IEP) for children 5–21, it left to state discretion early intervention. Part B of P.L. 99-457 in 1986 mandated a FAPE for children 3–21 by 1991 and provided incentives for states to provide services for those ages 0–3. In 1991, the 1975 and 1986 legislation were combined in P.L. 101-476, the Individuals with Disabilities Education Act which was subsequently reauthorized as the Individuals with Disabilities Improvement Act of 2004 or IDEA 2004. With this legislation, early intervention was provided through an individualized family service plan (IFSP) for children with disabilities or at risk 0–3 years and special education through an IEP for children with disabilities ages 3–21. In addition to educational services in a least restrictive environment, related and transition services should be provided, and families have the right to due process. The Americans with Disabilities Act amended in 2008 guarantees through Section 504 that children with disabilities are not discriminated against in the general education classroom and may receive accommodations and modifications to their curriculum based on their disability. Part C of IDEA 2004 requires states to develop "child find" systems to identify those needing to be evaluated for qualification for an IFSP or IEP. The NICU follow-up clinic furnishes at least part of that bridge.

It seems clear that as neonatal-perinatal medicine was beginning to question the acceptability of the outcomes of its intervention in the educational and psychological literature, there was a movement away from Gesell's theories of primacy of maturation to an acceptance of the possibility of change through behavioral and educational interventions. The optimism was not unbridled, however. In the Infant Health and Development Project, the effectiveness of early intervention seemed more robust for the heavier low birthweight infants than the lighter, i.e., <2000 g, low birthweight infants; Blair and Ramey [14] state "it is the co-occurrence of LBW and environmental disadvantage that places the LBW infant at highest risk for developmental delay… *Within* high- or low-risk groups, environmental factors account for the majority of variance in outcome. In *between*-group comparisons, however, an interaction may exist in which environmental factors become less important with increased severity of perinatal status." Orton et al. [15] in a meta-analysis of early intervention studies found little evidence of effectiveness on motor outcome at infant and school age. The improved cognitive outcomes in infancy and preschool were not sustained to school age. Nevertheless, the experience gleaned from the actions of the twentieth century, medical, educational, and legislative, permitted those interested in follow-up to hope for at least the partial amelioration of deficits. These efforts at amelioration included focus on both the major and increasingly more commonly recognized minor sequelae associated with NICU stays. The optimism in the treatment of all children with disabilities, including those who have a history of care in the NICU, is further reinforced by the perspective of the WHO disability terminology which places emphasis on functional interaction rather than pathophysiology (Fig. 3).

An Educational Role for the Follow-Up Clinic

The Accreditation Council for Graduate Medical Education (ACGME) has recognized that there is a need to follow those infants with an NICU history. In the 2013 revision of their 2006 requirements for graduate medical education in neonatal-perinatal medicine, the requirement for an NICU follow-up clinic is clearly stated [16].

VIII.C.6 A sufficient number of discharged infants must be available to assure appropriate outpatient experience for each fellow.

VIII.C.6a) This should occur in a NICU follow-up clinic.

VIII.C.6b) The clinic must have staff with expertise in performing developmental assessments, as well as skilled neonatal or pediatric faculty as teachers.

VIII.Cc) These experiences should enable fellows to understand the relationship between neonatal illnesses and later health and development, and to become aware of the socioeconomic impact and psychosocial stress that such infants may place on a family.

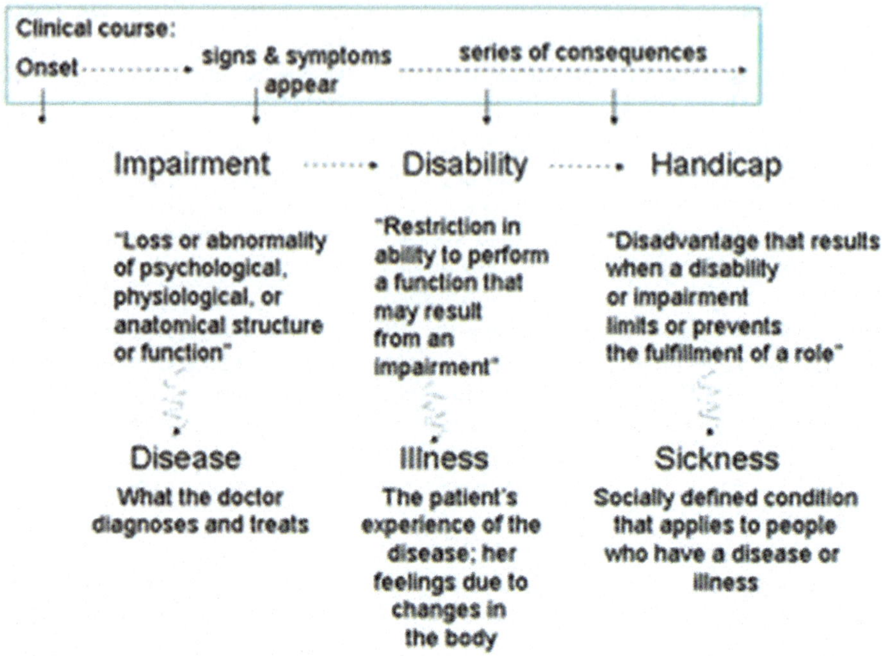

Fig. 3 WHO disabilty terminology

Relation to the Medical Home

An NICHD workshop [17] in 2004 met to examine guidelines for follow-up clinics. Other than discussions regarding growth, the manuscript primarily uses developmental outcomes as the measure to be used to assess interventions in the nursery. In the clinics, the recommendation for further care is "refer for diagnostic or intervention services as needed." There was little focus on the effectiveness of the interventions for delays, and, while neonatal diagnoses and interventions such as BPD, complex medical problems, and postnatal steroids are presented as risk categories, no mention is made of the appropriate venue for managing complex and chronic problems such as feeding or diuretics, steroids, and oxygen. Because few of these clinics are part of a medical home, while the expertise in the medical management of these problems is often found in the follow-up clinic, the management of these issues in the follow-up clinic can and often does lead to fragmentation of care in conflict with a medical home model. Other follow-up clinics will refer these complex patients to subspecialty clinics such as pulmonology or gastroenterology, and, in those cases, the relationship to the medical home will be dictated by the relation of the subspecialty clinic to the patient's medical home. Certain subspecialty follow-up, such as ophthalmology, will always require special services often at other sites. Many follow-up clinics will limit their responsibility to neurodevelopmental evaluation and recommendation for intervention for these issues and defer other medical follow-up to the discretion of the primary care provider.

Questions Raised

While the need for comprehensive follow-up is clearly stated, there are numerous questions to be answered regarding the structure and administration of such a program. Who needs to be followed and at what intervals? What specialties comprise the optimal staffing? What testing should be done and how should it appropriately relate to individual state's early intervention programs? How should the program be financed, especially given the at-risk nature of the families involved and the expected high incidence of no shows? While the NICU follow-up clinic is a necessary bridge between the NICU and the early intervention system, providing both a quality control to the NICU and a "child find" apparatus to EI, there is no clear guide as to how to establish one of these clinics.

It is our hope that this monograph will give sufficient background to aid those who are privileged to care for and provide services to this special population. With this information, those personnel should be more able to advocate at their institutions in establishing and maintaining follow-up programs.

References

1. Lyon, J. (1985). *Playing god in the nursery*. Don Mills, ON: Stoddart.
2. Hess, J. H. (1953). Experiences gained in a thirty year study of prematurely born infants. *Pediatrics, 11*, 425–434.
3. Drillien, C. M. (1958). Growth and development in a group of children of very low birth weight. *Archives of Disease in Childhood, 33*, 10–18.
4. From an interview with William A. Silverman by Lawrence M Gartner in the Oral History Project in June 1997 through the Friends of Children Fund of the American Academy of Pediatrics.
5. American Academy of Pediatrics, Canadian Paediatric Society. (2002). Postnatal corticosteroids to treat or prevent chronic lung disease in preterm infants. *Pediatrics, 109*, 330–338.
6. American Academy of Pediatrics. (2010). Policy statement- Postnatal corticosteroids to prevent or treat bronchopulmonary dysplasia. *Pediatrics, 126*, 800–808.
7. American Academy of Pediatrics. (2014). Hypothermia and neonatal encephalopathy. *Pediatrics, 133*, 1146–1150.
8. White, R. D., Smith, J. A., Shepley, M. M., et al. (2013). Recommended Standards for Newborn ICU Design, eighth edition. *Journal of Perinatology, 33*, S2–S16.
9. Als, H., & McAnulty, G. B. (2011). The Newborn Individualized Developmental Care and Assessment Program (NIDCAP) with Kangaroo Mother Care (KMC): Comprehensive Care for Preterm Infants. *Current Women's Health Reviews, 7*(3), 288–301.
10. Robertson, A. F. (2003). Reflections on errors in neonatology: I. The 'hands–off years, 1920–1950. *Journal of Perinatology, 23*, 48–55.

11. Robertson, A. F. (2003). Reflections on errors in neonatology: II. The ‘heroic years, 1950–1970. *Journal of Perinatology, 23*, 154–161.
12. Robertson, A. F. (2003). Reflections on errors in neonatology: III. The “experienced” years, 1970–2000. *Journal of Perinatology, 23*(3), 240–249.
13. Shonkoff, J. P., & Meisels, S. J. (1990). Early childhood intervention: The evolution of a concept. In S. J. Meisels & J. P. Shonkoff (Eds.), *Handbook of early intervention*. New York, NY: Cambridge University Press.
14. Blair, C., & Ramey, C. T. (1997). Early intervention for low-birth-weight infants and the path to second-generation research. In M. J. Guralnick (Ed.), *The effectiveness of early intervention*. Baltimore, MD: Paul H Brookes Publishing.
15. Orton, J., Spittle, A., Doyle, L., et al. (2009). Do early intervention programmes improve cognitive and motor outcomes for preterm infants after discharge? A systematic review. *Developmental Medicine & Child Neurology, 51*, 851–859.
16. Accreditation Council for Graduate Medical Education. (2013). *ACGME Program Requirements for Graduate Medical Education in Neonatal-Perinatal Medicine*.
17. NICHHD, National Institute of Neurologic Disorders and Stroke, and CDC workshop. (2004). Follow-up care of hi-risk infants. *Pediatrics, 114*, 1377–1397.

Part II
NICU Background

Developmental Care in the Nursery

Sandra Willett

"Virtually every aspect of early human development, from the brain's evolving circuitry to the child's capacity for empathy, is affected by the ***environments*** *and* ***experiences*** *that are encountered in a cumulative fashion, beginning early in the prenatal period and extending throughout the early childhood years." From Neurons to Neighborhoods, Executive Summary, 2000 [1]*

Abstract Developmental care (DC) is a caregiving philosophy that focuses on minimizing stress and maximizing developmental opportunities for NICU infants and their families. The goal is to provide individualized, family-centered care in order to optimize immediate and long-term infant health and developmental outcomes. Developmental care guides broad NICU practices and policies. From daily bedside routines such as feeding and positioning to comprehensive protocols for staff training or environmental design, the approach considers the potential impact of every experience upon a developing, fragile infant. This chapter provides an overview of what is known about developmental care: the theoretical foundation; the core constructs that guide implementation; the strengths and limitations of current, common DC practices; and the existing evidence that supports or refutes these widely embraced principles.

Introduction

Babies born prematurely face numerous developmental challenges. Up to 50% will experience difficulties with motor, cognitive, attentional, and/or behavioral skills that impact school-age academic success and social participation [2, 3]. Deficits in working

S. Willett (✉)
Munroe-Meyer Institute of the University of Nebraska Medical Center, Omaha, NE, USA
e-mail: swillett@unmc.edu

H. Needelman, B. J. Jackson (eds.), *Follow-Up for NICU Graduates*,
https://doi.org/10.1007/978-3-319-73275-6_2

memory, visuomotor/perceptuomotor skills, attention, and executive function persist through adolescence contributing to learning disabilities and lower IQ scores [4, 5]. While factors such as severity of medical condition, suboptimal nutrition, and necessary medical interventions during a period of biological fragility contribute to adverse preterm infant developmental outcomes, the neonatal intensive care unit (NICU) *environment* and *experience* are increasingly scrutinized as a significant stressor altering infant/family health and well-being [6]. NICU "developmental care" focuses on developmentally appropriate infant and family-centered care experiences. It is "defined by efforts in unit design, equipment selection, policies, care protocols, and staff training to maintain the basic physical, sensory, and interpersonal needs of the preterm infant while minimizing exposure to noxious stimuli" [7]. The hope of such caregiving is to optimize immediate and long-term infant cognitive and psychosocial development and to empower the families who parent NICU graduates [8].

From a public health perspective, developmental care implies a continuum from preconceptual maternal healthcare to prenatal care to postnatal care through late childhood/early adolescence [9, 10]. Any adversity along this continuum such as trauma, medical complications, social instability, or prematurity potentially alters outcome trajectories. Developmental care (DC) as a NICU philosophy emerged in the late 1970s and early 1980s as advancements in medical technology allowed survival of the smallest, sickest infants [11, 12]. As infant mortality decreased, prematurity-related developmental sequelae increased. Heidelise Als, the founder of NICU developmental culture, recognized the importance of comprehensive, individualized medical and developmental care in optimizing infant outcomes. With her profound insight and influence, the NICU care paradigm shifted from technology driven to relationship-based, from an environment focused on the needs of medical care providers to the needs of infants/families, and from hard-wired age or stage-related guidelines to individualized, family-centered clinical decision-making. Developmental care is a comprehensive, transdisciplinary approach. The movement is associated with various monikers ("neuroprotective care" [13, 14], "developmentally supportive care" [14], and most recently "trauma-informed, age-appropriate care" [15]) or with specific approaches ("neonatal individualized developmental care and assessment program" [16] and "neonatal integrative developmental care model" [17]). Semantics aside, developmental care acknowledges that premature babies, and their families, are "under construction." All inputs, social, physical, emotional, and environmental, potentially strengthen or compromise foundational neuro-circuity. Three basic, broad constructs are fundamental to developmental care:

Dr. Heidelise Als, Ph.D., founder of Neonatal Individualized Developmental Care and Assessment Plan(NIDCAP). This approach revolutionized developmental care and emphasizes the need for "tailoring" the environment to match the individual needs of a given infant

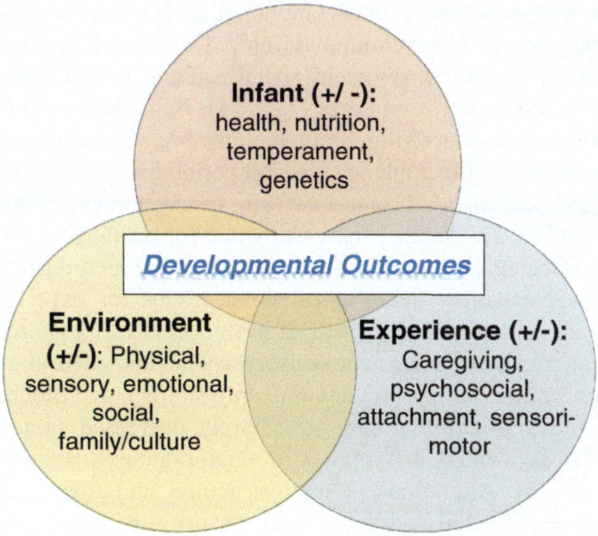

Fig. 1 Development is a complex, interactive dynamic between infant, environment, and experience. All three realms are dramatically altered with premature birth. Developmental outcomes, in all domains, social, emotional, motor, cognitive, and behavioral, are the cumulative result of positive and negative influences over time

1. At any given point in time, each infant/family is unique. Developmental care recognizes the individual strengths, needs, and vulnerabilities of each infant/family unit.
2. Infants and families experience incredible stress during NICU hospitalization. NICU caregiving prioritizes respect and protection of infant/caregiver dyads during a sensitive period of neurobehavioral development and emotional attachment.
3. Environment and experience have a profound impact upon the premature infant's developing neurosensory systems. Cumulative positive and negative inputs determine eventual infant and family outcomes [8, 18, 19] (Fig. 1).

This chapter will explore developmental interventions and assessments for premature infants hospitalized in the NICU. What do we know? What is still in question? And to maximize immediate and long-term outcomes, which developmental practices should be "standard" best-practice?

Developmental Vulnerability: Rationale for Developmental Care

Stress and the NICU Environment

Development is the acquisition of new capabilities over time as a result of physical maturation and experiential learning. Although milestones appear to be predictable and time-ordered, passage of time alone does not guarantee sequential change [20].

Experience, as well as physical, social, emotional, and cultural environment, matters for determining how and when children develop. Developmental plasticity refers to cumulative "shaping" of behaviors in social, affective, and cognitive domains through positive and negative influences [11, 21]. Behavioral changes both reflect and drive changes in underlying brain architecture. Conception through early infancy is a profoundly vulnerable yet resilient period for developmental neuroplasticity, for in this window rapid central nervous system growth, differentiation, and integration occur [12, 22]. While babies are "wired for learning" at birth with innate, experience-expectant capabilities [1], healthy caregiving relationships and supportive environments establish neurological connectivity for experience-dependent social, emotional, and cognitive growth. If birth occurs early, neurodevelopmental processes are altered by atypical neurosensory, social, and environmental stressors. Changes in central nervous system anatomy, physiology, and biochemistry result. In long-term studies of children born prematurely, decreased white and deep gray matter volumes as well as differences in synaptogenesis and myelination are documented [4, 23]. Alterations in microstructural and functional connectivity affecting regions associated with motor and sensory function, auditory and visual processing, memory, attention, and language [2, 24, 25] correlate with clinically significant cognitive and language delays that persist through adolescence [4].

The premature population demonstrates long-term anatomic, neurochemical, and hormonal changes known to be associated with early life stress and cortisol production [2, 26]. In animal studies, early life stressors "can distort rodents' approach to problem-solving, interfere with their ability to successfully orient their behavior toward a goal, and reduce their ability to engage with new or uncertain conditions" [27]. These reported behavioral responses implicate regions of the CNS involved with self-regulation, memory, and executive function and are strikingly similar to those exhibited in children with attention deficit disorder or learning disabilities, disorders known to be common among children born prematurely [11, 14, 28]. In humans, such learning difficulties are manifest early. With preterm infants born at less than 30 weeks of gestation, Lobo et al. examined manual exploration of objects, an important precursor for cognitive development, during the first 6 months of life. Infant exploratory motor behaviors such as fingering, transferring, and squeezing were coded during interaction with seven different, standardized infant toys. Compared to term infants, preterm infants demonstrated decreased overall object exploration, decreased visual-haptic multimodal exploration, impaired bimanual exploration, and reduced variability in exploratory behaviors that reflect severity of prematurity-associated risk [29].

Extrauterine environmental and physical demands acutely stress the fragile, premature infant's physiological capabilities. Such stress, if intense or prolonged, becomes "toxic" causing chronic hyperstimulation of the autonomic system [15, 27]. Consider a fetus, in utero, at 28 weeks of gestation. Every physiological need, from gas exchange to temperature regulation, is modulated by the uterine environment. Premature birth disrupts this symbiosis. The dimly lit, sound-protected, movement-contained, and predictable maternal rhythmicity is replaced by the high-tech NICU. Bright lights, noisy equipment, high activity levels, painful interventions, unpredictable care patterns, and parental separation prevail. Parental separation alone is known to disrupt infant physiologic stability [8, 30], hippocampal synaptogenesis [31], and social responsivity [18]. Routine but necessary NICU caregiving practices, diaper changes, bathing, or weighing precipitates behavioral and physiological responses indicative of stress [11, 15, 19]. Smith et al. prospectively followed 44 infants born prior to 30 weeks of gestation reporting an average of 11, with a range of 4–18 stressful procedures daily in the first 14 days of a preterm infant's life [23]. Stress is problematic in the NICU environment. Though complete elimination of prematurity-associated stress is unlikely, developmental care aims to minimize it.

Developmental Care Constructs

Developmental care (DC) in the NICU begins the moment a baby enters the world, regardless of gestational age. Nurses, physicians, occupational therapists, physical therapists, respiratory therapists, developmental specialists, speech language pathologists, and psychologists make up the professional caregiving team. But, undeniably, the most critical and consistent members of the NICU infant care team are the *parents*. Broadly, DC encompasses appropriate modulation of external sensory information (tactile, vestibular, visual, gustatory, proprioceptive, and auditory), sleep protection, energy conservation through clustering of cares and positioning, and parental engagement [17, 18, 32]. In a large, multicenter project (Neonatal Adequate Care for Quality of Life), Montirosso et al. [33] assessed implementation of developmental care using the Quality of Care Checklist, a survey probing care practices, parent policies, environmental controls, and infant pain management. Infant neurobehavioral outcomes at discharge as measured by the Neonatal Network Neurobehavioral Scale (NNNS) were improved in NICUs with higher infant-centered care, as measured by parental involvement with cares and other developmental practices, and pain management survey scores.

Specific elements of DC are "bundled" into "approaches" in an attempt to standardize practice patterns. These include Neonatal Individualized Developmental Care and Assessment Plan (NIDCAP), Maternal-Infant Transaction Program (MITP), Creating Opportunities for Parental Empowerment (COPE), Hospital to Home: Optimizing Infant's Environment (H-HOPE), Family Nurture Intervention (FNI), and Neonatal Integrative Developmental Care Model. None of these is as widely acknowledged or has had as universal an impact upon NICU culture as Heidelise Als' Neonatal Individualized Developmental Care and Assessment Plan (NIDCAP). In fact, the very foundations of all developmental care and the approaches listed above are centered upon Als' synactive theory of development and understanding infant neurobehavioral cues.

Understanding Infant Neurobehavioral Cues

Als, a developmental psychologist renowned for her studies of premature infant behavior and brain development, transformed NICU care culture. With meticulous detail and eloquent thought, she forever altered how premature infants were perceived through application of T. Berry Brazelton's pioneering work with newborn neurobehavior. Brazelton, "impressed from the beginning by the newborn infants' ability to interact with the environment and by their capacity to deal selectively with environmental stimuli" (Brazelton and Nugent, 2011, p. 3) [34], challenged the notion that infants were predominantly reflexive, incapable of learning or social interaction, and merely passive recipients of sensory stimulation. Instead, he, with newborns, and Als, with preterm infants, described environmentally and socially responsive infant *communication* strategies built on four interdependent subsystems: autonomic/physiologic, motor, state, and attention/social interaction [16, 34, 35]. Als' synactive theory of neurobehavioral development (Fig. 2) posits these four interdependent subsystems work "synactively" with environmental and caregiver influences to promote infant self-regulation.

Self-regulation is the pinnacle of infant neurobehavior. When the four subsystems interact harmoniously, babies self-regulate, actively coping with environmental demands and interacting for social engagement [35, 36].

Als' conceptual framework demonstrates characteristics of both a hierarchical and dynamical systems model [37]. Subsystems evolve and differentiate in a fairly defined maturational sequence with physiological systems at the "core" or taking

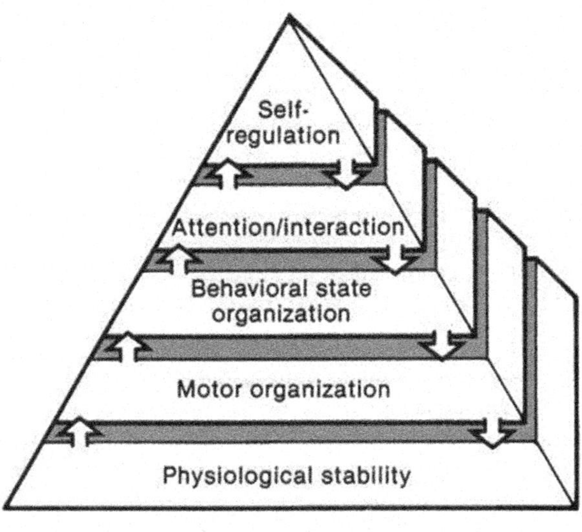

Fig. 2 Als' synactive theory of development proposes that infant self-regulation is dependent upon stability within four interrelated subsystems: attention/interaction, behavioral state/organization, motor organization, and physiological stability. These subsystems emerge in a hierarchical manner developmentally with physiologic needs laying the foundation. But, at any given moment, subsystems interact dynamically, supporting or compromising each other

precedence (hierarchical); yet, systems interact with each other, the environment, and caregivers in such a way that any one may support or compromise another (dynamical system). For example, if a baby's physiologic foundation is challenged by a medical condition such as respiratory distress, the infant suppresses motor activity, enters a protective sleep or drowsy state, and "tunes out" social interaction to preserve physiological stability. This innate, protective mechanism occurs regardless of gestational age or developmental stage and serves a useful purpose: it conserves energy for growth. It is a manner of coping. Conversely, with this same struggling infant, if social/interaction or sensory information is of inappropriate timing or intensity, a cascade of physiologic compromise occurs. Decreasing environmental neurosensory demands or providing motor system support through swaddling/containment, in this instance, serves to promote physiologic stability. The self-regulated infant maintains balance within and between subsystems during periods of stress without external assistance [36]. More importantly, the self-regulated infant has the capacity to explore within and learn from the physical and social environment.

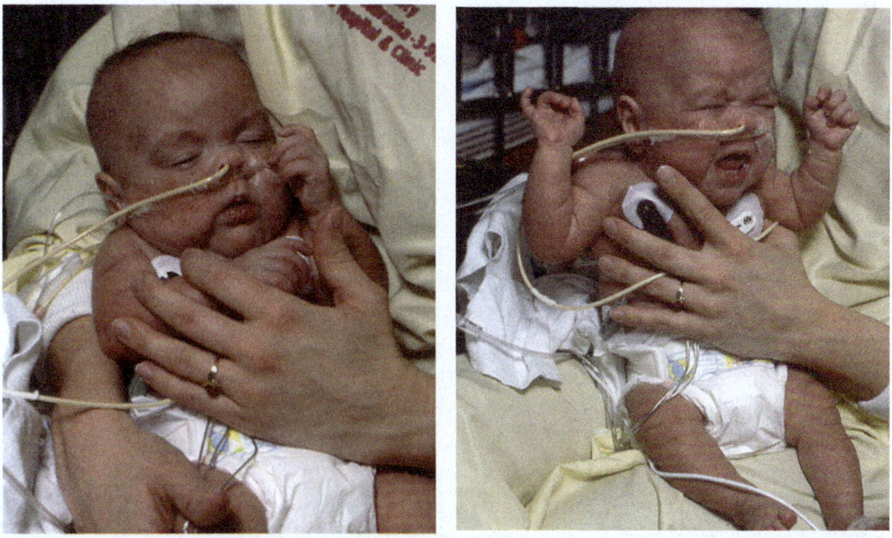

Term equivalent age preterm infant with bronchopulmonary dysplasia and feeding difficulties exhibits stress cues and poor regulation when transitioned to caregivers lap (left). Motor containment and a rest break to recover are necessary to maintain physiologic stability. She drifts into protective sleep after the stressful interaction (right)

Within each subsystem, specific behaviors or subtle physical cues may be "read" to understand if an infant is expressing "coping/readiness" or "stress" or approach or avoidance [38]. Autonomic or physiologic cues involve basic bodily functions. Overt signs of autonomic stress include changes in color, heart rate, respiratory rate or effort, and oxygen saturation levels. Within context, startles, hiccups, sneezes, gags, emesis, or gastrointestinal motility may also herald physiologic instability. Motor stress is displayed by changes in posture or movement, increased extension, increased or decreased resting tone and activity levels, increased tremulousness, or movement disorganization. Flexion, hand to mouth or hands to midline, sucking, and grasping

	Stress Signs	Coping Signs
Physiologic	Tachypnea/Apnea Tachycardia/Bradycardia Mottling/Cyanosis/Ruddiness Hiccups Gagging/poor feeding Yawning, sighing, sneezing Oxygen desaturation	Sucking, coordinated feeding Healthy color (pink) Oxygen levels, heart rate, respiratory rate remain within acceptable ranges
Motor	Hypo or hypertonicity Extreme postures Increased extension Tremors or startles Motor agitation/squirming	Flexed postures Movement variability Appropriate tone Smooth movements Hand to face or hand to mouth
State	Irritability, excessive crying Inconsolability Difficulty with state transition Difficulty awakening or staying awake	Achieves and maintains a quiet, alert state Transitions between states Consoles easily
Attention/Interaction	Grimace Gaze Aversion Glassy eyes or locked stare 'Gape' face	Bright eyes Open, engaged facial expression Visual attention

Fig. 3 Potential indicators of infant stress/coping within the four subsystems of Als' synactive model

reflect coping. State, quantified numerically along a continuum from one (deep sleep) to six (highly agitated, crying), reflects degree of infant arousal. State alone may not indicate coping or stress. Instead, the baby's ability to maintain state organization, robustness, and range of, as well as transition between, states during caregiving or interaction is of interest [37, 38]. Babies who are awake, alert, and attending to social or environmental stimuli are considered state four, an optimal state for learning. Babies engaged in interaction who transition into sleep or agitated states may be communicating stress or fatigue. Finally, social/interactive cues are last to emerge developmentally and perhaps most difficult to read in preterm infants [39]. Stress cues in this subsystem share similarities with the subtle body language cues that even adults display: grimaces, gaze aversion, looking away, inattention, panicked alertness, irritability, or inconsolability [38]. Ongoing observation of neurobehavioral cues across the four subsystems allows caregivers and families to monitor an infant's adaptation to routine cares or therapeutic interventions and to adjust environmental and sensory inputs accordingly (Fig. 3).

Environment and Experience: What Matters and Why

Developmental Care Practice Themes

Developmental care interventions begin with the basic premise that environment and experience matter and can be manipulated to protect physiologically fragile NICU infants. But which elements of developmental care are considered standard?

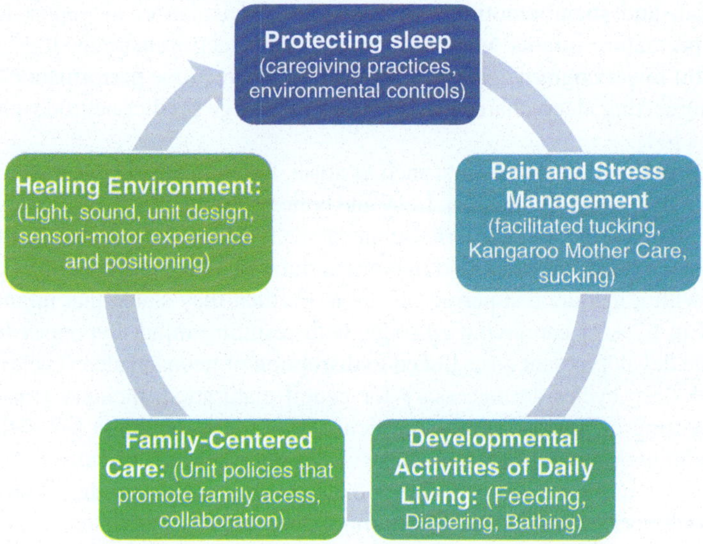

Fig. 4 Core constructs or basic elements of developmental care with selected examples of interventions include protected sleep, pain and stress management, developmental activities of daily living, family-centered care, and the "healing environment"

This next section will explore a broad range of developmental care strategies from specific practices like kangaroo mother care to NICU-wide environmental modifications like noise and light control. These care practices are not implemented to mediate acute medical conditions. Instead they guide baby and family friendly care, foster infant/mother attachment, and promote age- or "stage"-appropriate learning experiences to optimize infant outcomes. Coughlin and colleagues, acknowledging that DC was universally embraced but inconsistently implemented, reviewed the literature to define core constructs [40]. They described five: protected sleep, pain and stress assessment and management, developmental activities of daily living (feeding, positioning, skin care), family-centered care, and the healing environment. The Neonatal Integrative Neuroprotective Care Model™ (Fig. 4: Altimier, 2015; Phillips, 2015) using slightly different semantics expands this to seven with the addition of protecting skin and optimizing nutrition [14, 41].

In practice, individual DC constructs are tightly interrelated and overlapping: addressing one ultimately influences others.

First and Foremost Sleep: The Foundation for Growth and Well-Being

Sleep is crucial for infant health and well-being [42, 43]. Babies born prematurely demonstrate disorganized sleep, shorter sleep duration, greater indeterminate sleep, and disrupted sleep-wake cyclicity [42, 44, 45]. Such alterations affect cognitive, behavioral, and social-emotional development. Weisman et al. report low birthweight; premature infants who demonstrate poor sleep transitions at 37 weeks of age exhibit lower cognitive, verbal, symbolic, and executive performance at 5 years of age suggesting sleep characteristics upon discharge predict neurodevelopmental outcomes [44].

NICU environmental stressors, such as noise, lighting, or poorly timed caregiving, disrupt neonatal sleep cycles [43]. Levy and colleagues, using video and polysomnography, studied a group of near-term infants in the NICU to determine if hands-on care affected sleep [46]. They reported maximum durations between handling episodes at 50 min, with a median rest period of 2 min. Further, they stated that infant arousals occurred in 57% of caregiving episodes with resulting infant hypopnea in 28% of those. The developmental costs linked to disruption of neonatal sleep cycles are high. Patterned sleep cycles are necessary for neural maturation, memory consolidation, and processing of sensory stimuli, particularly visual information [47–49]. Preterm infants with poorly regulated sleep-wake cycles demonstrate difficulty with gaze regulation at 4 months of age and higher distractibility at 18 months. Both traits are strong predictors of attentional and learning deficits [49]. Disrupted sleep is further associated with extrauterine growth restriction. Common in premature infants, this growth pattern occurs when an infant expends essential calories during stress-induced wakefulness, calories that, during restful sleep, would be available for growth and tissue healing [42, 50]. An adverse developmental trajectory ensues: decreased growth impacts brain development during a period of rapid differentiation; decreased brain growth impacts cognitive development which in turn impacts performance at school age and potential for adult educational achievement. Finally, sleep deprivation alters a baby's ability to allocate attention and maintain self-regulation [48, 50]. The sleep-deprived infant spends less time alert, awake, and attending, which impacts early social interactions and bonding. The attention-socialization-stress cycle is linked to adverse mental health outcomes [51]. It is no surprise that DC focuses on modifiable NICU environmental factors known to protect sleep and reduce infant stress: noise, light, and caregiving practices.

The Auditory Environment/Experience

Noise and Auditory Development in the NICU

For premature babies, the very equipment that sustains life potentially disrupts sleep and elevates physiological stress by creating background "noise." 1974 EPA guidelines, acknowledging health risks associated with noise-induced stress, state continuous noise should never exceed 100 dB nor impulse noise 140 dB; in hospitals, 45 dB is acceptable during daytime but 35 dB recommended at night [52, 53]. Routine NICU sound levels vary from 50 to 90 dB with intermittent noise as high as 120 dB [53–55]. Graven reports both increased risk for sensorineural hearing loss and substantial sleep interference for babies exposed to sustained sound levels above 60 dB [43].

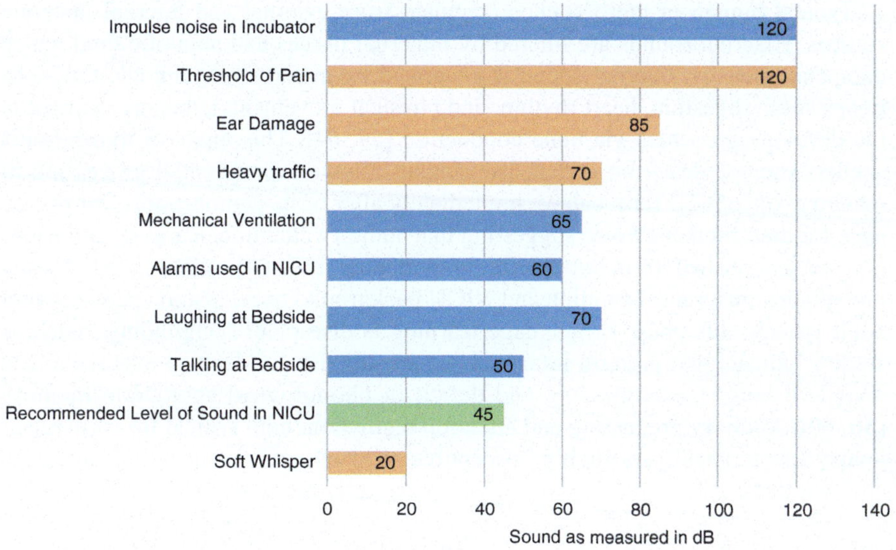

Fig. 5 Common environmental and NICU sound levels

Feeding pumps, for infants too immature to eat, and physiological monitors that alert caregivers out of visual range of potentially life-threatening events alarm at 60–80 dB [53]. The human factor also contributes. Bedside conversations register sound levels of 50 dB and laughter 70 dB [53]. Lasky and Williams report that NICU newborns in open cribs breathing room air achieved American Academy of Pediatrics' recommended sound level exposure (45 dB) only 36% of their stay, babies with nasal cannulas 21% of their stay, and extremely low birthweight babies only 6% of their stay [56]. They conclude that in most "NICU care circumstances, the AAP recommended sound levels were almost never achieved" [56] (Fig. 5).

In addition to disrupting sleep and causing physiologic stress, NICU noise affects auditory, speech language, and social development. In the womb, the fetus can perceive and respond to sound as early as 26 weeks [57, 58]. Differentiation of complex sounds and varying speech phonemes is possible by 30 weeks, marking the onset of speech/language development [59, 60]. Changes in amount and type of fetal acoustic experience impact auditory structure and function at multiple levels. Animal studies establish prenatal exposure to intense low-frequency sound damages hair cells in the developing cochlea as well as CNS auditory processing connections [43]. Chang and Merzenich report rat pups raised in continuous noise fail to recognize certain sound frequencies or patterns, basic benchmarks of auditory processing, until they are three to four times older than peers raised in a noise-free setting [61]. Chicks exposed to excessive noise during the fetal period demonstrate decreased birthweight, decreased brain weight/size, and changes in auditory system neuronal size and structure [62]. In utero monkeys and rats exhibit an increase of circulating stress hormones in response to noise and subsequent abnormal social behavior after birth [63]. Clearly, early sound experiences have wide-reaching influence.

Typical human in utero sounds originate from external and internal maternal sources. External sounds are filtered by maternal tissues and amniotic fluid which dampens intensity. Internal sound is generated by maternal physiologic functions (respiration, digestion, heart rhythm, and physical movements), then transmitted to the developing cochlea via bone conduction [59, 64]. This mode of transmission shelters the cochlea from certain frequencies allowing preferential processing of maternal voice [64]. Term babies, immediately after birth, demonstrate clear preference for their mother's voice suggesting that auditory attention, learning, and memory are established from intrauterine experiences [59, 64]. When a NICU stay precipitates maternal separation and NICU background noise distorts perception of adult speech, this phase of language learning is disrupted. Longitudinal outcome studies indicate that preterm infants are at greater risk for delayed expressive and receptive language processing and deficit in phonological short-term memory [50, 60]. Auditory processing and attention regulation, both critical for speech/language development, are highly susceptible to damage from altered fetal sound environments.

Strategies to Optimize Auditory Experience

Determining optimal auditory experiences for the developing infant in a noisy NICU is difficult. Obviously, noise abatement strategies are the first line of defense and will be discussed further in the next section. In conjunction with "noise" management, Caskey and colleagues suggest that for auditory input it is not only "how much" or "how loud" but "who" provides it that matters. Studying NICU infants as young as 32 weeks, they report (1) hourly adult word counts increased by more than 380% at 32 weeks and 220% at 36 weeks when a *parent* was visiting, (2) child vocalizations per hour increased when a *parent* was present by as much as 129%, (3) infant/adult conversational turns increased by 520% at 32 weeks and 160% at 36 weeks when a *parent* was present, and (4) infant vocalizations and "conversations" were greater during *parent* feedings than nurse feedings [60]. In 2014, this same group demonstrated that higher adult word counts in the NICU are positively correlated with higher Bayley III language composite scores at 7 and 18 months of age [65]. Clearly, having a parent present changes the amount of language a preterm infant is exposed to and the amount of reciprocal social interactions a baby experiences, both important precursors for speech/language and social development. The power of parental presence will be fully detailed in a later section.

NICU music therapy (MT), the intentional and mindful use of soothing sound, has been promoted as a developmentally appropriate source of auditory input. If used *mindfully*, music can mask environmental noise, promote calming, and reduce stress for both infant and parents. MT has been applied and studied for a diverse range of NICU purposes including analgesia during painful procedures, soothing during withdrawal in neonatal abstinence syndrome, feeding and sucking entrainment, overall stress reduction, growth and behavioral enhancement, promotion of attachment behaviors, and development of social, language, or cognitive skills [66]. No long-term follow-up studies substantiate the impact of MT on infant developmental outcomes, but Standley, in a 2012 meta-analysis, found strong effect sizes (Cohen's d of 0.70 or greater) favoring MT study groups for behavioral and physiological measures of infant stress reduction, improved sucking and feeding abilities, and decreased length of hospital stay [67]. Standley goes on to report that infants aged 28–35 weeks demonstrated the greatest benefit, females were more responsive than males, and intervention effectiveness decreased as birthweight increased [67]. Bieleninik, Ghetti, and Gold conducted a meta-analysis of 14 randomized controlled MT trials that included 964 preterm infants and 266 parent participants [68]. They established that MT reduced infant respiratory rate by 3.91 breaths per minute, a statistically significant and clinically relevant indicator of relaxation response, and MT decreased measures of maternal anxiety significantly, shifting mean anxiety scores from clinical to subclinical levels [68]. While further research is needed with regard to standard "dosing" and acceptable volume parameters, *live maternal* singing is favored over adult speaking or recorded music for improving infant behavioral state and reducing stress measures [69, 70]. As with any intervention, appropriate timing of MT is critical as is monitoring an individual infant's response before, during, and after music exposure to prevent overstimulation, physiological compromise, and interference with sleep cycles.

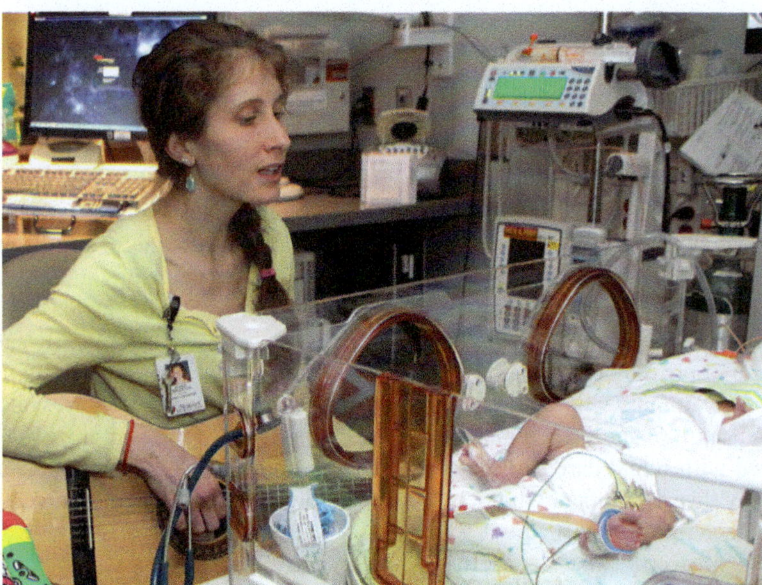

Appropriately timed live singing may reduce stress and improve physiologic stability in medically fragile, preterm infants. Infant response must be monitored carefully before, during, and after music exposure. Photo compliments of Nebraska Medicine

Noise Abatement Strategies

To reduce stress and optimize adult speech exposure, NICU noise abatement is and continues to be a focus of developmentally supportive intervention. Graven in 2003 recommended standard sound abatement strategies that incorporate both staff awareness and systematic monitoring [43]. Increased awareness alone decreases ambient nursery noise [55, 71]. Strauch et al. report a simple "quiet hour" per shift significantly drops sound levels and increases the percentage of time babies spend in light or deep sleep [50, 72]. Current NICUs incorporate multiple levels of sound control. Ambient sound monitors, intelligent or vibrating alarm systems, enclosed staff workstations, structural sound-absorbing design elements, and traffic patterns for noise containment are standard [55, 71, 73]. Ongoing programs to increase noise awareness are and should be routine policy. Families, visitors, and staff complete compliance modules that address safe sound practices. Single family room (SFR) designs are standard, replacing the noisy open-bay infant care areas [55, 71, 73]. SFR design drops noise levels on average 12 dB below the AAP's recommended safe level of 45 dB [74].

Single Family Room Designs: The Ideal Solution?

The rationale for SFR is much more comprehensive than sound control [51, 75, 76]. Environmental control, sleep protection, infant stress reduction, and family involvement drove the design shift; all constructs are intimately associated with developmental care. Indeed, SFR is associated with multiple advantages. For the infant, these include greater rate of weight gain, decreased episodes of apnea, improved sleep and behavioral regulation, reduced infection rates and mortality, decreased length of hospitalization, fewer medical procedures, quicker transition to enteral feeds, fewer rehospitalizations, and better neurodevelopmental outcomes at term and 18 months of age [76, 77]. SFR shifts focus from infant-centered care to the infant/family dyad. Associated family outcomes include increased parental visitation, increased family participation in infant cares, improved psychosocial indices, greater family satisfaction, and improved breastfeeding outcomes [73, 74]. Lester, in an 18-month follow-up study of infants who graduated from two different nursery designs, concluded that SFR "jump-starts" high maternal engagement [77]. Mothers in the SFR group engaged with infant care early and sustained high-level involvement compared to open-bay design mothers who demonstrated slow ramping toward involvement as infant discharge approached.

Yet controversy exists. Concerns regarding staffing patterns, ease of patient care/accessibility, and costs of redesign remain. Shepley et al. argue that SFR construction cost is recuperated in the first 12 months of operation [78]. Lester and colleagues report less overall stress, more personal accomplishment, and less emotional exhaustion among nurses practicing in a SFR environment [76]. These same nurses indicate that SFR design facilitates a more professional environment and greater family-centered care. Pineda et al. consecutively assigned 136 infants born prior to 30 weeks of gestation to either SFR or open bay over a 3-year period (2007–2010) [79]. Measures of cerebral maturation were lower in SFR infants at term age; Bayley Scales of Infant Development language scores were lower, and Infant Toddler Social-Emotional Assessment externalizing behavior scores were higher among the SFR group at 2 years of age. Although the authors conclude that SFR may not be as beneficial as suggested, limitations to their study abound. Small sample size for cerebral maturation measures (approximately 20 for each group), non-randomized room assignment, greater Medicaid-insured families among the SFR group, and an urban NICU population with historically low parental visitation rates impedes generalizability.

The Visual Environment/Experience

Light and Visual Environment in the NICU

Premature infants exposed to light of inappropriate intensity or duration exhibit physiological signs of stress and adverse clinical outcomes such as poor weight gain or behavioral and sleep disturbances [80]. Exposure to atypical light and visual sensory experiences during fetal development impacts the structure of neurosensory systems and functionally contributes to deficits in visual acuity, depth perception, visuospatial memory, visual attention, and visuomotor control/perception [81–83]. In utero, baby's eyes are continuously shielded from bright light and patterned images; and circadian rhythms, influenced by light cycling after birth, are driven by changes in maternal temperature, activity, and hormone production [42]. The visual system develops prenatally in an environment devoid of direct light and patterned visual information. Anatomical protective mechanisms, pupillary constriction, eyelid thickness, and motor behaviors for eye shielding that modulate extrauterine visual inputs at birth are poorly developed prior to 40 weeks of gestation [47]. The mismatch between preterm endogenous visual capabilities and exogenous extrauterine stimulation impacts cell production, neural interconnectivity, and synaptogenesis in the visual cortex [47]. In animal studies, inappropriately timed onset of visual input interferes or competes with other developing sensory systems. Lickliter manipulated prenatal tactile, visual, and vestibular inputs in quail to determine the effects of "developmentally" disordered sensory augmentation [84, 85]. Chicks exposed to patterned light prenatally did not exhibit a preference for their species-specific maternal call after hatching; and increased tactile or vestibular information altered perceptual responsiveness in both the visual and auditory systems. Such findings foster the belief that sensory systems are temporally constrained during early development [84–86]. Any alterations in timing, amount, or type of sensory experience profoundly affect perceptual skills. As a result of this theory, dim lighting, protection from direct light, and minimal visual stimulation prior to 32 weeks are a developmental practice standard in the NICU. After 32 weeks, cycled or natural lighting to promote diurnal rhythmicity is individualized based on infant physiologic responses and caregiving needs; but the value and impact of early visual "stimulation" with images in NICU infants continue to be uncertain.

Visual attention in infancy is closely linked with cognitive processes and requires allocation of energy reserves. As such, visual or multimodal sensory demands or "stimulation" may contribute to physiological compromise. Yet, early visual experience and exploration are critical for cognitive, motor, and social development. As early as a few weeks post-term, gaze shifting, focus, and visual following behaviors as well as "look" durations are measureable indicators of infant attention and interest [87]; and visual attention is a predictor of later information processing abilities. Rose et al. report 5-month-old infants (corrected age) born prematurely demonstrate differences in visual attention [88]. These babies display "more off-task behavior, longer look durations, and slower visual shift rates" when engaged in visual motor tasks. These same visuomotor and visuoperceptual skills are needed for eventual academic success with reading and writing.

Understanding Appropriate Visual Experiences in the NICU

So what types of visual experiences should be encouraged for preterm infants? First, consider that gestational age alone does not guarantee readiness for visual experiences or the social reciprocity that accompany them. A medically stable infant at 36 weeks may demonstrate robust readiness for visual interactions, while a 42-week infant with chronic lung disease still struggles. An infant's physiological capability for interaction and state regulation must be monitored. Second, does the visual information carry social and contextual relevance? The most important source of visual information for any preterm infant is not high-contrast, simple images but the maternal face. Programmed for animate and mobile objects, infant visual attention is drawn to the ever-changing intersensory (visual, auditory, affective, and social-emotional) canvas a mother provides. Babies already know mom's scent and voice prior to "seeing" her. This "knowledge" creates sensory redundancy, an important mechanism for information processing and consolidation of learning [86, 89]. Lickliter [86] indicates that selective attention during early infancy is "biased toward stimulus properties that are common or redundant across sensory modalities" (Lickliter, 2011, p. 3) and such redundancy guides and shapes perception, learning, and memory. Experiences that promote sensory redundancy have not been studied specifically in the preterm human population. But the contrast between the contingency-based, intersensory uterine environment, provided, modulated, and contextually wrapped by a mother, and the NICU is apparent.

The Caregiving Environment/Experience

Protect, Prevent, and Preserve

Caregiving practices are necessary tasks associated with routine baby-related care as well as medical procedures related to premature birth. Coughlin defines tasks related to personal care (bathing, eating, dressing, repositioning, toileting) as "age-appropriate activities of daily living" and emphasizes that such tasks not only meet basic hygiene and health needs but build social-emotional relationships [15]. Yet for preterm infants, these same routine caregiving tasks are known to cause pain, disrupt sleep, and increase physiologic and behavioral signs of stress. Levy and Hassan (2016) recorded caregiving in the NICU and reported that (1) hands-on care occurred during 27% of a 4 h observation period, (2) 44% of these contacts were related to medical cares, (3) cares were initiated across all behavioral states, and (4) in only half of the observations were infants given sufficient time to complete a full 60 min sleep-wake cycle [46]. Within 60 s of these hands-on caregiving episodes, infant hypopnea (16%), apnea (8%), and oxygen desaturation (20%) events were documented. What can be done to reduce the stress related to caregiving and promote "age-appropriate" activities of daily living? While it must begin with reading infant stress and readiness cues, the developmental caregiving environment embraces three basic premises: protect sleep, prevent pain, and preserve the integrity of fragile, developing sensory, musculoskeletal, and social-emotional systems.

Care Timing: To or Not to Cluster?

The importance of sleep and the influence of physical environment (space, sound, and light) upon sleep have been discussed previously. Human caregiving is another factor to consider in the infant sleep equation. Clustering of necessary cares according to infant sleep-wake cycles is promoted as a DC method to protect sleep, minimize stress, and conserve energy through planned infant rest and recovery periods. While this appears logical and simple, it is not. For very preterm infants, sleep cycles are indeterminate, and essential cares may be too time dependent to be deferred [42]. For medically stable premature infants, clustered care and protected rest periods in the NICU are associated with improved weight gain, decreased apnea, and longer periods of sleep [90]. Yet for infants who are extremely premature or medically fragile, clustered cares may induce too great an energy cost, and the deeper sleep that follows may actually indicate exhaustion. The argument for individualized caregiving schedules according to infant stress and readiness cues remains for procedural caregiving that is not medically urgent.

Pain Reduction Strategies for Developmental Care

Pain is a largely subjective, individualized experience with both physical and emotional consequences. Without direct biological markers, pain measurement in neonates relies on facial expression or stress-related behavioral and physiological correlates. Marchant, in a review of evidence regarding neonatal pain, concludes (1) cortical, subcortical, peripheral, and hormonal elements required for sensation of noxious stimuli and pain are present by 20 weeks of gestation; (2) stereotypical pain-related hemodynamic, behavioral, and physiologic responses are alleviated in neonates after administration of analgesics suggesting pain responses are indeed present at an early age; and (3) exposure to pain in the neonatal period alters neuronal networks resulting in lasting tactile hypersensitivity (hyperalgesia and allodynia) [92], white and gray matter loss, and changes in pain-related CNS regions associated with sensory, affective, and cognitive processing [91]. Babies in the NICU are subjected to approximately 10–15 invasive, painful procedures daily [93]. Frequent procedural pain, or pain associated with routine cares such as heel sticks or eye examinations, in combination with stressful social (maternal separation) and physical (handling and sensory input) environments is known to compromise growth and negatively alter infant's developmental, cognitive, affective, and behavioral health [93, 94].

Simple, cost-effective means of procedural pain reduction in the NICU may be considered maternal-specific such as breastfeeding, or skin-to-skin holding, or may be administered by any NICU caregiver: facilitated tucking, nonnutritive sucking, holding/rocking, and oral administration of sucrose [94–96]. Although the evidence is mixed and more research indicated, a Cochrane review by Riddell and colleagues concludes that of the studied non-pharmacologic pain management strategies, nonnutritive sucking, swaddling/facilitated tucking, and rocking/holding yield the strongest evidence as indicated by measures of standardized mean difference between study groups and controls [94]. These three strategies may be used safely to decrease pain reactivity and decrease behavioral and physiological indicators of pain in neonates and young infants. Although skin-to-skin holding is associated with lower scores on the Premature Infant Pain Profile, physiological indicators such as heart rate were not different between skin-to-skin and other pain-relief strategies. The use of oral sucrose for analgesia is perhaps the most studied non-pharmacologic pain management strategy for NICU infants. Stevens et al. (2013) indicated 57 studies with 4730 infants met criteria for inclusion in their Cochrane Review of sucrose [93]. Meta-analysis revealed that compared to controls, infants receiving sucrose had significantly lower Premature Infant Pain Profile scores and reduced total crying time post-heel stick. However, these findings did not carry-over to eye examinations. Long-term effects of sucrose from repeated administration and among infants born prior to 27 weeks, of extremely low birthweight, or medically unstable/ventilated are uncertain, and appropriate dosing for preterms is still debated [92, 97]. Naughton (2013) in an integrative review of the synergistic effects of sucrose and nonnutritive sucking concludes the combination is safe and effective and provides clinically significant relief from procedural pain in preterm neonates [97].

Positioning for Energy Conservation and Stress Reduction

Recreation of a flexed, contained fetal position for the NICU infant promotes physiological stability, preserves energy for growth, and provides increased opportunities for motor-based coping behaviors in an environment biased by gravity. Babies born preterm miss the last trimester of contained "exercise" within the intrauterine environment. Lacking intrauterine-induced flexor bias and adequate muscle power, the preterm infant presents with a gravity-dependent, "flattened" posture: external rotation and abduction of the hips, external rotation and retraction of the shoulders, flattened spinal and pelvic alignment, and forced lateral rotation of the head and neck [15]. These postures, in combination with prematurity-related muscle weakness, interfere with infant-initiated, active mobility impacting flexion attempts, midline orientation of limbs and head, and hand-to-mouth "coping" behaviors [98]. In the early hours after birth, infant head position is a primary concern as extreme lateral rotation is associated with changes in cerebral hemodynamics in babies born prior to 32 weeks of gestation.

Postural alignment and support in the NICU is provided by a variety of "positioning aids" which are as complicated as fluid-filled, moldable cushions or as simple as a rolled up blanket. When and how to use these devices depends upon infant behavioral and physiological responses as well as medical acuity. A healthy 34-week infant who is thermally regulated, growing well, able to get hands to mouth for self-soothing, and physiologically stable during position changes or cares may not need postural supports. In fact, encouraging active movement is important for exploration, communicating with caregivers, and promoting development of strength/mobility. However, a 38-week infant with significant lung disease, who struggles with work of breathing and physiological compromise during cares, benefits from the energy conservation that postural supports provide. "Individualized" positioning is key.

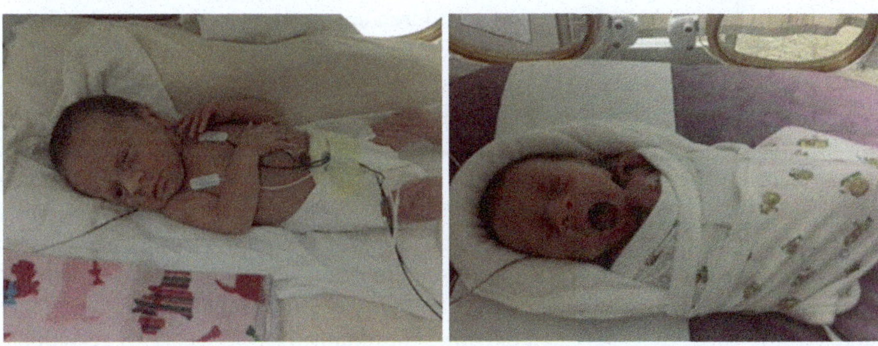

Unsupported supine positioning may cause increased energy expenditure and physiologic compromise. This infant, at 30 weeks of gestation is nestled in a gel cushion on the left. Note her loosely extended legs, her difficulty with midline head orientation, and her inability to get her hands near her face for self-soothing. On the right, when swaddled in a nesting device, her head is at midline, her arms contained near her face, and her legs maintained in flexion. This assisted containment reduces energy expenditure and promotes self-regulation

Prone Positioning: The Pros and Cons

Among the core developmental care constructs originating with Als' NIDCAP model are the importance of (1) midline, flexed positioning in all postures to promote coping behaviors and physiological stability and (2) positioning to decrease stress, conserve energy, and improve sleep and cardiorespiratory function [38]. Grenier et al. retrospectively reviewed videotapes of 15 NICU infants with a mean age of 32 weeks of gestation to evaluate which infant resting positions were associated with motor-based stress or self-regulatory behaviors [99]. Their findings, based on a lower ratio of motor-based self-regulatory to stress behavior cues, suggest that stable, preterm infants placed in prone (nested or un-nested) and side-lying nested positions are less stressed and more capable of self-regulatory strategies than when in other positions [99]. Babies in prone actually demonstrated fewer self-regulatory attempts which the authors interpret as decreased overall stress reducing the need for self-regulation, a finding that supports the energy conservation advantages of prone. Prone positioning with preterm, hospitalized infants is associated with reduced salivary cortisol levels, decreased energy expenditure, improved Brazelton sleep scores, improved chest wall synchrony during respiration, increased time in quiet sleep, and decreased time in wakeful states [98, 100, 101], all factors associated with reduced stress and improved growth parameters. Gilles, Wells, and Bhandari, in a 2012 Cochrane review of 24 studies examining the value of positioning for acute respiratory distress in a total of 581 hospitalized infants, of which 60% were preterm and 70% ventilated, conclude that (1) blood oxygenation levels are slightly higher (2% increase) prone as compared to supine; (2) on average, respiratory rate is four breaths/minute lower in prone; and (3) no adverse effects of prone position were identified [102]. However, the authors include a warning statement: "It is important to remember that these children were hospitalized. Therefore, given the association of the prone position with sudden infant death syndrome (SIDS), the prone position should NOT be used for children unless they are in hospital and their breathing is constantly monitored" (Gillies et al., 2012, p. 2).

Balancing the advantages of prone positioning with modeling of safe sleep practice remains a controversial issue for NICU developmental practice. Preterm and low birthweight babies are at a higher risk for sudden infant death syndrome (SIDS); yet, upon discharge from the NICU, they are more likely to be placed in non-supine sleep positions than term infants [103, 104]. To allow infants to become accustomed to back sleeping and to model safe sleeping practices for families prior to discharge, the American Academy of Pediatrics recommends that infants be placed supine as early as 32 weeks of gestational age. As recently as 2010, a random crib audit in a large tertiary care NICU revealed only 39% of infants over 32 weeks of gestational age were placed supine for sleep, and only 23% of parents surveyed post-discharge reported full compliance with safe sleep practices [103]. Gelfer and colleagues developed an algorithm to assist with problem-solving timing of supine sleeping practices with NICU infants. In their clinical decision-making framework, infant weight ($\geq$ 1500 g), medical factors, respiratory status, crib environment, and anticipated discharge date are factors for consideration.

Positioning for Musculoskeletal Integrity/Mobility

Positioning to preserve integrity of a rapidly developing, plastic musculoskeletal system is another focus of developmental caregiving. Differentiation of muscle tissue type, formation of articular structures, and change in active and passive muscular capacity are driven by intrinsic muscular activity and external forces during fetal and early infant development [98]. For babies born early, gravitational force and a flat extrauterine caregiving environment combined with the infant's weak, immature musculature markedly inhibits spontaneous activity. These factors are attenuated by the increased amount of extrauterine time between birth and onset of adequate antigravity muscle control (i.e., a baby born at 39 weeks develops adequate extrauterine head and limb control for a variety of antigravity movements within 3–6 weeks; a baby born at 32 weeks of gestation will be exposed to gravitational forces and dependent resting postures for twice as long before this same control develops). Without attention to varied, supportive resting positions, deformational forces on the premature infant's spine, skull, and extremities, as well as imbalances in developing muscle groups, result [105, 106]. Historically, preterm birth is associated with an increased risk for muscle tightness in neck extension, shoulder adduction/retraction, hip external rotation/abduction, and ankle eversion. All of these factors are thought to impact early eye-hand coordination and mobility skills [105, 106]. Anecdotal evidence suggests that changes in respiratory care and positioning practices appear to have reduced these postural risks. But, plagiocephaly and scaphocephaly remain of concern for premature babies. At term equivalent age, prevalence of deformational plagiocephaly is reported to be as high as 38% and scaphocephaly as high as 73% in very preterm infants [107]. Throughout an infant's hospitalization, diligent repositioning, kangaroo care, tummy time, and a variety of resting head and holding positions are key for preventing skull deformities. Babies' skulls change rapidly, growing as much as 0.79 cm in a day [20]. Even a 24 h period of inattention to resting position during a growth spurt may detrimentally influence head shape with an infant too weak to actively turn their head (Fig. 6).

Sweeney and Gutierrez detail best-practice guidelines for postural support and musculoskeletal development in the NICU. They emphasize (1) varied positions, prone, supine, supported upright, and side-lying on either side, for varying gravitational forces and resulting muscle activity patterns; (2) support of the head and trunk in neutral or slight flexion when supine or side-lying, alternating head turn, and degree of turn when prone; (3) containment of upper extremities to prevent shoulder retraction and orient arms/hands toward body/face and midline; and (4) flexion of lower extremities with neutral hip rotation/abduction [105]. Nesting supports and swaddling are methods of "containment" routinely used to accomplish these goals. Swaddling, or wrapping an infant in such a way that limbs are supported within a blanket and hands are close to face/mouth, promotes self-regulation, improves sleep, and conserves energy for focused activities such as feeding or social interaction [106].

While limb containment for energy conservation and musculoskeletal alignment is a recommended practice, complete immobilization is not. When awake and

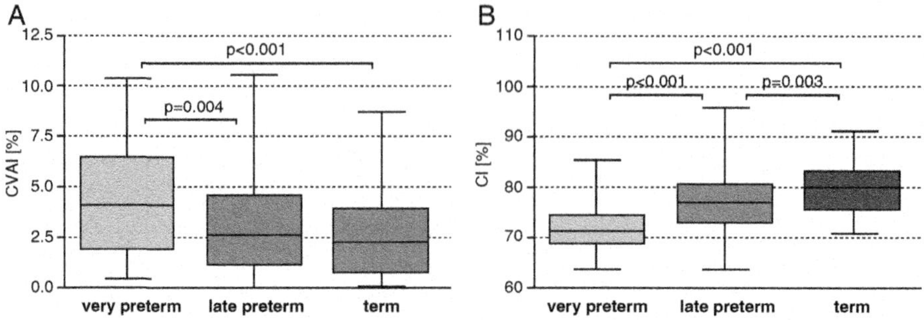

Fig. 6 Ifflaender et al. (2013) demonstrated that risk of plagiocephaly and scaphocephaly, respectively quantified by measures of cranial vault asymmetry (CVAI) and cranial index (CI), is directly related to degree of prematurity. CI is a ratio calculated by dividing the medial-lateral cranial width in mm by anterior-posterior skull width and multiplying by 100 (ML/AP × 100); CVAI is a ratio calculated by dividing the measures obtained from diagonal lines bisecting the skull 30 degrees from midline (|Diagonal A–Diagonal B × 100 ÷ Diagonal A or B (whichever is longer)). For more information regarding plagiocephaly measurement and severity classification, see https://pediatricapta.org/special-interest-groups/HB/ORTH_961942_PlagiocephalyScale_BWInfo.pdf

physically able, all babies need opportunities to spontaneously stretch, kick, and move to gain strength as well as perceptual awareness of body in space. In utero, babies demonstrate active exploration of their own body and the uterine environment through movement. They breathe, step, turn somersaults, bring their hands to their face, suck their fingers, reach for and grasp the umbilical cord, and demonstrate complex facial expressions. This spontaneous exploratory movement, though seemingly random, allows the infant to discover new tasks and solutions, laying the foundation for learning [89]. Preterm infants miss this period of prolonged practice in a contained environment which may impact not only physical strength and motor skills but cognitive development. Heathcock et al., using a mobile kicking paradigm to quantify associative learning and memory, demonstrated that 3-month-old infants born prior to 33 weeks of gestation did not learn the contingency between their own kicking and mobile movements during a 6-week intervention period [108]. In contrast, term infants demonstrate learning and memory within a single intervention session. Such findings demonstrate that subtle but significant prematurity-related learning impairments associated with movement are present at a very early age.

Passive movements, or range of motion exercises, for preventing osteopenia related to inactivity have been investigated in the premature population [109, 110]. Schulzke et al. reviewed "physical activity" findings from 8 small clinical trials including 214 infants between 26 and 34 weeks of age [110]. Physical activity groups receiving range of motion exercises and joint compression/loading were compared to controls receiving holding or tactile stimulation. While moderate short-term effects on bone mineralization upon program completion did not persist at 12 months of adjusted age, small positive effects on daily weight gain and linear growth were reported in the physical activity group. Yet long-term benefits and risks could not be evaluated due to methodological limitations and failure to include infants who were at high baseline risk of osteopenia. The authors conclude that "current evidence does not support the routine use of physical activity programs in preterm infants" [110]. In fact, preterm infants are vulnerable to musculoskeletal injury from passive movement or joint stresses. Weakness, ligamentous laxity, and increased malleability of connective tissue prior to term increase the risk of joint effusion, subluxation, or fracture from improperly applied forces [105, 106]. Passive movements of premature infants are not recommended. Instead, active, antigravity infant movement should be encouraged in a variety of positions during holding or daily cares.

Beyond Positioning: Should "Stimulation" Be Provided?

The fluid-filled intrauterine environment is rich with intersensory stimuli: proprioceptive, vestibular, tactile, gustatory, and auditory. Without this input, does the premature infant in the NICU benefit from enriched sensory experiences or stimulation? It depends on who provides the information, what type and intensity is provided, and most importantly how the infant responds physiologically and behaviorally. Random, unpredictable inputs by unfamiliar individuals carry no contingency or structured social predictability from which infants learn. Yet these same sensory experiences, provided during meaningful caregiving interactions with trusted social partners, typically Mother, may increase infant alertness and state control. The goal is not to enhance sensory inputs but to build social reciprocity and attachment. The Hospital to Home: Optimizing Infant's Environment (H-Hope) studies compared immediate and 6-week adjusted age outcomes between two maternal-infant cohorts: one receiving a standard education program and the other a twice daily, maternally administered sensory program commencing at 32 weeks of gestational age [111, 112]. The infant-directed sensory program, modulated according to infant behavioral cues, incorporated 10 min of Auditory input with "motherese," Tactile input with stroking or massage, and eye to eye maternal-infant Visual interaction, followed by 5 min of horizontal rocking or Vestibular input (ATVV). NICU infants receiving the ATVV program exhibited increased alertness, more rapid progression from gavage to complete oral feedings, better weight gain, and earlier hospital discharge [112, 113]. At 6 weeks of adjusted age, maternal-infant dyads from the ATVV group demonstrated higher mutual responsiveness; and infants demonstrated higher clarity of cues [112]. Although such findings early in development may be associated with improved social, communication, and literacy skills, generalizability from this particular study is limited as only dyads with two social risk factors were included and all participants were either African American or Latina. The Mother-Infant Transaction Program (MITP) combined parent training for reading infant cues with a 9-week NICU program of enhanced sensory input: touch, movement, massage, and KMC [114, 115]. Videotaped maternal-infant interactions at term equivalent age revealed MITP mothers demonstrated more positive affect, scored higher maternal-infant synchrony indices, and responded more appropriately to infant cues. Infants of these mothers exhibited fewer stress cues and gained weight more rapidly than control infants. Follow-up at 6 months of age displayed differences in infant communication scores favoring MITP. Long-term follow-up of cognitive outcomes has been mixed [114]. The role and effects of enhanced sensory input for hospitalized NICU infants remain unclear. Graven and colleagues, whose research has focused on the physical and developmental environment of the at-risk infant, have developed guidelines for regulation of sensory stimulation in the NICU [43, 47] (for more information regarding the highly respected, annual Graven's conference, visit www.cme.hsc.usf.edu).

Infant Massage in the NICU

An ancient parenting practice embraced by many eastern cultures to promote infant well-being, massage in the NICU is an interventional strategy proposed for reducing infant stress and improving physiologic regulation [116–118]. A range of prescriptive protocols and credentialing programs exist. All include some type of tactile input ranging from prescriptive, systematic stroking with or without kinesthetic input to light, comforting touch. In the preterm infant, positive touch is espoused for wide-ranging health benefits. Proponents cite improved circulation, bone density, and gut motility, reduction in biomarkers associated with stress such as cortisol, and increased secretion of hormones necessary for growth, immune, and gastrointestinal function [119, 120]. Yet little evidence to support these claims exists. Badr et al. [118], in a 2015 meta-analysis of preterm infant massage, reviewed 34 studies, 30 of which were randomized, controlled trials. These studies, spanning four continents, examined a variety of outcome variables including neurodevelopmental indices, length of infant stay, mother and infant behaviors, breastfeeding rates, pain, caloric intake, and measures of infant physiologic well-being. The strongest finding based on aggregate analysis was that massaged infants demonstrated "modest" but statistically significant improvements in daily weight gain (13 studies included). Results from four studies suggested statistically higher scores on the Bayley Scales Mental Developmental Index (7.89 points), but age of testing varied. Despite individual studies reporting beneficial effects or "positive trends" between massaged infants and specific outcome variables mentioned above, none of these findings reached statistical significance with meta-analysis as a result of moderate to high study heterogeneity, poor methodologic quality, and high risk of bias. Overall, evidence supporting use of massage to benefit preterm infant physiologic function, stress reduction, or positive neurodevelopmental outcomes is weak [116–118]. However, when massage is delivered by the mother and an infant's physiologic responses are closely monitored, maternal-infant attachment, responsivity, and early interactional capabilities may be strengthened. These potential effects have yet to be investigated.

Developmentally Supportive Feeding (contributed by Deb Egan, RN, BSN)

Full oral feeding is the most complex and challenging skill the NICU infant must master before discharge home [121–123]. Transition from enteral tube to oral feeding for infants born preterm is not as simple as a neuro-maturational or specific age-delineated milestone. Individual infant differences in muscle strength and tone, state regulation, endurance, and suck-swallow-breathe coordination determine "readiness" and safety for oral feeds [123]. Feeding difficulty, which may be defined as oral-motor compromise or inadequate oral intake for growth, is the most common reason for prolonged NICU hospitalization; and these difficulties often persist well beyond discharge. The incidence of post-NICU feeding problems ranges from 19 to 80% with the highest risk among babies born before 28 weeks or less than 1000 g [121]. In a recent observational cohort of 378 Dutch infants born between 24 and 36 weeks of gestation, Hoogewerf et al. reported an overall 20% incidence of parent-reported infant feeding difficulties 1–2 years post-NICU care with the Montreal Children's Hospital Feeding Scale [124]. Such feeding difficulties may include feeding refusals, prolonged feeding times, poor acceptance of textures, coughing/gagging or choking, and/or poor chewing abilities [121]. For infants born greater than 32 weeks of gestation, prolonged tube feedings (greater than 30 days) are associated with feeding concerns post-NICU discharge and for infants less than 32 weeks, lower birthweight or SGA [122].

Extremely preterm, low birthweight, or medically complex infants bring the most medical and/or developmental comorbidities to the feeding task [121]. Considering these fragile infants begin oral feeding when their physiology, anatomy, and central nervous system are still immature, the task must be approached cautiously and individualized to the infant's neurodevelopmental capabilities. Cue-based, infant-led, or co-regulated feeding practices are considered more developmentally supportive and less likely to precipitate adverse feeding outcomes than volume- or schedule-driven feeding [125]. Watson, in a 2016 Cochrane review, reported low level evidence that responsive feeding practices lead to slower weight gain in premature infants but reduced transition time from enteral tube to oral feeds [125]. As both infant expression and caregiver interpretation of infant cues (neurobehavioral, readiness, and hunger) are critical to feeding success, developmental practices that maximize parental interaction are an essential component of feeding progression. The infant must show a desire to eat or "cue" by awakening, alerting, or demonstrating motor signs of hunger (stirring, hand-to-mouth, rooting behaviors) for feeding engagement and then must maintain an alert state as well as sustain motor competence (strength, endurance, suck-swallow-breathe coordination) to complete adequate oral intake for growth [121, 125]. This "co-regulated" dynamic is challenging in the NICU caregiving environment.

Experiences before starting oral feeds may also impact feeding outcomes. If an infant has repeated negative oral experiences, he may resist eating by refusing to open his mouth, refusing to suck/swallow, or even gagging with any oral stimulus.

Multiple intubations, long-term feeding tube use, gastroesophageal reflux, or thrush also may interfere with a baby's desire to eat. Alterations in central nervous system development as a result of chronic desaturations, intraventricular hemorrhage, or congenital anomalies impact immediate and long-term feeding success. If the baby has a cleft or groove in the palate or lip, compression and suction of the nipple are reduced. If the fluid bolus is too large, the infant runs the risk of aspiration or micro aspiration, which may cause coughing, gagging, or chronic respiratory problems. The infant with bronchopulmonary dysplasia struggles to coordinate suck/swallow/breathe and may avoid feeding altogether because the physiological need to breathe exceeds the drive to eat. If feeding is difficult, tiring, or uncomfortable, an infant may choose to minimize intake and, when transition to solids is expected developmentally, may resist or even avoid textured food.

Like any complex developmental milestone, learning to feed orally takes practice. In utero, the baby has months of swallowing and sucking experience before the actual need to coordinate suck-swallow-breathe for nutritional intake. The term infant practices swallowing colostrum, a thicker fluid, at the breast for several days before breastmilk "comes in" and breastfeeding commences. When an infant is born early, or is ill at birth, the natural progression is interrupted and practice delayed. Nonnutritive sucking (NNS) during NICU tube feedings is an opportunity to practice and strengthen premature infant sucking. Meta-analysis of existing evidence by Foster and colleagues suggests that NNS decreases length of transition between gavage to full oral feeds and between start of oral feeds to full feeds and decreases overall length of NICU stay [126].

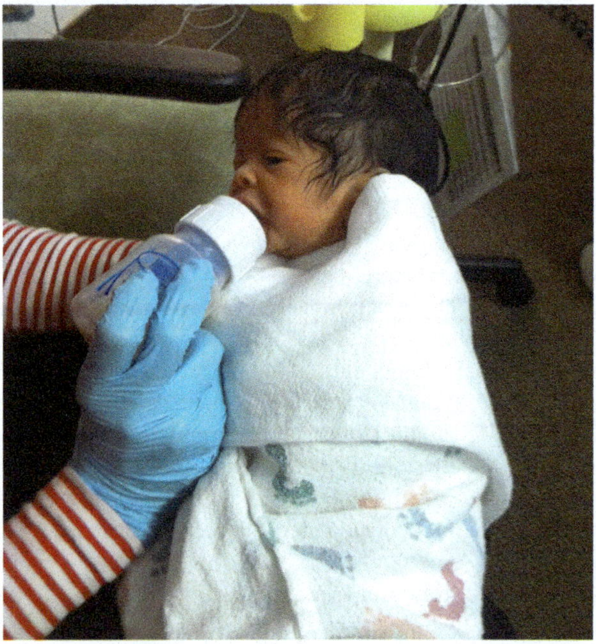

36-week-old infant in side-lying and using a low flow nipple. Transition to oral feeding is a complex process that must be guided by infant neurobehavioral and developmental "readiness"

Interventions in the NICU that may scaffold feeding success are firmly grounded in developmental care constructs. These include:

1. *Reduce stress*. Infants in the NICU experience many repetitive negative or stressful experiences. Mitigating pain or stress with developmental or pharmacological support is important, and utilizing parents for alleviation of pain or stress is ideal.
2. *Maximize relationships*. Feeding is a social, relationship-based activity. An infant needs consistent auditory, proprioceptive, and tactile input during routine caregiving interactions such as feeding to build trust and attachment. This reinforces social reciprocity and communication as much as feeding behaviors. Infants in the NICU have multiple caregivers, so maximizing family involvement is crucial for building relationships.
3. *Support breastfeeding*. From a nutritional standpoint, research overwhelmingly establishes breastmilk as optimal, but so is the act of breastfeeding [127].One major reason for this is because breastfeeding optimizes the infant's developmental environment through close proximity to the mother. When bottles are introduced, mothers do not "need" to be at the bedside as others can feed the baby. The result of maternal-infant separation is that preterm infants may be less likely to breastfeed after discharge from the NICU. There is also a common misconception that bottle feeding hastens discharge. Though milk transfer may take longer to achieve, the breast is a safer place to learn to eat for the premature baby. This is because the breast stops flowing when baby stops sucking, making breathing easier and safer. The artificial nipple continues to flow even when the baby pauses to breathe, making aspiration more likely.
4. *Respect infant "cues."* As discussed in the neurobehavioral section of this chapter, infants communicate clearly through their state, attention, autonomic, and motor behaviors. It is important to attend to and respect these cues before, during, and after oral feeding.
5. *If using bottles, initiate feeds with low flow nipples*. Excess flow increases risk of aspiration. Other important considerations for nipple choice include bottle venting, nipple firmness, temperature of the milk, bottle angle during feeding, and feeding position. Each of these variables must be individualized to promote a given infant's success.
6. *Empower parents to be primary NICU caregivers*. Although every family situation is different, individualizing the infant's feeding plan from the beginning to maximize family involvement increases the likelihood of feeding success.
7. *Utilize the developmental support staff*. Lactation, speech pathology, OT, and PT further enhance opportunities to scaffold feeding behaviors for both the parent and the infant in the NICU setting.

Family-Centered Care

Prematurity and Its Impact Upon Parents

For parents of premature infants, the unexpected timing of birth and stressful NICU hospitalization markedly increases the risk of depression, anxiety, and post-traumatic stress symptoms. Morey indicates 87% of mothers who give birth prematurely report increased arousal, avoidance, and "reexperiencing" of traumatic NICU events [128]. Adverse parenting outcomes associated with preterm birth include poor parental mental health, dysfunctional parenting, and negative parent-infant interactions which persist well beyond NICU discharge into early school years [128–130]. Parental confidence is profoundly shaken by the preterm birth experience. Early high-quality parent-infant interactions positively influence both cognitive and social infant development [131, 132]. High parenting efficacy, defined as the self-perceived capability to execute and organize tasks related to child care, is known to buffer adverse developmental effects associated with stress, depression, or compromised child development [133–135]. Parenting self-efficacy rating scales reflect actual long-term parenting competency and subsequently a parent's ability to provide a nurturing, healthy environment [134, 135]. Parents of preterm infants tend to score lower on self-efficacy rating scales, and these scores do not increase over time or with experience. While increased stress and anxiety contribute to parental self-efficacy concerns, a premature infant's temperament also plays a role. Premature infants are less likely to seek social interaction, are harder to read with regard to social cues, and demonstrate greater irritability and inconsolability and less positive emotional response [132].

Family-Centered Intervention Programs in the NICU

Infant-family relationships lay the foundation for social, emotional, and developmental well-being. Spittle et al., in their Cochrane review of early intervention programs for premature infants, state that interventions which "focus on parent-infant relationships have a greater impact on cognitive outcomes at infant and pre-school ages than interventions that focus on infant development or parent support alone" [136]. Creating Opportunities for Parent Empowerment (COPE), a NICU parental educational program that begins within a week of admission and offers family support up to 6 months following discharge demonstrated decreased infant hospital stays, decreased parental stress levels, and increased parenting self-efficacy as measured by development of critical parenting skills [129, 135]. The Family Nurture Intervention studies, which focus on building positive parent-infant interaction during hospitalization, evaluated cognitive, language, and motor infant outcomes at 18 months of corrected age [137]. Findings from parent-focused programs such as these indicate significantly increased cognitive and language scores. Clearly, fostering early family engagement is critical for promoting positive infant immediate and long-term outcomes.

Power of Parental Presence: Dual Benefits

Parental presence in the NICU when infants are smallest and sickest is strongly associated with better infant regulation, attention, and physiologic stability, attributes favorable for early learning [57, 138, 139]. For example, during kangaroo mother care, babies demonstrate better heart rate stability, increased alertness/attention, and improved patterns of deep sleep, all indicators of reduced stress and improved state regulation [140–142]. Mothers who provide kangaroo mother care or infant massage demonstrate less depressive symptoms, more attachment behaviors, and shorter latencies to joint attention with their infants [140–142]. Hence, maternal-infant engagement has a positive impact upon maternal well-being and infant learning/receptiveness. Caskey et al. reports preterm infants fed by parents instead of nursing staff exhibit greater infant social interactivity and vocalization [60]. Holditch-Davis theorizes any interventions' potential for *sustained* developmental impact *depends* upon dynamic, symbiotic parent-infant dyad change [113].

Compared with term peers, babies born prematurely have six times the odds of experiencing disorganized attachment at 36 months of age after controlling for SES and medical risk factors. This increases the risk for later psychopathology, neurodevelopmental problems, and affective disturbances [30]. For premature infants, parental proximity that fosters attachment is disrupted by infant medical fragility, the NICU environment, altered parental roles in the NICU, negative emotional responses to preterm birth, and prolonged physical separation [143]. Family-centered care in the NICU recognizes the infant develops within the context of family relationships, and these relationships must be protected and strengthened throughout hospitalization. The American Academy of Pediatrics redefined Patient and Family-Centered Care in 2012 stating that family collaboration is critical for clinical decision-making, parents are integral partners in health care, and families should be present and participatory in all aspects of care [144] (Fig. 7).

When planning and delivering health care services, honor racial, ethnic, cultural, socioeconomic and family experiences/preferences. ***Listen*** *to and* ***respect*** *each child/family.*
Ensure ***flexibility*** *in organizational policies, procedures and provider practices in order to facilitate choice and tailor services according to the needs, beliefs and cultural values of each child/ family.*
Complete, honest and unbiased information should be shared with patients and families so they might effectively participate in ***decision-making*** *at the level that they choose. Cultural and linguistic diversity should be considered when sharing health literacy information.*
Provide and ensure formal and informal for the child and family at every stage of the child's life.
Collaborate *with patients and families at all levels of health care: in the delivery of care, in professional education, policy making, program development, implementation and evaluation, in health care facility design. Include patients/families in advisory councils, committees, task forces, research, and quality improvement initiatives as leaders or co-leaders.*
Recognize and build on the ***individual strengths*** *of children and families.* ***Empower*** *them to discover their own strengths, to build confidence and participate in choices and decisions about healthcare.*

Fig. 7 Six core constructs that guide patient- and family-centered care as defined by the American Academy of Pediatrics (2012) [144]

European countries are embracing the concept of "couplet care," or admission of the family unit for the entire length of infant stay, as the next generation of family-centered care [19]. Such practices in the USA are limited by nursery design and governmental family leave policies.

Kangaroo Mother Care

Kangaroo mother care facilitates infant temperature and physiologic regulation, reduces energy expenditure, promotes growth, and improves sleep. Though typically associated with holding by mothers, fathers, too, play an important role with this skin-to-skin bonding experience in the NICU

The optimal healing environment for any infant is his mother. She provides physiological modulation, co-regulation of state and behavior, social-emotional support, and ideal, developmentally appropriate sensorimotor experiences, conditions which cannot be reconstructed in the NICU setting. Skin-to-skin, kangaroo mother care (KMC) accomplishes all these goals in one simple act. Kangarooed babies sleep better, exhibit fewer stress cues, and demonstrate better physiologic regulation and attention and fewer apnea/bradycardia episodes [12, 73]. When lying on the mother's chest, the infant has optimal developmental exposure to maternal language as well as rich intersensory and sensorimotor experiences. Frequency of parent visitation is associated with increased episodes of kangaroo care, improved breastfeeding rates, decreased length of hospitalization, and improved maternal mental health [73]. Overall, KMC is the "gold" developmental care standard improving both immediate and long-term health outcomes in preterm infants [139].

Originally developed as an "alternative approach" to traditional medical care and used primarily in "low-income" facilities with limited resources, KMC originated as a comprehensive care measure in the 1970s in Columbia [142, 145, 146]. An approximate 40% drop in Columbian infant mortality rates resulted [141]. Well-designed studies of skin-to-skin mother care demonstrate that babies as young as 28 weeks of gestation maintain thermoregulation; have fewer apnea/bradycardia episodes; demonstrate lower respiratory rates, stress hormone levels, and behavioral pain indicators; and sleep more deeply with KMC than those in incubators [140, 142, 146]. These same studies find no greater incidence of infection [146, 147]. In fact, it is hypothesized that exposure to natural maternal flora strengthens a baby's immune system [147]. Bounds et al., in a 2016 meta-analysis of 124 international KMC studies, summarize key outcomes in low birthweight, preterm infants linked to KMC: (1) 36% lower mortality; (2) decreased risk of neonatal sepsis, hypothermia, hypoglycemia, and hospital readmission; and (3) increased exclusive breastfeeding upon discharge [148]. KMC also increases measures of maternal attachment and responsiveness, decreases maternal depression indices, and increases self-reported measures of parenting competency, all important predictors for improved long-term infant developmental outcomes. With regard to international standards, the First European and Seventh International Workshop on Kangaroo Mother Care (2010) adopted this position statement: "Kangaroo Mother Care should begin as soon as possible after birth, be applied as continuous skin-to-skin contact to the extent that this is possible and appropriate and continue for as long as appropriate" [141].

Despite international recognition of KMC benefits, teasing out which outcome factors are purely related to skin-to-skin holding is difficult. Implementation of this practice, and how it is specifically defined, is still inconsistent in western, technologically advanced settings [149]. KMC practice limitations based on infant medical acuity vary and may be unrelated to developmental stage and/or gestational age of the infant. Instead, individualized indices of infant medical stability must be considered, family-readiness respected, and infant physiological cost-benefit related to transfer/positioning weighed. While definitive guidelines and high-quality evidence are still needed, KMC is considered safe and efficacious. In general, KMC should be individualized according to infant-family needs and expectations, offered and encouraged as an option for all families, initiated within the first week of life if the infant/family is medically able, and considered valuable regardless of dose or frequency.

Developmental Approaches

"Developmental approaches" combine specific elements of developmental care into standardized practice protocols. Regardless of the protocol, such approaches share common themes which are grounded in Als' synactive theories: reading infant cues, minimizing stress, modifying the environment, engaging parents, and providing developmentally appropriate sensory and learning experiences. Various authors/

entrepreneurs "wrap" best-practice NICU constructs into trademarked "packages" which, unfortunately, may be linked to "financial gain." The Neonatal Individualized Developmental Care and Assessment Program (NIDCAP) set the standard for developmental care. It is the most extensively studied and internationally promoted NICU developmental approach. Because Als, the founder of NIDCAP, laid the theoretical foundation for developmental care and because most other "approaches" borrow heavily from her tenets, this section will focus on NIDCAP.

NIDCAP Federation International, a nonprofit, incorporated, and international professional membership organization, bears responsibility for quality NIDCAP developmental care training and education [150]. While the primary focus is on naturalistic observation of infant behaviors and how this information can guide individualized care, the program mandates "all or nothing," system-wide policy, and practice implementation to achieve status as a credentialed site [36, 150–152]. Seven key components, phased in over a 5–7-year period, are required: (1) two trained and NIDCAP-certified professionals from two different disciplines, (2) two full-time equivalent salaried positions dedicated for these certified individuals, (3) a NIDCAP-trained and NIDCAP-certified multidisciplinary leadership support team, (4) a core group of NIDCAP-trained and NIDCAP-certified nursing staff from all shifts, (5) a parent council, (6) reliable and regular continuing education supporting NIDCAP efforts, and (7) development and administration of a NIDCAP Nursery Certification Program which oversees ongoing recertification processes [150–152]. Each NIDCAP-certified professional completes a mandatory 12-month training program for competency with the Assessment of Preterm Infant Behavior (APIB). The APIB is systematically used every 7–10 days to evaluate each NICU infant's behavior before, during, and after routine cares. Assessment results guide recommendations for individualized environmental modifications, positioning, and sensory experiences. The financial investment in this process, both in time and monetary resources, is substantial and often a limiting factor.

For institutions to invest in this practice and certification effort, solid, irrefutable evidence of its superiority to any other caregiving approach is necessary. Individual studies, many from invested stakeholders in the movement, report shortened length of stay, improved weight gain, and enhanced neurobehavioral and developmental outcomes that persist through early childhood [150]. Yet authors who have reviewed and analyzed the body of existing evidence argue that these studies are of mixed methodological quality as a result of small sample sizes, low statistical power, high drop-out rates, high risk of contamination or crossover bias, and compromised internal validity. Symington, in a Cochrane review of general developmental care trials, many of which were termed NIDCAP, reported mixed evidence for NIDCAP as follows: limited evidence of long-term, positive effects on behavior and movement, no evidence of effect on cognition, statistically significant effect on reduced incidence and severity of lung disease, no effect on feeding and growth parameters, and a positive effect on hospitalization cost [32]. Wallin's systematic review of NIDCAP acknowledged similar findings but concluded that solid scientific evidence was limited [36]. The most recent meta-analysis by Ohlsson and Jacobs reports on 11 primary and 7 secondary NIDCAP studies of which 2 were high quality [153].

They state there were no statistically significant differences between NIDCAP and control groups for either short-term medical or long-term neurodevelopmental outcomes. While the value of developmentally supportive care and the tremendous influence of Als and NIDCAP is widely acknowledged and embraced, the need for credentialing remains in question. Until evidence is clearly established, the packaging of specific care elements, the dosing/frequency of those elements, and the standardized manner of application may be less important than the culture shift toward family-centered, individualized, and age-appropriate, developmentally supportive care.

Developmental Testing in the NICU: Is It of Value?

Developmental assessments in the NICU may be used for a variety of reasons: (1) determining individual infant neurobehavioral strengths and vulnerabilities; (2) educating families regarding infant neurobehavioral characteristics or developmental status; (3) determining behavioral, functional, or neuromotor status of the infant; (4) assessing specific interactional or functional skills; (5) understanding infant development at a given time point or progress over time; and (6) determining the immediate need for or response to intervention [154]. The goals of assessment and physiological status of the infant typically predicate the choice of testing tool. However, clinical utility of the test is also a factor. Cost of testing materials, time to administer, training or certification requirements, and psychometric soundness (reliability, validity, sensitivity, and specificity) are important to consider. The majority of NICU developmental or motor tests require physical handling and/or position changes to quantify infant responses. Accordingly, cost/benefit of testing must be weighed relative to an infant's physiological tolerance. Infant response during testing, too, must be carefully monitored and adjusted according to infant neurobehavioral cues. Figure 8 summarizes the most common tests and measures used in the NICU for neurobehavioral or neuromotor assessment.

For fragile infants, observational, hands-off tools of spontaneous behaviors such as the assessment of preterm infant behavior may be used to understand infant response to environmental inputs/stressors or Prechtl's general movement assessment to examine neurologic integrity as a function of motor activity. For near-term, medically stable infants, tests that incorporate changes of position or movement through space like the Test of Infant Motor Performance are appropriate to determine developmental status or identify need for referral to community-based intervention programs. Many of these neonatal/early infancy tests place a heavy emphasis on tone, reflexes, and stimulus-response handling relationships: outdated constructs with regard to developmental outcomes. This type of "neuro-maturational" information, though perhaps relevant for assessment of current neurological status, correlates poorly with motor learning capabilities and acquisition of functional movement. Other limitations of existing tests include (1) most lack normative data from preterm samples which limits interpretation of results and (2), with the exception of the TIMP and GMA, most lack strong psychometric testing [154]. The majority

Name of Test	Purpose	Subscales or Categories of Assessment	Age ranges for administration	Strengths/ Limitations
Assessment of Preterm Infant Behavior (APIB)[35, 154, 156] www.nidcap.org	Discriminative; Assesses Neurobehavioral function or behavioral organization of preterm infant	Self-regulatory behavior, state, reflexes, threshold of disorganization (tactile and vestibular demands)	28 to 48 weeks gestational age	Naturalistic observation, identifies infant cues, strengths/vulnerabilities and educates families; lengthy to administer (30-60 minutes); extensive training required for administration and scoring.
Neonatal Behavioral Assessment Scale (NBAS)[34, 157] www.brazelton-institute.com	Discriminative and predictive; Assesses infant neurobehavioral function, identifies areas of infant strength/ deficit; builds relationship between infant and caregivers	Behavioral responses and reflexes; infant responsiveness, motor tone and activity, state, self-regulation, interactive capabilities; examiner support necessary for infant stability.	36 weeks to 2 months PMA.	Identifies infant cues, educates families regarding infant strengths/vulnerabilities. Flexible administration, may be incorporated into daily routines. Requires training/certification.
NICU Network Neurobehavioral Scale (NNNS)[158]	Discriminative; determines neurobehavioral function of preterm, high-risk, or drug-exposed infants	Passive/active tone and primitive reflexes; state, sensory and interactive responses; behavioral indices of stress/abstinence	30 – 46 weeks PCA	Scores at 44 weeks PMA correlate with motor outcomes at 12 to 36 months; 115 test items, 45 of which require physical manipulation of the infant; Requires certification.
General Movement Assessment (GMA) [159, 160, 161]	Discriminative and predictive; Qualitative assessment of spontaneous movement; used to evaluate early brain function	Movement quality: variety, fluidity, complexity	36 weeks PCA to 4 months	Observation of 30 to 60 minutes of videotaped movement, non-invasive. Cramped, synchronous movements at 37 to 46 weeks PMA predictive of CP. Training/certification required.
Dubowitz or Hammersmith Infant Neurological Examination (HINE)[162]	Discriminative and predictive; Assesses neurological status of infant; identifies neurological abnormalities	Tone, tone patterns, reflexes, movement, abnormal signs and behavior.	30 weeks to 4 months of age	Quick (15 minutes), practical and requires no formal training or proprietary forms for scoring; prognosticates severity of motor outcomes.
Test of Infant Motor Performance (TIMP)[163, 164] www.timp.com	Discriminative, evaluative, and predictive; Assesses functional motor behaviors; identifies risk for motor impairment and/or Cerebral Palsy; evaluates change over time or in response to interventioin	Spontaneous movement, orientation, postural and selective motor control; response to positioning/ handling, visual and auditory stimuli	34 weeks through 4 months of (adjusted) age	Shortened, screening version available (TIMPSI), Test items correlate with routine daily caregiving demands, Scores at 12 weeks PMA predictive of motor outcomes at 12 months. Self-study training or formal training required, 25 to 30 minutes to administer.

Fig. 8 A summary of developmental, neurobehavioral, or related tests and measures for NICU infants [147, 148]

of these tests, if administered prior to term or even during the first 12 weeks of early infancy, have limited predictive and discriminative validity. A high rate of false-positive findings is characteristic which results in infants being over-identified as atypical. Assessment findings in the NICU typically guide short-term treatment plans or family education about a baby's current strengths and needs, but caution regarding prognostication is imperative. National Institute of Neurological Disorders and Stroke (NINDS) strongly recommends routine use of the TIMP and GMA for early identification of movement disorders in at-risk NICU infants and the HINE for predicting severity of motor involvement [165].

Conclusion

Life course theory states that risk and protective factors combine for a dynamic health trajectory [166]. For infants in the NICU, developmental risk cannot be eliminated; however, protective factors can be employed. Lester and colleagues outline several best-practice developmental care suggestions, all of which emphasize the influence of family or mother-infant bonding [73]. First, NICU care should emphasize family-focused care. This philosophy recognizes the pivotal role of family, seeks to partner with families for individualized family-child NICU care plans, and empowers parents to advocate for their infant's unique needs. Of all the evidence regarding developmental care, one theme is robust: the NICU team that fosters family involvement fosters positive infant outcomes. Infants who experience early nurturing and high-quality social-emotional relationships are more likely to realize optimal immediate and long-term cognitive and behavioral development. Next, regardless of the specific approach, developmental NICU care seeks to minimize environmental stressors and to maximize a "healing environment" for the fragile preterm infant. Limiting noise and light, promoting energy conservation through swaddling, protecting sleep, and respecting infant neurobehavioral cues for feeding and/or interaction readiness are crucial. To once again quote *From Neurons to Neighborhoods*, a thought-provoking text regarding the effects of environment and experience upon our most valuable resource, children: *"What happens during the first months and years of life matters a lot, not because this period of development provides an indelible blueprint for adult well-being, but because it sets either a sturdy or fragile stage for what follows"* [1].

References

1. "Executive Summary." Institute of Medicine. (2000). *From neurons to neighborhoods: The science of early childhood development*. Washington, DC: The National Academies Press. https://doi.org/10.17226/9824.
2. Pitcher, J. B., Smith, A. E., Clow, A., Kenaway, D. J., & Ridding, M. C. (2012). Physiological evidence consistent with reduced neuroplasticity in human adolescents born preterm. *The Journal of Neuroscience, 32*(46), 16410–16416.

3. D'Agata, A. L., Young, E. E., Cong, X., Grasso, D. J., & McGrath, J. M. (2016). Infant medical trauma in the Neonatal Intensive Care Unit (IMTN): A proposed concept for science and practice. *Advances in Neonatal Care, 16*(4), 289–297.
4. Myers, E., & Ment, L. R. (2009). Long-term outcome of preterm infants and the role of neuroimaging. *Clinics in Perinatology, 36*, 773–789. https://doi.org/10.1016/j.clp.2009.07.008.
5. Jongbloed-Pereboom, M., Janssen, A. J., Steenbergen, B., & Nijhuis-van der Sanden, M. W. G. (2012). Motor learning and working memory in children born preterm: A systematic review. *Neuroscience and Biobehavioral Reviews, 36*, 1314–1330.
6. Spittle, A. J., Ferretti, C., Anderson, P. G., Orton, J., Eeles, A., Bates, L., Boyd, R. N., Inder, T. E., & Doyle, L. W. (2009). Improving the outcome of infants born at <30 weeks gestation: A randomized controlled trial of preventative care at home. *BMC Pediatrics, 3*(9), 73. https://doi.org/10.1186/1471-2431-9-73.
7. National Perinatal Association. *NICU Developmental Care: Position Paper*. 2008. Retrieved from www.nationalperinatal.org.
8. Craig, J. W., Glick, C., Phillips, R., Hall, S. L., Smith, J., & Browne, J. (2015). Recommendations for involving the family in developmental care of the NICU baby. *Journal of Perinatology, 35*, S5–S8.
9. Heckman, J. J. (2008). Schools, skills, and synapses. *Economic Inquiry, 46*(3), 289–324.
10. Doyle, O., Harmon, C. P., Heckman, J. J., & Tremblay, R. E. (2009). Investing in early human development: Timing and economic efficiency. *Economics and Human Biology, 7*(1), 1–6.
11. Browne, J. V. (2011). Developmental care for high-risk newborns: Emerging science, clinical application, and continuity from Newborn Intensive Care Unit to community. *Clinics in Perinatology, 38*, 719–729.
12. Browne, J. V.. New perspectives on premature infants and their parents. *Zero to Three*. 2003. 1–12. Retrieved from www.zerotothree.org.
13. McGrath, J. M., Cone, S., & Samra, H. A. (2011). Neuroprotection in the preterm infant: Further understanding of the short- and long-term implications for brain development. *Newborn and Infant Nursing Reviews, 11*, 109–112.
14. Altimier, L. B. (2015). Neuroprotective core measure 1: The healing NICU environment. *Newborn and Infant Nursing Reviews, 15*, 91–96.
15. Coughlin, M. E. (2014). *Transformative nursing in the NICU: Trauma informed age appropriate care*. New York, NY: Springer, ISBN: 978-0-8261-9657-6.
16. Als, H. (1982). Toward a synactive theory of development: Promise for the assessment of infant individuality. *Infant Mental Health Journal, 3*, 229–243.
17. Altimier, L. B., Kenner, C., & Damus, K. (2015). The Wee Care Neuroprotective NICU Program (Wee Care): The effect of a comprehensive developmental care training program on seven neuroprotective core measures for family-centered developmental care of premature neonates. *Newborn and Infant Nursing Reviews, 15*(3), 91–96.
18. Eichel, M., Coughlin, M. E., Fran, P. F., & Moreno, L. (2007). *Developmentally supportive care: An evidence-based self-study module*. Murrysville, PA: Children's Medical Ventures.
19. Westrup, B. (2015). Family-centered developmentally supportive care: The Swedish example. *Archives de Pédiatrie, 22*, 1086–1091.
20. Adolph, K. E., & Robinson, S. R.. (2015). Motor development. *Handbook of Child Psychology and Developmental Science*. 2:4:1–45.
21. Harden, B. J. (2015). *Services for families of infants and toddlers experiencing trauma: A research-to-practice brief.* Brief prepared for the Office of Planning, Research and Evaluation, Administration for Children and Families, U.S. Department of Health and Human Services.
22. Spittle, A. J., Thompson, D. K., Brown, N. C., Treyvaud, K., Cheong, J. L. Y., Lee, K. J., Pace, C. C., Olsen, J., et al. (2014). Neurobehaviour between birth and 40 weeks gestation in infants born < 30 weeks gestation and parental psychological wellbeing: Predictors of brain development and child outcomes. *BMC Pediatrics, 14*, 111, http://www.biomedcentral.com/1471-2431/14/111.
23. Smith, G. C., Gutovich, J., Smyser, C., Pineda, R., Newnham, C., Tjoeng, T. H., Vavasseur, C., Wallendorf, M., Neil, J., & Inder, T. (2011). Neonatal intensive care unit stress is associated with brain development in preterm infants. *Annals of Neurology, 70*, 541–549.

24. Smyser, C. D., Inder, T. E., Shimony, J. S., Hill, J. E., Degnan, A. J., Snyder, A. Z., & Neil, J. J. (2010). Longitudinal analysis of neural network development in preterm infants. *Cerebral Cortex, 20*, 2852–2862.
25. Smyser, C. D., Snyder, A. Z., & Neil, J. J. (2011). Functional connectivity MRI in infants: Exploration of the functional organization of the developing brain. *NeuroImage, 56*, 1437–1452.
26. Fernandez, E. F., Montman, R., & Watterberg, K. L. (2008). ACTH and cortisol response to critical illness in term and late preterm newborns. *Journal of Perinatology, 28*, 797–802.
27. Hamoudi, A., Murray, D. W., Sorensen, L., & Fontaine, A. (2015). *Self-regulation and toxic stress: A review of ecological, biological, and developmental studies of self-regulation and stress*. OPRE Report # 2015–30, Washington, DC: Office of Planning, Research and Evaluation, Administration for Children and Families, U.S. Department of Health and Human Services.
28. Maroney, D. I. (2003). Recognizing the potential effect of stress and trauma on premature infants in the NICU: How outcomes are affected? *Journal of Perinatology, 23*, 679–683.
29. Lobo, M. A., Kokkoni, E., Cunha, A. B., & Galloway, J. C. (2015). Infants born preterm demonstrate impaired object exploration behaviors throughout infancy and toddlerhood. *Physical Therapy, 95*(1), 51–64.
30. Pennestri, M. H., Gaudreaua, H., Bouvette-Turcot, A. A., Mosss, E., Lecompte, V., Atkinson, L., Lydon, J., Steiner, M., & Meaney, M. J. (2015). Attachment disorganization among children in neonatal intensive care unit: Preliminary results. *Early Human Development, 91*, 601–606.
31. Kiechl-Kohlendorfer, U., Merkle, U., Deufert, D., Neubauer, V., Peglow, U. P., & Griesmaier, E. (2015). Effect of developmental care for very premature infants on neurodevelopmental outcome at 2 years of age. *Infant Behavior & Development, 39*, 166–172.
32. Symington, A. J., & Pinelli, J. (2006). Developmental care for promoting development and preventing morbidity in preterm infants. *Cochrane Database of Systematic Reviews, 2*, CD001814. https://doi.org/10.1002/14651858.CD001814.pub2.
33. Montirosso, R., Del Prete, A., Bellu, R., Tronick, E., & Borgatti, R. (2012). Level of NICU quality of developmental care and neurobehavioral performance in very preterm infants. *Pediatrics, 129*(5), e1129–e1137.
34. Brazelton, T. B., & Nugent, J. K. (2011). *Neonatal behavioral assessment scale*. London: Mac Keith Press.
35. Als, H., Butler, S., Kosta, S., & McAnulty, G. (2005). The Assessment of Preterm Infants' Behavior (APIB): Furthering the understanding and measurement of neurodevelopmental competence in preterm and full-term infants. *Mental Retardation and Developmental Disabilities Research Reviews, 11*(1), 94–102. https://doi.org/10.1002/mrdd.20053.
36. Wallin, L., & Eriksson, M. (2009). Newborn Individual Development Care and Assessment Program (NIDCAP): A systematic review of the literature. *Worldviews on Evidence-Based Nursing, 6*(2), 54–69.
37. Sweeney, J. K., Heriza, C. B., Blanchard, Y., & Dusing, S. C. (2010). Neonatal physical therapy. Part II: Practice frameworks and evidence-based practice guidelines. *Pediatric Physical Therapy, 22*, 2–11.
38. Lowman, L. B., Stone, L. L., & Cole, J. G. (2006). Using developmental assessments in the NICU to empower families. *Neonatal Network, 25*(3), 177–185.
39. Skene, C., Franck, L., Curtis, P., & Gerrish, K. (2012). Parental involvement in neonatal comfort care. *Journal of Obstetric, Gynecologic, & Neonatal Nursing, 41*, 786–797. https://doi.org/10.1111/j.1552-6909.2012.01393.x.
40. Coughlin, M., Gibbins, S., & Hoath, S. (2009). Core measures for developmentally supportive care in neonatal intensive care units: Theory, precedence and practice. *Journal of Advanced Nursing, 65*, 2239–2248. https://doi.org/10.1111/j.1365-2648.2009.05052.x.
41. Phillips, R. M. (2015). Seven core measures of neuroprotective family-centered developmental care: Creating an infrastructure for implementation. *Newborn and Infant Nursing Reviews, 15*, 87–90.

42. White, R. D. (2015). Neuroprotective core measure 4: Safeguarding sleep—Its value in neuroprotection of the newborn. *Newborn and Infant Nursing Reviews, 15*, 114–115.
43. Graven, S. N. (2000). The full-term and premature newborn: Sound and the developing infant in the NICU: Conclusions and recommendations for care. *Journal of Perinatology, 20*, S88–S93.
44. Weisman, O., Magori-Cohen, R., Louzoun, Y., Eidelman, A. I., & Feldman, R. (2011). Sleep-wake transitions in premature neonates predict early development. *Pediatrics, 128*, 706. https://doi.org/10.1542/peds.2011-0047.
45. Calciolari, G., & Montirosso, R. (2011). The sleep protection in the preterm infants. *The Journal of Maternal-Fetal and Neonatal Medicine, 24*(S(1)), 12–14.
46. Levy, J., Hassan, F., Plegue, M. A., Sokoloff, M. D., Kushwaha, J. S., Chervin, R. D., Barks, J. D. E., & Shellhaas, R. A. (2016). Impact of hands-on care on infant sleep in the Neonatal Intensive Care Unit. *Pediatric Pulmonology*, 1–7. https://doi.org/10.1002/ppul.23513.
47. Graven, S. N. (2011). Early visual development: Implications for the neonatal intensive care unit and care. *Clinics in Perinatology, 38*, 671–683.
48. Tarullo, A. R., Balsam, P. D., & Fifer, W. P. (2011). Sleep and infant learning. *Infant and Child Development, 20*, 35–46.
49. Geva R, Yaron H and Kuint J. Neonatal sleep predicts attention orienting and distractibility. *Journal of Attention Disorders*. Published online 26 July 2013. https://doi.org/10.1177/1087054713491493. Retrieved from http://jad.sagepub.com/content/early/2013/07/26/1087054713491493.
50. Brown, G. (2009). NICU noise and the preterm infant. *Neonatal Network, 28*(3), 165–173.
51. Shahheidari, M., & Homer, C. (2012). Impact of the design of Neonatal Intensive Care units on neonates, staff and families: A systematic literature review. *The Journal of Perinatal & Neonatal Nursing, 26*(3), 260–266.
52. EPA.gov[Internet].UnitedStatesGovernment,WashingtonDC:[updated2014May9;cited2014 Nov 15]. *EPA identifies noise levels affecting health and welfare*; [one screen]. Retrieved from http://www2.epa.gov/aboutepa/epa-identifies-noise-levels-affecting-health-and-welfare.
53. Bremmer, P., Byers, J. F., & Kiehl, E. (2003). Noise and the premature infant: Physiological effects and practice implications. *Journal of Obstetric, Gynecologic, and Neonatal Nursing, 32*, 447–454.
54. Freudenthal, A., van Stuijvenberg, M., & van Goudoever, J. B. (2013). A quiet NICU for improved infants' health, development and well-being: A systems approach to reducing noise and auditory alarms. *Cognition, Technology & Work, 15*, 329–345.
55. Laubach, V., Wilhelm, P., & Carter, K. (2014). Shhhh....I'm growing: Noise in the NICU. *Nursing Clinics of North America, 49*, 329–344.
56. Lasky, R. E., & Williams, A. L. (2009). Noise and light exposures for extremely low birth weight newborns during their stay in the neonatal intensive care unit. *Pediatrics, 123*, 540–546.
57. Picciolini, O., Porro, M., Meazza, A., Gianni, M. L., Rivoli, C., Lucco, G., et al. (2014). Early exposure to maternal voice: Effects on preterm infants' development. *Early Human Development, 90*, 287–292.
58. Stromswold, K., & Sheffield, E. (2004). *Neonatal intensive care unit noise and language development*. Rutgers University Center for Cognitive Science Technical Report. 1–15.
59. McMahon, E., Wintermark, P., & Lahav, A. (2012). Auditory brain development in premature infants: The importance of early experience. *Annals of the New York Academy of Sciences, 1252*, 17–24.
60. Caskey, M., Stephens, B., Tucker, R., & Vohr, B. (2011). Importance of parent talk on the development of preterm infant vocalizations. *Pediatrics, 128*, 910–916.
61. Chang, E. F., & Merzenich, M. M. (2003). Environmental noise retards auditory cortical development. *Science, 300*, 498–502.
62. Kesar, A. G. (2014). Effect of prenatal chronic noise exposure on the growth and development of the body and brain of chick embryo. *International Journal of Applied and Basic Medical Research, 4*(1), 3–6.

63. American Academy of Pediatrics, Committee on Environmental Health. (1997). Noise: A hazard for the fetus and newborn. *Pediatrics, 100*(4), 724–727.
64. Rand, K., & Lahav, A. (2014). Maternal sounds elicit lower heart rate in preterm newborns in the first month of life. *Early Human Development, 90*, 679–683.
65. Caskey, M., Stephens, B., Tucker, R., & Vohr, B. (2014). Adult talk in the NICU with preterm infants and developmental outcomes. *Pediatrics, 133*, e578–e584. www.pediatrics.org/cgi/doi/10.1542/peds.2013-0104.
66. Loewy, J., Steward, K., Dassler, A. M., Telsey, A., & Homel, P. (2013). The effects of music therapy on vital signs, feeding, and sleep in premature infants. *Pediatrics, 131*(5), 902–918. www.pediatrics.org/cgi/doi/10.1542/peds.2012-1367.
67. Standley, J. J. (2012). Music therapy research in the NICU: An updated meta-analysis. *Neonatal Network, 31*(5), 311–316.
68. Bieleninik, L., Ghetti, C., & Gold, C. (2016). Music therapy for preterm infants and their parents: A meta-analysis. *Pediatrics, 138*(3), 1–17.
69. Courtnage, A., Chawla, H., Loewy, J., & Nolan, P. (2002). Effects of live, infant-directed singing on oxygen saturation, heart rates and respiratory rates of infants in the neonatal intensive care unit. *Pediatric Research, 2346*(51), 403A.
70. Nakata, T., & Trehub, S. E. (2004). Infants' responsiveness to maternal speech and singing. *Infant Behavior & Development, 27*, 455–456.
71. Domanico, R., Davis, D. K., Coleman, F., & Davis, B. O. (2011). Documenting the NICU design dilemma: Comparative patient progress in open-ward and single family room units. *Journal of Perinatology, 31*, 281–288.
72. Strauch, C., Brandt, S., & Edwards-Beckett, J. (1993). Implementation of a quiet hour: Effect on noise levels and infant sleep states. *Neonatal Network, 12*(2), 31–35.
73. Lester, B. M., Miller, R. J., Hawes, K., Salisbury, A., Bigsby, R., Sullivan, M. C., & Padbury, J. F. (2011). Infant neurobehavioral development. *Seminars in Perinatology, 35*, 8–19.
74. Goines, L. (2008). The importance of quiet in the home: Teaching noise awareness before the infant is discharged from the NICU. *Neonatal Network, 27*(3), 171–176.
75. Stevens, D., Thompson, P., Helseth, C., & Pottala, J. (2015). Mounting evidence favoring single-family room neonatal intensive care. *Journal of Neonatal-Perinatal Medicine, 8*, 177–178.
76. Lester, B. M., Hawes, K., Abar, B., Sullivan, M., Miller, R., Bigsby, R., et al. (2014). Single-family room care and neurobehavioral and medical outcomes in preterm infants. *Pediatrics, 134*(4), 754–760.
77. Lester, B. M., Salisbury, A. L., Hawes, K., Dansereau, L. M., Bigsby, R., Laptook, A., Taub, M., Lagasse, L. L., Vohr, B. R., & Padbury, J. F. (2016). 18-month follow-up of infants cared for in a single-family room Neonatal Intensive Care Unit. *Journal of Pediatrics, 177*, 84–89.
78. Shepley, M., Smith, J. A., Sadler, B. L., & White, R. D. (2014). The business case for building better neonatal intensive care units. *Journal of Perinatology, 34*(11), 811–815.
79. Pineda, R. G., Neil, J., Dierker, D., Smyser, C. D., Wallendorf, M., Kidokoro, H., et al. (2014). Alterations in brain structure and neurodevelopmental outcome in preterm infants hospitalized in different neonatal intensive care unit environments. *The Journal of Pediatrics, 164*(1), 52–60. e2.
80. Rodriguez, R. G., & Pattini, A. E. (2016). Neonatal intensive care unit lighting: Update and recommendations. *Archivos Argentinos de Pediatría, 114*(4), 361–367.
81. Taylor, N. M., Jakobson, L. S., Maurer, D., & Lewis, T. L. (2009). Differential vulnerability of global motion, global form, and biological motion processing in full-term and preterm children. *Neuropsychologia, 47*(13), 2766–2778.
82. Tremblay, E., Vannasing, P., Roy, M.-S., Lefebvre, F., Kombate, D., et al. (2014). Delayed early primary visual pathway development in premature infants: High density electrophysiological evidence. *PLoS One, 9*(9), e107992. https://doi.org/10.1371/journal.pone.0107992.
83. Di Rosa, G., Cavallaro, T., Alibrandi, A., Marseglia, L., Lamberti, M., Giaimo, E., Nicotera, A., Bonsignore, M., & Gagliano, A. (2016). Predictive role of early milestones-related

psychomotor profiles and long-term neurodevelopmental pitfalls in preterm infants. *Early Human Development, 101*, 49–55.
84. Lickliter, R. (1990). Premature visual experience accelerates intersensory functioning in bobwhite quail neonates. *Developmental Psychobiology, 23*, 15–27.
85. Lickliter, R. (1993). Timing and the development of perinatal perceptual organization. In G. Turkewitz & D. A. Devenny (Eds.), *Developmental time and timing* (pp. 105–123). Hillsdale, NJ: Erlbaum.
86. Lickliter, R. (2011). The integrated development of sensory organization. *Clinics in Perinatology, 38*(4), 591–603.
87. Von Hofsten, C., Gredeback, G., & Johnson, S. (2010). Eye tracking in infancy research. *Developmental Neuropsychology, 35*(1), 1–19.
88. Rose, S. A., Feldman, J. F., & Jankowski, J. J. (2001). Attention and recognition memory in the 1st year of life: A longitudinal study of preterm and full-term infants. *Developmental Psychology, 31*, 135–151.
89. Smith, L., & Gasser, M. (2005). The development of embodied cognition: Six lessons from babies. *Artificial Life, 11*(1–2), 13–29.
90. Holsti, L., Grunau, R. E., Oberlander, T. F., & Whitfield, M. F. (2005). Prior pain induces heightened motor response during clustered care in preterm infants in the NICU. *Early Human Development, 81*, 293–302.
91. Marchant, A. (2014). 'Neonates do not feel pain': A critical review of the evidence. *Bioscience Horizons, 7*, hzu006. https://doi.org/10.1093/biohorizons/hzu006.
92. Murki, S. & Subramanian, S.. Sucrose for analgesia in newborn infants undergoing painful procedures: RHL commentary (last revised: 1 June 2011). *The WHO Reproductive Health Library*; Geneva: World Health Organization.
93. Stevens, B., McGrath, P., Gibbins, S., et al. (2003). Procedural pain in newborns at risk for neurology impairment. *Pain, 105*(1–2), 27–35.
94. Pillai Riddell, R. R., Racine, N. M., Gennis, H. G., Turcotte, K., Uman, L. S., Horton, R. E., Ahola Kohut, S., Hillgrove Stuart, J., Stevens, B., & Lisi, D. M.. Non-pharmacological management of infant and young child procedural pain. *Cochrane Database of Systematic Reviews*. 2015, Issue 12. No.: CD006275. https://doi.org/10.1002/14651858.CD006275.pub3.
95. Harrison, D., Reszel, J., Wilding, J., Abdulla, K., Bueno, M., Campbell-Yeo, M., Dunn, S., Harrold, J., Nicholls, S., Squires, J., & Stevens, B. (2015). Neuroprotective core measure 5: Minimizing stress and pain—Neonatal pain management practices during heel lance and venipuncture in Ontario, Canada. *Newborn and Infant Nursing Reviews, 15*, 116–123.
96. Johnston, C., Campbell-Yeo, M., Fernandes, A., Inglis, D., Streiner, D., & Zee R. (2014). Skin-to-skin care for procedural pain in neonates. *Cochrane Database of Systematic Reviews*. Issue 1. Art. No.: CD008435. https://doi.org/10.1002/14651858.CD008435.pub2.
97. Naughton, K. A. (2013). The combined use of sucrose and nonnutritive sucking for procedural pain in both term and preterm neonates: An integrative review of the literature. *Advances in Neonatal Care, 13*(1), 9–19.
98. Monterosso, L., Kristjanson, L., & Cole, J. (2002). Neuromotor development and the physiological effects of positioning in very low birth weight infants. *Journal of Obstetric, Gynecologic, & Neonatal Nursing, 31*, 138–146.
99. Grenier, I. R., Bigsby, R., Vergara, E. R., & Lester, B. M. (2003). Comparison of motor self-regulatory and stress behaviors of preterm infants across body positions. *American Journal of Occupational Therapy, 57*, 289–297.
100. Jarusa, T., Bart, O., Rabinovich, G., Sadeh, A., Bloch, L., Dolfin, T., & Litmanovitz, I. (2011). Effects of prone and supine positions on sleep state and stress responses in preterm infants. *Infant Behavior & Development, 34*, 257–263.
101. Masterson, J., Zucker, C., & Schulze, K. (1987). Prone and supine positioning effects on energy expenditure and behavior of low birth weight neonates. *Pediatrics, 80*(5), 689–692.
102. Gillies, D., Wells, D., & Bhandari, A. P. (2012). Positioning for acute respiratory distress in hospitalised infants and children. *Cochrane Database of Systematic Reviews, 7*, CD003645. https://doi.org/10.1002/14651858.CD003645.pub3.

103. Gelfer, P., Cameron, R., Masters, K., & Kennedy, K. A. (2013). Integrating "back to sleep" recommendations into Neonatal ICU practice. *Pediatrics, 131*(4), e1264–e1270. https://doi.org/10.1542/peds.2012-1857.
104. American Academy of Pediatrics. (2011). Technical Report. SIDS and other sleep-related infant deaths: Expansion of recommendations for a safe infant sleeping environment. *Pediatrics, 128*, e1341–e1367, www.pediatrics.org/cgi/doi/10.1542/peds.2011-2285.
105. Sweeney, J. K., & Gutierrez, T. (2002). Musculoskeletal implications of preterm infant positioning in the NICU. *The Journal of Perinatal & Neonatal Nursing, 16*(1), 58–70.
106. Byrne, E., & Garber, J. (2013). Physical therapy intervention in the neonatal intensive care unit. *Physical & Occupational Therapy in Pediatrics, 33*(1), 75–110. https://doi.org/10.3109/01942638.2012.750870.
107. Ifflaender, S., Rudiger, M., Konstantelos, D., Wahls, K., & Burkhardt, W. (2013). Prevalence of head deformities in preterm infants at term. *Early Human Development, 89*, 1041–1047.
108. Heathcock, J., Bhat, A., Lobo, M., & Galloway, J. (2004). The performance of infants born preterm and full-term in the mobile paradigm: Learning and memory. *Physical Therapy, 84*(9), 808–821.
109. Erdem, E., Tosun, Ö., Bayat, M., et al. (2015). Daily physical activity in low risk extremely low birth weight preterm infants: Positive impact onbone mineral density and anthropometric measurement. *Journal of Bone and Mineral Metabolism, 33*, 329. https://doi.org/10.1007/s00774-014-0594-6.
110. Schulzke, S. M., Trachsel, D., & Patole, S. K.. (2007). Physical activity programs for promoting bone mineralization and growth in preterm infants. *Cochrane Database of Systematic Reviews*, Issue 2. Art. No.: CD005387. https://doi.org/10.1002/14651858.CD005387.pub2.
111. White-Traut, R. C., Norr, K. F., Fabiyi, C., Rankin, K. M., Zhyouing, L., & Liu, L. (2013). Mother-infant interaction improves with a developmental intervention for mother-preterm infant dyads. *Infant Behavior & Development, 36*(4), 694–706. https://doi.org/10.1016/j.infbeh.2013.07.004.
112. White-Traut, R. C., Rankin, K. M., Yoder, J. C., Liu, L., Vasa, R., Geraldo, V., & Norr, K. F. (2015). Influence of H-HOPE intervention for premature infants on growth, feeding progression, and length of stay during initial hospitalization. *Journal of Perinatology, 35*(8), 636–641. https://doi.org/10.1038/jp.2015.11.
113. Holditch-Davis, D., White-Traut, R. C., Levy, J. A., O'Shea, M., Geraldo, V., & David, R. J. (2014). Maternally administered interventions for preterm infants in the NICU: Effects on maternal psychological distress and mother–infant relationship. *Infant Behavior & Development, 37*, 695–710.
114. Hauglann, L., Handegaard, B. H., Ulvund, S. E., Nordhov, M., Ronning, J. A., & Kaaresen, P. I. (2015). Cognitive outcome of early intervention in preterms at 7 and 9 years of age: A randomized controlled trial. *Archives of Disease in Childhood. Fetal and Neonatal Edition, 100*, F11–F16. https://doi.org/10.1136/archdischild-2014-306496.
115. Milgrom, J., Newnham, C., Martin, P. R., Anderson, P. J., Doyle, L. W., Hunt, R. W., Achenbach, T. M., Ferretti, C., Holt, C. H., Inder, T. E., & Gemmill, A. W. (2013). Early communication in preterm infants following intervention in the NICU. *Early Human Development, 89*, 755–762.
116. Vickers, A., Ohlsson, A., Lacy, J. B., & Horsley, A. (2004). Massage for promoting growth and development of preterm and/or low birth-weight infants. *Cochrane Database of Systematic Reviews, 2*, CD00390.
117. Wang, L., He, J. L., & Zhang, X. H. (2013). The efficacy of massage on preterm infants: A meta-analysis. *American Journal of Perinatology, 30*(9), 731–738.
118. Badr, K., Abduallah, B., & Kahale, L. (2015). A meta-analysis of pre-term infant massage: An ancient practice with contemporary applications. *Maternal Child Nursing, 40*(6), 344–358.
119. Smith, J. R. (2012). Comforting touch in the very preterm hospitalized infant: An integrative review. *Advances in Neonatal Care, 12*(6), 349–365.
120. Choi, H. J., Kim, S. J., Oh, J., Lee, M. N., Kim, S. H., & Kang, K. A. (2016). The effects of massage therapy on physical growth and gastrointestinal function in premature infants: A pilot study. *Journal of Child Health Care, 20*(3), 394–404.

121. Shaker, C. S. (2017). Infant guided, co-regulated feeding in the Neonatal Intensive Care Unit. Part I: Theoretical underpinnings for neuroprotection and safety. *Seminars in Speech and Language, 38*, 96–105.
122. Bell, E. F., Stark, A. R., Adamkin, D. H., et al. (2008). American Academy of Pediatrics Committee on Fetus and Newborn. Hospital discharge of the high-risk neonate. *Pediatrics, 122*(5), 1119–1126.
123. Lubbe, W. (2017). Clinician guide for cue-based transition to oral feeding in preterm infants: An easy-to-use clinical guide. *Journal of Evaluation in Clinical Practice.* 1–9. Retrieved from wileyonlinlibrary.com/journal/jep.
124. Hoogewerf, M., ter Horst, H. J., Groen, H., Nieuwenhuis, T., Box, A. F., & van Dijk, M. W. G. (2017). The prevalence of feeding problems in children formerly treated in a neonatal intensive care unit. *Journal of Perinatology, 37*, 578–584.
125. Watson J., & McGuire, W. (2016). Responsive versus scheduled feeding for preterm infants. *Cochrane Database of Systematic Reviews.* Issue 8. Art. No.: CD005255. https://doi.org/10.1002/14651858.CD005255.pub5.
126. Foster, J. P., Psaila, K., & Patterson, T.(2016). Non-nutritive sucking for increasing physiologic stability and nutrition in preterm infants. *Cochrane Database of Systematic Reviews*, Issue 10. Art. No.: CD001071. https://doi.org/10.1002/14651858.CD001071.pub3.
127. Eidelman, A. I., et al. (2012). Policy statement: Breastfeeding and the use of human milk. *Pediatrics, 129*, e287. https://doi.org/10.1542/peds.2011-3552.
128. Morey, J. A., & Gregory, K. (2012). Nurse-led education mitigates maternal stress and enhances knowledge in the NICU. *Maternal Child Nursing, 37*(3), 182–191.
129. Melnyk, B. M., Crean, H. F., Feinstein, N. F., & Fairbanks, E. (2008). Maternal anxiety and depression following a premature infants' discharge from the NICU: Explanatory effects of the COPE program. *Nursing Research, 57*(6), 383–394.
130. Winstanley, A., Sperotto, R. G., Putnick, D. L., Cheriand, S., Bornstein, M. H., & Gattis, M. (2014). Consistency of maternal cognitions and principles across the first five months following preterm and term deliveries. *Infant Behavior & Development, 37*, 760–771.
131. Wade, M., Madigan, S., Akbari, E., & Jenkins, J. M. (2015). Cumulative biomedical risk and social cognition in the second year of life: Prediction and moderation by responsive parenting. *Frontiers in Psychology, 6*, 354. https://doi.org/10.3389/fpsyg.2015.00354.
132. Schwartz, M. K. (2005). Parenting pre-term infants: A meta-analysis. *Maternal Child Nursing, 30*(2), 115–120.
133. Pennella, C., Whittingham, K., Boyd, R., Sanders, M., & Colditz, P. (2012). Prematurity and parental self-efficacy: The preterm parenting & self-efficacy checklist. *Infant Behavior & Development, 35*, 678–688.
134. Melnyk, B. M., Oswalt, K. L., & Sidora-Arcoleo, K. (2014). Validation and psychometric properties of the neonatal intensive care unit parental beliefs scale. *Nursing Research, 63*(2), 105–115.
135. Melnyk, B. M., Crean, H. F., Feinstein, N. F., & Fairbanks, E. (2007). Testing the theoretical framework of the COPE program for mothers of critically ill children: An integrative model of young children's post-hospital adjustment behaviors. *Journal of Pediatric Psychology, 32*(4), 463–474.
136. Spittle, A., Orton, J., Anderson, P., Boyd, R., & Doyle, L. W. (2012). Early developmental intervention programmes post-hospital discharge to prevent motor and cognitive impairments in preterm infants. *Cochrane Database of Systematic Reviews, 12*, 1–65.
137. Welch, M. G., Firestein, M. R., Austin, J., Hane, A. A., Stark, R. I., Hofer, M. A., Garland, M., Glickstein, S. B., Brunelli, S. A., Ludwig, R. J., & Myers, M. M. (2015). Nurture intervention in the neonatal intensive care unit improves social-relatedness, attention, and neurodevelopment of preterm infants at 18 months in a randomized controlled trial. *Journal of Child Psychology and Psychiatry, 56*, 1202–1211.
138. Dearn, T., & Shoemark, H. (2014). The effect of maternal presence on premature infant response to recorded music. *Journal of Obstetric, Gynecologic, & Neonatal Nursing, 43*, 341–350.

139. Engmann, C., Wall, S., Darmstadt, G., Valsangkar, B., & Claeson, M.. (2013). Consensus on kangaroo mother care acceleration. *The Lancet* [Internet]. Nov [cited 2014 Sept 28]; 382: [1 p]. Retrieved from https://doi.org/10.1016/S0140-6736(13)62293-X.
140. Conde-Agudelo, A., & Díaz-Rossello, J. L. (2014). Kangaroo mother care to reduce morbidity and mortality in low birthweight infants. *Cochrane Database of Systematic Reviews*, Issue 4. Art. No.: CD002771. https://doi.org/10.1002/14651858.CD002771.pub3.
141. Nyqvist, K. H., Anderson, G. C., Bergman, N., Cattaneo, A., Charpak, N., Davanzo, R., et al. (2010). Towards universal Kangaroo Mother Care: Recommendations and report from the First European conference and Seventh International Workshop on Kangaroo Mother Care. *Acta Paediatrica, 99*, 820–826.
142. Salam, R. A., Mansoor, T., Mallick, D., Lassi, Z. S., Das, J. K., & Bhutta, Z. A. (2014). Essential childbirth and post-natal interventions for improved maternal and neonatal health. *Reprod Health* [Internet] [cited 2014 Sept 28]; 11(Suppl 1): [17 pages]. Retrieved from http://www.reproductive-health-journal.com/content/11/S1/S3.
143. Vanderveen, J. A., Bassler, D., Roberson, C. M. T., & Kirpalani, H. (2009). Early interventions involving parents to improve neurodevelopmental outcomes of premature infants: A meta-analysis. *Journal of Perinatology, 29*, 343–351.
144. American Academy of Pediatrics, Policy Statement. (2012). Patient- and family-centered care and the pediatrician's role. *Pediatrics, 129*, 394–404.
145. Bergman, N. J. (2015). Neuroprotective core measures 1-7: Neuroprotection of skin-to-skin contact (SSC). *Newborn and Infant Nursing Reviews, 15*, 142–146.
146. Davanzo, R., Brovedani, P., Travan, L., Kennedy, J., Crocetta, A., Sanesi, C., Strajn, T., & De Cunto, A. (2013). Intermittent kangaroo mother care: A NICU protocol. *Journal of Human Lactation, 29*, 332–338.
147. Park, H., Choi, B. S., Lee, S. J., Son, I., Seol, I., & Lee, H. J. (2014). Practical application of kangaroo mother care in preterm infants: Clinical characteristics and safety of kangaroo mother care. *Journal of Perinatal Medicine, 42*(2), 239–245.
148. Boundy, E. O., Dastjerdi, R., Spiegelman, D., Fawzi, W. W., Missmer, S. A., Lieberman, E., Kajeepeta, S., Wall, S., & Chan, G. J. (2016). Kangaroo mother care and neonatal outcomes: A meta-analysis. *Pediatrics, 137*(1), e20152238.
149. Chan, G. J., Valsangkar, B., Kajeepeta, S., Boundy, E. O., & Wall, S. (2016). What is kangaroo mother care? Systematic review of the literature. *Journal of Global Health, 6*(1), 010701.
150. NIDCAP Federation International. Retrieved from www.nidcap.org/about.aspx.
151. Lawhon, G., & Hedlund, R. E. (2008). Newborn individualized developmental care and assessment program training and education. *The Journal of Perinatal & Neonatal Nursing, 22*(2), 133–144.
152. Smith, K., Buehler, D., Hedlund, R., Kosta, S. (2011). NIDCAP Nursery Certification Program (NNCP). *A guide to preparation, application and implementation of NIDCAP Nursery Certification.* NIDCAP Federation International. http://50.28.28.219/wp-content/uploads/2013/12/NNCP-Guide-7Jul11.pdf.
153. Ohlsson, A., & Jacobs, S. E. (2013). NIDCAP: A systematic review and meta-analyses of randomized controlled trials. *Pediatrics, 131*, e881–e893.
154. Noble, Y., & Boyd, R. (2012). Neonatal assessments for the preterm infant up to 4 months corrected age: A systematic review. *Developmental Medicine and Child Neurology, 54*, 129–139.
155. El-Dib, M., Massaro, A. N., Glass, P., & Aly, H. (2011). Neurodevelopmental assessment of the newborn: An opportunity for prediction of outcome. *Brain & Development, 33*, 95–105.
156. Als, H., Lester, B. M., Tronick, E. Z., & Brazelton, T. B. (1982). Manual for the assessment of preterm infant's behavior (APIB). In H. E. Fitzgerald, B. M. Lester, & M. W. Yogman (Eds.), *Theory and research in behavioral pediatrics* (pp. 35–63). New York: Plenum Press.
157. Bartram, S. C., Barlow, J., Wolke, D. (2015). The Neonatal Behavioral Assessment Scale (NBAS) and Newborn Behavioral Observations system (NBO) for supporting caregivers and improving outcomes in caregivers and their infants. *Cochrane Database of Systematic Reviews*, Issue 6. Art. No.: CD011754. https://doi.org/10.1002/14651858.CD011754.

158. Lester, B. M., Andreozzi-Fontaine, L., Tronick, E., & Bigsby, R. (2014). Assessment and evaluation of the high risk neonate: The NICU network neurobehavioral scale. *Journal of Visualized Experiments, 90*, e3368. https://doi.org/10.3791/3368.
159. Einspieler, C., Prechtl, H. F. R., Bos, A., et al. (2004). *The qualitative assessment of general movement in preterm, term and young infants*. Mac Keith: London, UK.
160. Einspieler, C., HFR, P., Ferrari, R., et al. (1997). The qualitative assessment of general movement in preterm, term and young infants review of the methodology. *Early Human Development, 50*, 47–60.
161. National Institute of Neurological Disorders and Stroke (NINDS). Retrieved April 29, 2017, from https://www.commondataelements.ninds.nih.gov/Doc/NOC/Prechtls_Assessment_of_General_Movements_NOC_Link.pdf.
162. Maitre, N., Chorna, O., Romeo, D. M., & Guzzetta, A. (2016). Implementation of the Hammersmith infant neurological examination in a high-risk infant follow-up program. *Pediatric Neurology, 65*, 31–38.
163. Campbell, S. K. (2005). *The test of infant motor performance. Test user's manual, version 2.0*. Chicago: Infant Motor Performance Scales, LLC.
164. National Institute of Neurological Disorders and Stroke (NINDS). Retrieved April 29, 2017, from https://www.commondataelements.ninds.nih.gov/Doc/NOC/Test_of_Infant_Motor_Performance_NOC_Link.pdf.
165. National Institute of Neurological Disorders and Stroke (NINDS). *Common Data Elements for Cerebral Palsy*. Retrieved April 29, 2017, from https://www.commondataelements.ninds.nih.gov/CP.aspx#tab=Data_Standards.
166. Fine, A., & Kotelchuck, M. (2010). *Rethinking MCH: The life course model as an organizing framework*. U.S. Department of Health and Human Services Health Resources and Services Administration Maternal and Child Health Bureau. [Internet]. Retrieved from http://www.hrsa.gov/ourstories/mchb75th/images/rethinkingmch.pdf.

Utilizing Neonatal Brain Imaging to Predict Neurodevelopmental Outcomes

Eric S. Peeples

Abstract For 40 years, clinicians have sought to utilize neonatal neuroimaging results to not only aid in diagnosis and support clinical decision making, but also attempt to predict neurodevelopmental outcomes in high-risk infants. This effort has been supported by rapidly advancing imaging technology and a wealth of research correlating neuroimaging results with short-term outcomes. The univariate nature of imaging and dynamic pathophysiology of neonatal brain injury, however, have resulted in widely variable sensitivity and specificity of neuroimaging in predicting development outside of the first 2 years of life. This chapter reviews the history and current state of brain imaging for predicting later neurodevelopmental outcomes in high-risk neonates. Imaging modalities that are discussed include magnetic resonance imaging (MRI), ultrasound, computed tomography, as well as more recent magnetic resonance modalities such as magnetic resonance spectroscopy, arterial spin-labelling perfusion MRI, and diffusion tensor imaging. Currently, the predictive capabilities of isolated neuroimaging results do not adequately allow for either the exclusion of high-risk infants from structured follow-up or the selection of infants for additional therapies. Combining modes and timing of imaging, however, may allow for improved predictive values. As neuroimaging research evolves, it will be important to consider how investigators may utilize neuroimaging to improve outcomes for neonates rather than solely anticipating outcomes.

In 1977, Drs. Volpe and Pasternak used technetium scans to demonstrate increased radionuclide uptake in the parasagittal cortex of three infants with hypoxic-ischemic encephalopathy (HIE). Although their imaging findings initially correlated well with the proximal limb weakness observed on physical exam, the weakness gradually improved over the first few months of life, bringing into question the value of their initial association [1]. Although the technology involved in neonatal neuroimaging

E. S. Peeples (✉)
University of Nebraska Medical Center, Omaha, NE, USA
e-mail: epeeples@childrensomaha.org

H. Needelman, B. J. Jackson (eds.), *Follow-Up for NICU Graduates*,
https://doi.org/10.1007/978-3-319-73275-6_3

has developed significantly since 1977, the correlation of neuroimaging results with immediate or short-term outcomes, but not with long-term outcomes, continues to be widespread in the literature today.

In an editorial accompanying a seminal 1978 study, Dr. Volpe suggested that one of the primary goals of neuroimaging in preterm infants with brain injury was to "identify those with a hopeless prognosis" [2]. In doing so, physicians would have the ability to make clinical decisions and appropriately counsel families regarding prognosis early in the hospital course. At the time, correctly identifying infants with a poor prognosis was challenging due to the restraints of the imaging modalities. As neuroimaging has evolved, accurately identifying neuropathology has become more sophisticated, but with the ease of diagnosis has come an increasingly complex and diverse spectrum of neonatal brain injury, which has in some ways made outcome prediction via imaging even less clear.

The univariate nature of imaging is one of the most significant limiting factors for its prognostic efficacy. Images alone are unable to account for the complex nature of perinatal and neonatal brain injuries and the many extracranial factors that are strongly correlated to development, including nutrition, social and economic factors, and access to appropriate therapies. Additionally, the dynamic nature of brain injury in this population underlies the importance of the timing of imaging and is likely a primary cause of the widely variable sensitivity and specificity of neuroimaging in predicting later development. With these limitations in mind, this chapter reviews the history and current state of brain imaging in high-risk neonates for the purpose of predicting later neurodevelopmental outcomes.

Ultrasound

Cranial ultrasound (US) is an easily performed bedside exam that does not expose infants to ionizing radiation. In the hands of an experienced sonographer, US can effectively detect many abnormalities from congenital structural changes to acquired intraventricular and periventricular hemorrhage. US is limited, however, in its ability to detect superficial cortical and watershed injuries, as well as subarachnoid hemorrhage, due to their proximity to the calvaria and the constraints of angling the US probe. Additionally, the timing of the US, experience of the sonographer, and equipment quality may have a considerable effect on the results, increasing the possibility of false-positive or false-negative results. While grayscale US images are not sensitive for diagnosing ischemic injury, Doppler is often added to ultrasound sequences to improve the detection of ischemia. In addition, the ease and relatively low cost of ultrasonography have made it a first-line exam for many indications.

Prematurity

In the mid-1970s, investigators began to use real-time cranial US to assess for intracranial hemorrhage in neonates [3]. Shortly after these initial studies, Papile et al. divided intraventricular hemorrhage (IVH) into four levels of severity (Fig. 1) [4]. Their grading system is still utilized today, with the exception that many clinicians now believe that grade IV IVH should be a separate entity, frequently referred to as periventricular hemorrhage (PVH). Over the next 10 years, the development of higher frequency transducers allowed for better visualization of the cortical white matter, and the ability to diagnose periventricular leukomalacia (PVL). Since then, many researchers have attempted to use US results to predict the outcomes of premature infants (Table 1).

Many early studies focused on the correlation between IVH diagnosed in the first 1–2 weeks of life and later developmental outcomes. Initial studies found that grade II or higher IVH was 79% sensitive for the development of cerebral palsy (CP) by 2 years of life [13], and the severity of hemorrhage correlated with worse motor and cognitive outcomes [14, 15]. Today, however, the risk of CP due to isolated IVH has been demonstrated to be much lower. More recent studies have estimated the rate of CP in isolated IVH to be between 9 and 17%, compared to 4–6% in infants with normal US results [16, 17]. In any case, using early CP as a binary variable in this population is a primitive metric, as many infants with IVH who demonstrate signs of CP at 1–2 years of life have minimal functional impairment and an overall intelligence similar to that of controls by the time they reach school age [18].

Although most early studies focused on IVH, injuries to the white matter such as PVL (Fig. 2) may be the strongest sonographic predictors of abnormal motor outcomes and CP in preterm infants [19, 20]. Signs of white matter injury such as

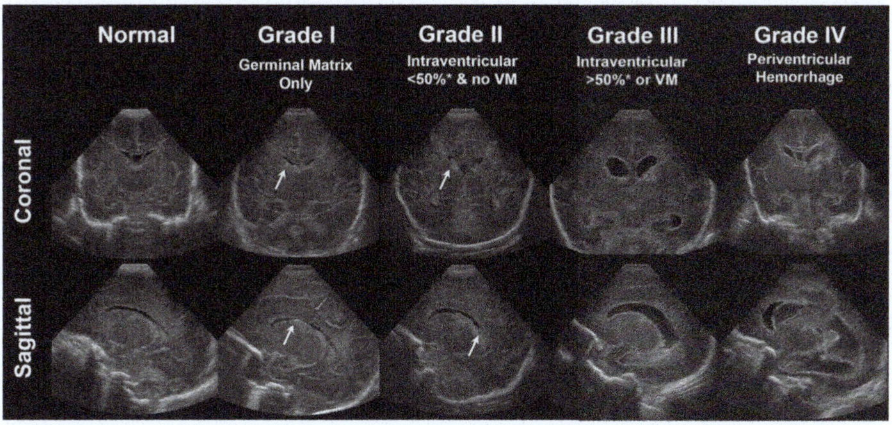

Fig. 1 Grading of intraventricular hemorrhage, as described initially by Papile et al. [4]. % volume of the lateral ventricle occupied by blood; *VM* ventriculomegaly

Table 1 Predictive value of different imaging modalities in extreme prematurity

	Sens (%)	Spec (%)	PPV (%)	NPV (%)	Outcome measured	Source
Term-equivalent MRI						
PVH, PVL, or infarct	100	79			CP at 18 months	[5]
White matter changes						
Moderate-severe	41–65	84–85			CP or severe cognitive or motor delay at 2 years	[6]
Moderate-severe			34	92	CP or MABC <5th percentile at 5 years	[7]
Any severity	88–94	30–31			CP or severe cognitive or motor delay at 2 years	[6]
Any severity			93	41	CP or MABC <5th percentile at 5 years	[7]
Diffuse cystic changes	33	94			Death or CP at 18–24 months	[8]
Gyral maturational delay	33	97			Death, CP, BSID III < 80, or vision or hearing loss at 18–24 months	[8]
Ultrasound						
PVH	67	53	11	95	CP at 18 months	[9]
Parenchymal echogenicity	54	96			Disabling CP at 2 years	[10]
Cystic PVL	67	96	62	97	CP at 18 months	[9]
Prolonged flare	17	85	9	92	CP at 18 months	[9]
Major lesions[a] <= 32 weeks	18–76	85–95	48	99	CP at 18-24 months	[5, 6, 11]
Major lesions[a] 33–36 weeks	86	99	83	99	CP at 2 years	[11]
Imaging at 1 week	16	99	75	85	CP or BSID II < 85 at 1 year	[12]
Imaging at 2 weeks	16	99	75	85	CP or BSID II < 85 at 1 year	[12]
Imaging at 3 weeks	37	99	87	87	CP or BSID II < 85 at 1 year	[12]
Imaging at 6 weeks	53	99	91	91	CP or BSID II < 85 at 1 year	[12]
Imaging at 40 weeks PMA	58	100	100	92	CP or BSID II < 85 at 1 year	[12]

[a]Grade III–IV IVH, cystic PVL, subcortical leukomalacia, basal ganglia lesions, or focal infarction
BSID Bayley Scales of Infant and Toddler Development, *CP* cerebral palsy, *MABC* Movement Assessment Battery for Children, *MRI* magnetic resonance imaging, *NPV* negative predictive value, *PPV* positive predictive value, *PVH* periventricular hemorrhage, *PVL* periventricular leukomalacia

parenchymal echogenicities or ventriculomegaly visualized in the first week of life have a 54% sensitivity but 96% specificity for disabling CP at 2 years of age [10]. Both the size and location of white matter lesions correlate with the severity of motor abnormalities seen in infancy [19–21]. Outside of infancy, 41–47% of infants with white matter lesions or ventriculomegaly develop psychiatric disorders or abnormalities in intelligence and adaptive functions at 6 years of life [22, 23].

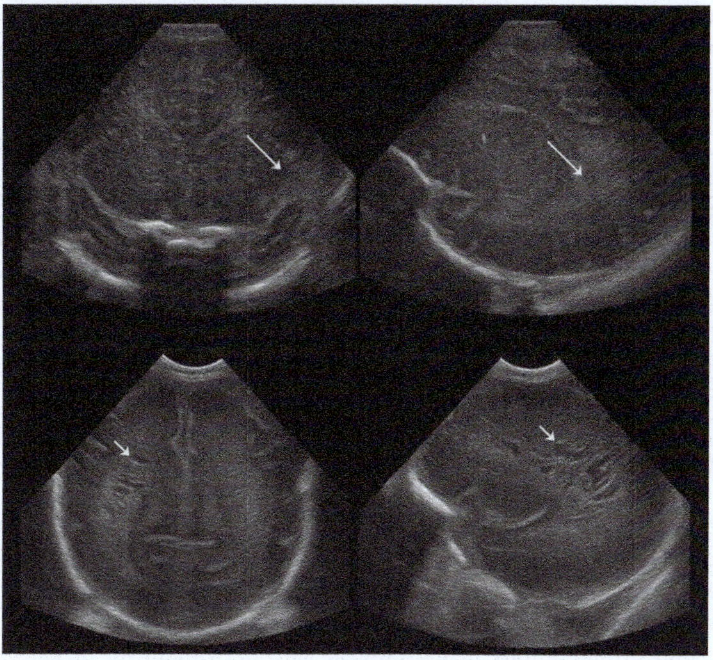

Fig. 2 Coronal (left) and sagittal (right) images demonstrating hyperechogenicity of the periventricular white matter consistent with non-cystic periventricular leukomalacia (top, long arrows) or cystic periventricular leukomalacia (bottom, short arrows). Images courtesy of Drs. Powers, Love, and Gollehon, Children's Hospital & Medical Center, Omaha, NE

More recently, cerebellar injury identified by US through the mastoid window has been discovered to be a significant determinant of later developmental impairment. In one study, 60% of infants with bilateral cerebellar hemorrhage developed CP, though the true prevalence in the study could not be determined because many infants did not have mastoid views performed [17]. A recent Neonatal Research Network trial of extremely preterm infants showed that in the units participating in the trial, only 46% of term-equivalent age US included mastoid views. Of those infants with mastoid views obtained, none demonstrated cerebellar lesions by US despite 15% of those infants having cerebellar lesions visualized on magnetic resonance imaging (MRI) [24]. One potential developmental confounder of these studies is the high frequency of concurrent IVH with cerebellar hemorrhage. In an attempt to define the risk of cerebellar hemorrhage alone, Limperopoulous et al. compared the 24-month outcomes of infants with isolated cerebellar injury, combined cerebellar and cerebral injury, and no injury diagnosed on US. They found that 66% of infants with isolated cerebellar hemorrhage later developed neurologic abnormalities versus 5% in the infants without injury. The neonates with both cerebellar and cerebral injury had a similar risk for developing neurologic abnormalities at 24 months of life but tended to have more severe outcomes [25].

Despite technological advances and high-frequency probes, the predictive value of US is still too poor to allow clinicians to provide accurate prognostic information regarding neurodevelopmental outcomes to families (Table 1). Thirty-four to 60% of infants who develop CP have no IVH, white matter cystic changes, or ventriculomegaly on US, and 6% have completely normal ultrasounds [17]. In an attempt to improve the negative predictive value (NPV) of US, some studies have grouped multiple types of injury into categories such as "major abnormalities," which may include grade III–IV IVH, cystic PVL, subcortical leukomalacia, basal ganglia lesions, and focal infarction. By improving the NPV, however, this method significantly decreases the positive predictive value (PPV) for cerebral palsy to only 48% [11]. Clinicians, therefore, face the dilemma of obtaining US results that may not add significantly to clinical decision making for the neonates.

Unfortunately, cranial US has yet to provide a robust biomarker for clinical outcomes in preterm infants. In fact, prediction models for extremely low-birth-weight infants using clinical variables alone may predict impairment better than those containing only US data, and adding the US results to the clinical models does not improve prediction of neurodevelopmental impairment at 18–22 months [26]. Because of the uncertain prognostic value, many physicians have voiced concern over the ubiquity of US imaging in this population, as well as the common perception that findings such as PVH and PVL carry universally poor outcomes, potentially leading to alterations in the clinical care or withdrawal of care [27]. For clinicians to make complex decisions such as withdrawal of support from US results, better models of developmental outcomes should be developed, and factors such as quality of life and familial impacts must be better explored [28].

Most current recommendations suggest routine screening cranial ultrasounds for all infants born at <30 weeks gestation [29], or to avoid missing the few cases that occur after 30 weeks some recommend <32 weeks [30]. The initial US should be performed between 7 and 14 days of life since the majority of IVH will have occurred by that time. A follow-up US is recommended at >36 weeks postmenstrual age and at least 6 weeks postnatal age to detect white matter changes which occur after the first several weeks of life (Fig. 3) [29, 30]. US obtained closer to 40 weeks corrected age also provide better prognostic accuracy than those obtained in the first few weeks of life [12]. The US sequences obtained should include not only sagittal and coronal views of the cerebrum but also mastoid views of the cerebellum.

Hypoxic-Ischemic Encephalopathy

After becoming a widely recognized modality in premature infants, the utility of cranial US in neonatal HIE began to be explored in the early 1980s [31]. Unlike the early studies in preterm infants that focused on the two findings of IVH and PVL, the two-dimensional US findings in neonatal HIE are broader but can be characterized as either central (basal ganglia, thalami, periventricular white matter, and brain stem) or peripheral (cortex and subcortical white matter) injury.

Extreme prematurity	
Early Ultrasound	7-14 days postnatal age
Late Ultrasound	36-40 weeks PMA AND ≥ 6 weeks postnatal age
MRI	Not currently recommended
Neonatal Encephalopathy	
Ultrasound	Upon admission
MRI	2-8 days post-injury
MRS	2-8 days post-injury
Congenital Heart Disease	
Ultrasound or MRI	Consider prior to repair
ECMO	
Ultrasound	Prior to and during ECMO
MRI or CT	Prior to discharge

Fig. 3 Suggested imaging and timing by etiology of injury. *CT* computed tomography, *ECMO* extracorporeal membrane oxygenation, *MRI* magnetic resonance imaging, *MRS* magnetic resonance spectroscopy, *PMA* postmenstrual age

The various injury patterns observed in infants with HIE have now been well defined by MRI and are discussed in greater detail in the MRI section of this chapter. Findings on US are less sensitive and specific than MRI but are similar in nature. The first pattern occurs after more prolonged and mild insults resulting in watershed injury at the intervascular boundaries, sparing the brain stem, cerebellum, and deep gray matter. Infants suffering from more severe acute hypoxic injury instead demonstrate changes primarily in the deep gray matter (basal ganglia, thalami, hippocampus), periventricular white matter, and perirolandic cortex. Due to the lower amount of metabolic activity in the remainder of the cortex, it is often spared except in cases of prolonged and severe exposure.

US performed in the first few days after injury demonstrates diffusely increased cerebral echogenicity (sometimes referred to as the "bright brain") and loss of the cerebral spinal fluid-containing spaces, which are signs of the development of cerebral edema and potentially early predictors of long-term injury. At 12 +/− 2 h of life, the "bright brain" has a sensitivity of 88% for death or severe disability in infants with HIE [32]. The loss of cerebral spinal fluid-containing spaces can also be found in the first few days of life in 9% of healthy newborns, and because of this the sizes of the ventricles, interhemispheric fissure, and subarachnoid space during that time do not correlate with the amount of injury or later developmental outcomes. After 7 days, however, smaller ventricles, interhemispheric fissure, and subarachnoid space are all associated with severe HIE as well as later death or neurodevelopmental impairment (Ilves, 2009).

The sensitivity and specificity of various US abnormalities in predicting outcomes for infants after HIE have varied widely among studies (Table 2). Studies performed with the early cranial ultrasound devices relied on significant abnormalities that could be visualized with the technology at the time. Because of this, some of the studies could demonstrate nearly 100% specificity in detecting death or moderate-to-severe

Table 2 Predictive value of examination and imaging modalities in term hypoxic-ischemic injury

	Sens (%)	Spec (%)	PPV (%)	NPV (%)	Outcome measured	Source
Structured physical exam						
0–9 days (med. 19 h)	100	31	53	100	Griffith's DQ < 70, CP, or hearing or vision loss	[33]
3–25 days (med. 7 days)	91	71	71	91	Griffith's DQ < 70, CP, or hearing or vision loss	[33]
MRI						
1–7 days	85	86			Death of moderate/severe disability (pooled)	[34]
8–30 days	99	53			Death of moderate/severe disability (pooled)	[34]
No abnormalities	61	92	92	59	Wechsler IQ > =70 at 6–7 years	[35]
PLIC signal intensity	48–90	93–100	88–100	63–87	Death, BSID II–III > 2 SD below mean, DQ < 85 or neurologic changes at 1–2 years	[36, 37]
Mod-Sev WM changes	25	95	84	54	Death or BSID II–III > 2 SD below mean	[36]
Bilateral BG change	84	42			Death or central motor deficit at 1 year	[38]
Low ADC (BG)	78	95	91	87	Death or Griffith's DQ < 85 at 18–46 months	[39]
Low ADC (thalamus)	48	98	93	74	Death or Griffith's DQ < 85 at 18–46 months	[39]
Focal infarct	32	89			Death or central motor deficit at 1 year	[38]
NICHD patterns 2B-3[a]	81	78	70	87	Death or Wechsler IQ < 70 at 6–7 years	[35]
MRS						
Lac/Cr	66	95	86	88	DQ < 85 at 1 year	[40]
Lac/NAA	96	61	67–89	72–94	Death or DQ < 85 at 18–46 months	[39, 41]
NAA/Cho	56	82	72–91	61–100	Death or DQ < 85 at 18–46 months	[39, 41]
Ultrasound						
Abnormal CBFV	57–59	88–100	93–94	73	Death, CP, or developmental delay at 1–2 years	[42–44]
Abnormal RI (ACA)	24–100	63–100	83–100	54–80	Death, CP, or developmental delay at 1–2 years	[43–45]
Abnormal RI (MCA)	82	89	93	73	Death, CP, or developmental delay at 1–2 years	[43]
Unilateral BG change	16	85			Death or central motor deficit at 1 year	[38]
Bilateral BG change	47	100			Death or central motor deficit at 1 year	[38]

(continued)

Table 2 (continued)

	Sens (%)	Spec (%)	PPV (%)	NPV (%)	Outcome measured	Source
Multiple cysts	32	100			Death or central motor deficit at 1 year	[38]
Computed tomography						
Decreased density	91	100	100	80	Death, CP, or developmental delay at 1–2 years	[43]

[a]National Institute of Child Health and Human Development (NICHD) patterns 2B = cerebral lesions with involvement of basal ganglia, thalamus, posterior or anterior limb of the internal capsule, or area of infarction; and 3 = cerebral hemispheric devastation

ACA anterior cerebral artery, *BG* basal ganglia, *BSID* Bayley Scales of Infant and Toddler Development, *CBFV* cerebral blood flow velocity, *CP* cerebral palsy, *DQ* developmental quotient, *IQ* intelligence quotient, *MCA* middle cerebral artery, *MRI* magnetic resonance imaging, *MRS* magnetic resonance spectroscopy, *NPV* negative predictive value, *PLIC* posterior limb of the internal capsule, *PPV* positive predictive value, *RI* resistive index

disability in infants with HIE, but sensitivities were as low as 6% [38]. Other studies have since demonstrated more moderate predictive values, estimating 42% sensitivity and 60% specificity in predicting negative outcomes when imaging within the first 6 h of life, corresponding to a PPV of 57% and NPV of 45% [45].

Due to the relatively low combined sensitivity and specificity of two-dimensional US to predict neurodevelopmental outcomes in infants after HIE, it is not currently recommended for use in diagnosis or prognosis. US should still be considered early in the hospital course, however, as a tool to help exclude other causes of neonatal encephalopathy that may mimic HIE.

Doppler

One main component of the pathophysiology of hypoxic-ischemic brain injury is the alteration of cerebral blood flow, in part due to vasoparalysis and subsequent lack of autoregulation [46]. As such, it has been hypothesized that abnormal cerebral blood flow should be strongly associated with outcomes. Since the mid-1980s [42], investigators have attempted to correlate neurodevelopmental outcomes with the cerebral blood flow velocity (CBFV) and resistive index (RI) of various cerebral blood vessels in infants who have suffered from hypoxic injury. The RI is an estimate of the resistance in a blood vessel and is derived from the equation (systolic velocity–diastolic velocity)/systolic velocity. A normal value in neonates is thought to be 0.75 +/− 0.1 [47]. Alterations in RI are seen in many patients with HIE, most likely due to increased resistance; however, it is important to also assess these patients for other etiologies of abnormal RI, such as cerebral edema, hemorrhage, patent ductus arteriosus, or cardiac dysfunction [47].

RI increases in infants with moderate HIE in the first 24 h of life, followed by a transition to low RI by the second day of life [48, 49]. After the initial elevated period, a value less than 0.55 has been determined as consistent with hypoxic-ischemic injury [42]. Reversal of this pattern has been observed in infants with the most severe injury, demonstrating low RI as early as 2–12 h after injury. This initial hyperperfusion may last up to 10 days after injury, and then gradually transitions to a prolonged hypoperfusion state with low CBFV and RI elevation that may last up to 5 months after injury [32]. A low RI and elevated CBFV after 12 h of life have consistently been associated with poor outcomes [32, 49, 50].

Doppler US possesses limitations which have so far kept it from becoming a universally relevant clinical tool. The first is that blood flow velocity is used as a surrogate for blood flow in these studies due to the inability to accurately measure vessel diameter, making it impossible to determine if elevated velocities are due to changes in vasomotor tone (i.e., vasoconstriction) or a true increase in blood flow. Also, after the initial hypoperfusion state in severe injury, cerebral blood flow is thought to dissociate from cerebral metabolism, resulting in "luxury perfusion" where the blood flow greatly exceeds the metabolic needs of the brain tissue. During this time, CBFV no longer reflects the condition of the injured brain, and other measurements that directly assess the cerebral metabolism, oxygenation, or electric activity of the brain would be more appropriate metrics. Lastly, since cardiovascular stability is affected in infants undergoing hypothermia, it is possible that hypothermia also affects cerebrovascular control, making the values derived from the studies of CBFV and RI performed on normothermic infants less applicable. Supporting this concern, the PPV of the traditional RI cutoff of less than 0.55 for poor outcome decreases from 84 to 60% in infants undergoing hypothermia [51].

Magnetic Resonance Imaging

In 1985, investigators began reporting the use of cerebral MRI in neonates [52], and shortly thereafter the first studies of MR imaging in neonatal encephalopathy were published [53, 54]. Despite the ease and ubiquity of cranial US, MRI has better sensitivity and specificity for brain injury in the neonates, especially in the thalami and basal ganglia, where over half of the injury may be missed by US [38]. The increased sensitivity for parenchymal injury has not proportionally translated into sensitivity for predicting outcomes, however. For instance, the same study that demonstrated superior diagnostic utility also showed a significant false-positive rate for neurodevelopmental outcomes, as each of the 15 infants with normal developmental outcomes had at least one abnormality visualized on MRI [38].

MRI utilizes electromagnetic radiation to excite protons within the brain tissue. The protons subsequently "relax" to an equilibrium state, and in doing so emit radiofrequency energy that is processed to generate images. The longitudinal and transverse components of that energy are termed T1 and T2 relaxation times, respectively, and differ depending on the tissue composition. By augmenting imaging parameters to maximize or minimize the T1 and T2 relaxation effects, one can produce images with certain contrasts. In general, a T2-weighted sequence is fluid sensitive, and tissues containing fluid (e.g., cerebrospinal fluid or interstitial fluid from parenchymal edema) have a high signal intensity and appear bright on imaging; conversely, fluid-containing tissues or structures have low signal intensity on T1-weighted sequences and appear dark. Blood is a fluid with complex and dynamic MRI signal characteristics due to changes in oxygenation over time that produce characteristic patterns of signal intensity on T1-weighted and T2-weighted sequences as the blood evolves.

MRI provides sensitive noninvasive imaging of the brain parenchyma without ionizing radiation. Unfortunately, its utility is often limited by the need to transfer potentially unstable infants from the intensive care unit to radiology, an issue complicated by the relatively long duration of the studies compared to other modalities such as computed tomography. Historically, many infants have also been sedated for MR imaging, adding additional risk; however, most MR protocols now attempt to feed and swaddle the infants in place of sedation.

Prematurity

In addition to detecting hemorrhage, which is also well visualized by US, MRI can differentiate between several common patterns of white matter changes in preterm infants. The first is prolonged immaturity and delayed myelination of the white matter, which presents as bands of low T2 intensity in the frontal periventricular white matter. The bands are thought to represent normal maturation by migrating glial cells and are not correlated with motor outcomes even when persisting to term-equivalent age (TEA) [5]. The second common injury pattern is diffuse white matter disease, often referred to as diffuse excessive high signal intensity (DEHSI). DEHSI is considered to represent a milder form of injury, and as such does not appear to significantly correlate with developmental outcomes [55, 56].

The remainder of MRI abnormalities described in preterm infants are more focal in nature. The most common focal changes are punctate lesions [57], thought to represent clusters of activated microglia. Since these lesions cannot be visualized with US, outcome data on this entity is very limited. Most likely, neonates with only a few lesions will develop normally, but those with lesions in the posterior limb of the internal capsule (PLIC) or more widespread lesions in the periventricular white matter may have abnormal motor development [58]. Although classic cystic PVL is the white matter lesion that has been the most thoroughly investigated in preterm infants, it represents only about 4% of the abnormalities seen on TEA MRI [57].

One of the first large studies to suggest obtaining routine MRI at TEA was by Woodward et al., which demonstrated significantly improved sensitivity of MRI over US in predicting a range of developmental impairment. Though improved, the sensitivity of moderate-to-severe white matter abnormalities on MRI for predicting cognitive delay at 2 years was still only 41%, and 65% for severe motor delay or CP, respectively. Additionally, the specificity for all outcomes assessed in their study was lower for MRI than US [6]. The strength of the association between white matter injury on MRI and motor outcomes also correlates with increasing injury severity [7].

One significant advantage of MR imaging is its superiority over US in imaging the posterior fossa. As mentioned earlier in this chapter, cerebellar hemorrhage may be visualized through US mastoid views and has been associated with significant risk of neurologic abnormalities in preterm infants [25, 59]. At 7 years of age, infants with cerebellar hemorrhage were found to be at higher risk for deficits in attention [60] but not memory or learning [61]. While few IVH are missed by routine US, cerebellar injury demonstrated on MRI is frequently missed by US [24, 62, 63]. A recent study demonstrated cerebellar hemorrhage on MRI in 10% of their cohort, whereas only 2% had hemorrhage detected on US. The difference in sensitivity for cerebellar injury may be translatable to improved prognostic ability, as those infants with hemorrhage only visible by MRI had a fivefold increased odds of abnormal neurologic outcome between 3 and 6 years of age over controls [64].

One of the challenges of the improved sensitivity of MRIs is that 10% of infants will demonstrate unexpected (i.e., not already diagnosed prenatally or postnatally by clinical signs or cranial US) abnormalities not acquired in the perinatal period [65]. In the adult population, 1–4% of brain MRI produce incidental findings, with up to one-third of those being neoplastic and many of the remainder considered benign without the need for follow-up [66]. Contrary to the adult literature, neonates tend to have a higher rate of incidental findings requiring follow-up. In the only study of incidental findings in the very-low-birth-weight population to date, only 25% of the findings were benign (e.g., small benign venous anomalies and arachnoid cysts) and required no further follow-up. The remaining 75% required further diagnostic follow-up and/or therapeutic intervention, including cortical tubers, significant dysmorphia of the brain stem or cerebellum, and ectopic pituitary [65]. The balance between revealing silent brain abnormalities of clinical significance and discovering findings of uncertain clinical significance that result in potentially unnecessary diagnostic follow-up has led to ongoing debate regarding the benefit and appropriateness of TEA MR imaging in this population.

Currently, there are several issues that must be overcome before TEA MRI can become a standard in preterm infants. Although MRI may have some prognostic value on its own, it does not significantly improve the ability to predict death or neurodevelopmental disability over the combination of early and late cranial US [24]. Another study showed that although white matter injury on MRI significantly increased the risk of MDI <70 or functional motor deficits at 12–18 months with a relative risk of 5.3, a structured neurological exam performed on the day of the MRI had a higher relative risk of 6.5 for the same outcomes [67]. Additionally, there are currently no data to support that any improved prognostic capabilities of MRI will translate into improved outcomes. Because of this, performing term-equivalent MRI for preterm infants was recently listed as one of the top five tests or treatments in newborn medicine that "cannot be adequately justified on the basis of efficacy, safety, or cost" [68].

Some have suggested that MRI should be obtained as standard for all very-low-birth-weight infants at 36–40 weeks postmenstrual age [69–71], though the most recent recommendations from the American Academy of Neurology do not suggest routine MRI for preterm infants, regardless of cranial US findings [29]. MRI may be helpful for better understanding more subtle developmental pathology, and performing MRI under the constraints of clinical research protocols may allow for the eventual development of methods to improve the outcome of these high-risk infants.

Hypoxic-Ischemic Encephalopathy

Two primary patterns of injury have been identified by MRI in term infants who have survived neonatal HIE, corresponding to regions of high metabolic need and therefore heightened vulnerability. The most common pattern, which is seen in ~45% of infants with HIE [72], presents after prolonged mild hypoxia and results primarily in injury to the white matter watershed areas of the brain, extending to the deep gray nuclei only in severe cases. The second primary pattern is seen in ~25% of infants with HIE [72], and is more often observed after severe acute hypoxia-ischemia. These infants develop predominantly bilateral central gray nuclei and perirolandic cortex injury, with occasional involvement of the hippocampus and brain stem (Fig. 4). Other types of injury that have been described in HIE include focal ischemic injury as well as periventricular white matter injury similar to the leukomalacia seen in preterm infants [73]. The latter injury is most common in infants born at lower gestational ages and results in a milder encephalopathy compared to the two primary patterns of injury [74].

Studies have attempted to associate particular injury patterns on MRI with developmental outcomes. A study performed in the pre-therapeutic hypothermia era assessed infants with HIE for abnormal signal intensity in the PLIC (Fig. 4),

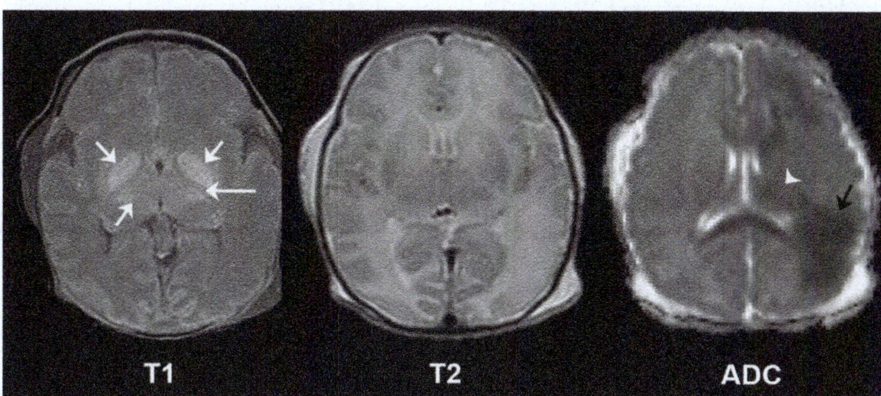

Fig. 4 Magnetic resonance imaging of an infant after hypoxic-ischemic encephalopathy demonstrating abnormal hyperintense T1 signal in the basal ganglia and thalamus (short white arrows) and lack of normal T1 signal in the posterior limb of the internal capsule (long arrow), as well as restricted diffusivity in the left thalamus (arrowhead) and white matter of the left hemisphere (black arrow). Images courtesy of Drs. Powers and Love, Children's Hospital & Medical Center, Omaha, NE

which can signify a loss of myelination. They found 90% sensitivity and 100% specificity for development of a low Griffith's Developmental Quotient or abnormal neurologic exam at 1 year of age in infants with moderate HIE [37]. They also demonstrated that the abnormal signal might not become apparent in the PLIC until about 4 days of life. Injury to the PLIC, as well as the basal ganglia and thalamus, is associated with more severe developmental phenotypes than watershed injury, with up to 94% of infants developing cognitive deficits, CP, and seizures at 1–2 years of life [37, 72, 75, 76]. The severity of the developmental phenotype is correlated with the degree of basal ganglia injury seen on MRI, especially when imaged between 7 and 14 days of life [75, 76]. As opposed to the motor deficits often seen with deep gray matter injury, infants with predominant watershed injury primarily develop cognitive impairments in the absence of motor deficits [72].

Due to overlap between injury patterns, MRI scoring systems have been developed to attempt to assess the effects of combined white and gray matter injury. One of the first scoring systems described by Barkovich et al. found that the combination of both basal ganglia and watershed injury was a better predictor of neuromotor and cognitive outcome at 12 months than either basal ganglia or watershed scores alone [77]. A more recent system derived from the National Institute of Child Health and Human Development (NICHD) Neonatal Research Network study on total body hypothermia came to a similar conclusion. In their study, the combination of level 2B and 3 (2B defined as the combination of both white and gray matter lesions and 3 defined as cerebral hemispheric devastation) had PPV of 70% and NPV of 87% for death or IQ < 70 at 6–7 years of age [35]. These scoring systems are vital for the standardization of reporting imaging results in clinical research, allowing for appropriate comparisons between studies. However, they are not commonly used in routine clinical care.

Although conventional T1- and T2-weighted MRI provides moderate sensitivity and specificity for predicting outcomes in neonates after HIE when performed toward the end of the first week post-injury [34, 39], the injury is frequently not apparent in the first few days of life. Diffusion-weighted MR imaging (DWI) allows for the calculation of a value termed the apparent diffusion coefficient (ADC), which is a measure of intracellular edema and may provide evidence of cerebral injury earlier than conventional MRI [78, 79]. ADC values decrease 1–2 days before injury appears on T1- and T2-weighted imaging, although false negatives are still common in the first 24 h after injury. The ADC values then pseudo-normalize 6–7 days after injury (or 10 days after injury in infants who have undergone hypothermia) and therefore may provide false-negative results if obtained after the first week (Fig. 5) [79–82]. Decreased ADC values have been associated with poor motor outcomes in infants up to 2 years of age [83, 84], and the combination of conventional MRI and ADC has a much stronger association with outcomes at 18–46 months than conventional MRI alone [39].

When correlating MRI findings to outcomes, special attention should be paid to the timing of developmental follow-up, as some infants may not demonstrate neurologic changes until after 1–2 years of age. In one cohort, 24% of children of infants who were considered developmentally normal at 2 years of age later demonstrated

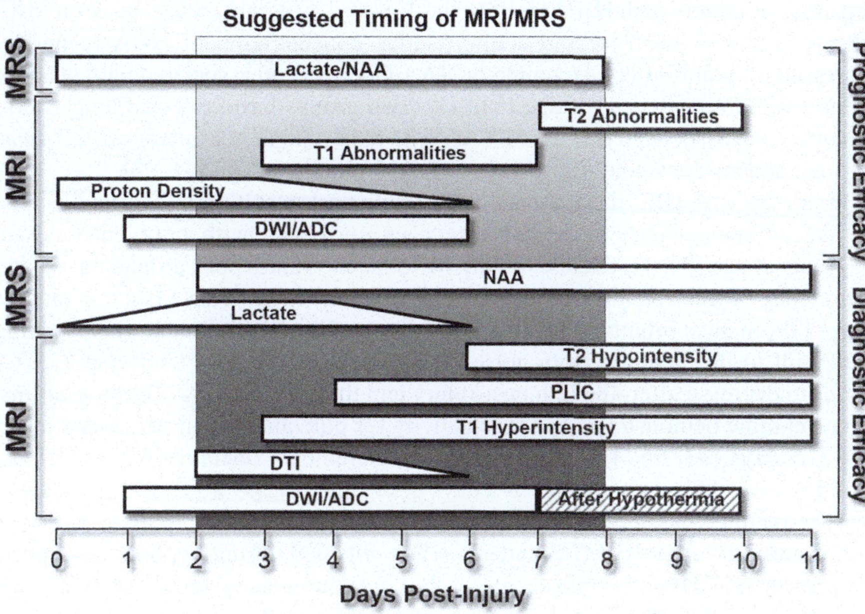

Fig. 5 Timing of diagnostic and prognostic utility of magnetic resonance imaging modalities

neurologic dysfunction at 5–6 years of life. Of those children, 80% had mild or moderate basal ganglia injury or marked white matter lesions on their neonatal MR imaging [85]. In another study, several subjects whose neonatal MRI demonstrated a watershed pattern of injury developed cognitive deficits at 30 months that were not apparent at 12 months in infants [72]. To improve the predictive ability of MRI to correlate with outcomes past the first few years of life, more advanced MR technology is under investigation.

Magnetic resonance spectroscopy (MRS) is a noninvasive method of measuring metabolites produced in the brain. Imaging with MRS may be beneficial in the first day after injury, at a time when conventional MRI changes are limited to edema, PLIC changes are not yet apparent, and DWI changes are still evolving. MRS also has the potential to reveal signs of other etiologies of encephalopathy, such as mitochondrial disorders and non-ketotic hyperglycinemia.

Initial studies into brain metabolism used phosphorus (^{31}P)-MRS and were integral in describing the concept of "secondary energy failure" that occurs 6–8 h after injury [86–89]. Several studies correlated an abnormal creatinine (Cr)/inorganic phosphate ratio to death or neurodevelopmental disability, with a PPV as high as 93% [87, 90, 91]. The development of proton (^{1}H)-MRS, however, which allows for significantly improved spatial resolution and visualization of a wider range of cerebral metabolites has kept ^{31}P–MRS from becoming widely utilized clinically.

A number of groups have used ^{1}H-MRS results to predict outcomes in infants after HIE [41, 92–94], and ^{1}H-MRS has been described as the strongest MR biomarker of outcomes in infants with HIE [34, 95]. The *N*-acetyl aspartate (NAA) peak on MRS reflects neuronal integrity, and a low relative concentration of NAA (generally expressed as a ratio of NAA/choline or NAA/Cr) has been associated with death or severe disability at 1 year of age [81, 94]. Two groups have assessed longer term outcomes and associated decreased NAA with abnormal motor outcomes, CP, hearing loss, and severe visual impairment at 24–30 months of age [41, 93].

In infants with HIE, the presence of a lactate peak resulting in elevated lactate/choline or lactate/NAA ratios has also been correlated with poor outcomes at 1–2 years of age [39, 41, 83, 93, 96–98]. MRS lactate values peak around the second or third day of life, and although lactate may continue to be detected up to 5 months of age [99] in most infants it rapidly diminishes by the end of the first week, making it difficult to differentiate from noise by 3–5 days post-injury. Conversely, NAA begins to decrease after 48–72 h post-injury and then remains low. Because of this, lactate/choline demonstrates better sensitivity for outcomes when performed in the first 3–5 days [93, 96, 100] while NAA measurements (lactate/NAA and NAA/choline) outperform lactate after 3–5 days post-injury [92].

Therapeutic hypothermia alters the energy homeostasis in the brain, resulting in significantly different ^{1}H-MRS patterns when obtained during hypothermia versus after rewarming. Hypothermia decreases the concentrations of key metabolites such as creatine, choline, glutamate, and NAA, and it may cause a small increase in phosphocreatine [101]. It is currently unclear whether these changes are enough to affect the ability of MRS to predict outcomes.

The most recent recommendations from the American Academy of Neurology suggest performing MRI between 2 and 8 days after hypoxic-ischemic brain injury (Fig. 5), including single-voxel MRS and DWI sequences if available.

Advanced MRI Modalities

As demonstrated by Doppler US studies, abnormal brain perfusion may be correlated with poor outcomes in infants after hypoxic-ischemic brain injury. Standard Doppler US technology, however, is limited in its ability to accurately measure blood flow velocity in deep tissues, especially at low velocities. Arterial spin-labelling (ASL) perfusion MRI is a noninvasive perfusion measure that can accurately quantify blood flow velocity deep in the brain and at low velocities. Higher ASL perfusion values, especially in the basal ganglia and thalami, are associated with death or CP at 9 or 18 months of age. The predictive performance increases further when combined with MRS Lac/NAA [102].

Another MRI sequence, diffusion tensor imaging (DTI) (Fig. 6), takes advantage of the tendency of water molecules to flow more easily along the path of an axon to allow visualization of neuronal tracts. The directional preference of water's path is termed anisotropy, and the fractional anisotropy (FA) is a value that can be derived from DTI. FA decreases when local tract structure is disrupted, signifying axonal or white matter bundle injury. Studies have demonstrated that DTI is feasible in neonates,

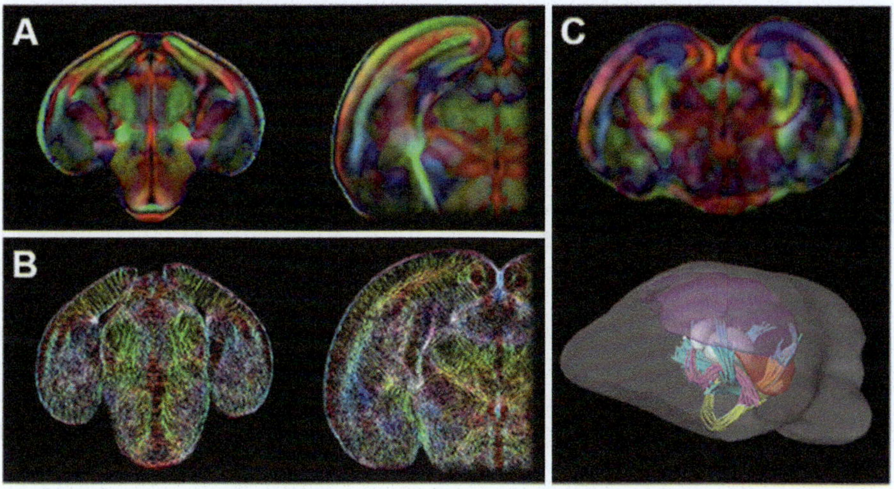

Fig. 6 (**a**) Coronal diffusion tensor (DTI) and (**b**) track density magnetic resonance images of embryonic mouse brains. (**c**) Coronal DTI color map and white matter tract reconstruction of an embryonic mouse brain. © 2016 Wu and Zhang CC-BY, version 4.0 [103]

and global FA in infants with HIE is improved after therapeutic hypothermia [104]. Decreased FA in the corpus callosum and corticospinal tracts has been associated with decreased BSID-II MDI and PDI, respectively, at 15 and 21 months after HIE [105]. In premature infants with PVH, an asymmetric DTI in the PLIC correctly predicted unilateral spastic CP in six of seven infants, and high FA asymmetry predicted spastic CP in all seven cases [106].

Other diffusion-related metrics such as diffusion kurtosis imaging, which measures the probability distribution of the water diffusion in DTI, may possess improved sensitivity to microstructural change over DTI [107], and additional techniques such as brain volumetrics, assessment of cortical folding, and resting-state functional MRI may also provide additional prognostic insight into the future. All of these techniques are currently early in their evaluation for use in the neonatal population.

Computed Tomography (CT)

In 1994, nearly 23% of infants with HIE were receiving at least one CT scan during their initial hospitalization despite concerns regarding exposing the immature brain to radiation, the ease of cranial ultrasound, and the superiority of MR imaging in this population [108]. However, as US and MRI techniques have continued to evolve and become more widely available, and more focus has been placed on neuroimaging techniques in the newborn period, CT imaging has been utilized less frequently. The main benefit of CT remains its rapid acquisition time obviating the need for sedation of the infant and better visualization of superficial structures over cranial US.

CT has been used in infants to demonstrate hemorrhage, generalized edema, and multicystic encephalomalacia in severe hypoxic injury [109, 110]. CT can also identify severe deep gray matter lesions with similar performance to MRI, and injuries to the thalami and basal ganglia on CT have been associated with death or major neurologic sequelae at 18 months of life [111]. CT does not perform well in the identification of white matter cortical injury or cerebellar injury, however [112, 113].

In addition to being less sensitive in the identification of hypoxic injury, CT has several disadvantages including the use of radiation. Although portable CT is available in a few institutions, most still require the infant to be transported to radiology for image acquisition. Due to the need for transport and radiation, CT is generally reserved for acute situations such as concern for hemorrhage when US is not feasible or informative and has not been widely utilized in prognostication for neonatal outcomes.

Imaging in Infants with Congenital Heart Disease (CHD)

Twenty-six to 59% of infants with CHD have evidence of structural brain abnormalities on brain imaging [114–116], and infants with CHD are at high risk for neurodevelopmental impairments. Imaging and neuropathology of infants with congenital heart disease demonstrate similar immaturity to that of preterm infants, with pathological studies most commonly showing focal or diffuse white matter injury, and less frequently gray matter damage [117].

Although infants with complex CHD are at added risk for adverse neurodevelopment due to their exposure to surgery, anesthesia, and cardiopulmonary bypass, much of the abnormal development of this population is likely due to abnormal in utero cerebral blood flow resulting in poor brain growth and development [118, 119]. In fact, the relative immaturity of their brain at term gestation may increase their susceptibility to white matter injury caused by hemodynamic changes in the perioperative period [120]. This theory is supported by the finding that infants with low preoperative brain maturity scores are at increased risk for postoperative white matter injury and that lower brain maturity is associated with more severe white matter injury [121].

Though frequently performed in the preoperative period, US has not been utilized clinically for predication of outcomes, but rather to assess for significant structural abnormalities that could complicate surgery. One study comparing US and MRI in a population of term infants with CHD found that MRI diagnosed brain injury in 26% of asymptomatic infants versus 3% with US, and 80% of the abnormal ultrasounds were considered to be false positives [116]. Additionally, preoperative US does not correlate with neurodevelopmental outcomes at 1 year of life [122]. US Doppler flow measures, however, have been used in the fetal and neonatal periods to assess cerebral blood flow velocity and have been correlated with abnormal neurodevelopmental outcomes [123, 124].

Although some have recommended an MRI in all infants with CHD that require early surgery due to the frequency of brain lesions in this population and the lower sensitivity and specificity of US [125], there is a paucity of data correlating outcomes to MRI in infants with congenital heart disease. In one of the few studies to assess the association between neuroimaging and development in this population, infants who had previously undergone the Fontan procedure for single ventricle physiology were followed up at 12–17 years of age for neuropsychological testing and MR imaging. The post-Fontan adolescents had lower Full-Scale IQ as well as lower Reading Composite scores and Mathematics Composite scores than population norms. The frequency of MRI abnormalities in those infants was 66%, approximately 11 times greater than the general population [126].

The American Heart Association suggests that "the indications for brain MRI for the asymptomatic child with CHD are poorly defined given the unclear prognostic value of abnormal findings and the lack of a consensus on the need for treatment of asymptomatic PVL. However, brain MRI may be a useful clinical adjunct … on a case-by-case basis, for the diagnosis and management of possible contributors to developmental delays" [127].

Imaging in Infants After Extracorporeal Membrane Oxygenation (ECMO)

Infants requiring extracorporeal membrane oxygenation (ECMO) are at very high risk of intracranial hemorrhage or infarction. These infants often receive serial neuroimaging to assess for complications such as hemorrhage or infarct, and published rates of abnormal neuroimaging after ECMO have ranged from 0 to 52% [128–130]. MRI has significantly better sensitivity than US, with up to 50% of infants with negative US demonstrating MRI abnormalities; however neither mode of imaging has been found to correlate with the Bayley Scales of Infant and Toddler Development (BSID) II or III or hearing or vision outcomes at 12–30 months of age [130].

The remaining literature assessing the association between neuroimaging and outcomes in infants after ECMO has been conflicting. Consistent with the above study, another group found no association between major brain lesions and neurodevelopmental outcome, though they did demonstrate that the presence of enlarged cerebrospinal fluid spaces (also referred to as benign extra-axial fluid of infancy) on MRI was associated with lower BSID II at 6 and 12 months of life [131]. Other studies have been more positive, finding significant associations between moderate-to-severe neuroimaging abnormality and death, CP, IQ scores, or abnormal neuromotor outcome [132–134].

Overall, the data regarding neurodevelopmental outcomes in infants with abnormal brain imaging post-ECMO does not support making clinical decisions based on imaging alone. To allow for identification of silent brain lesions, however, the Extracorporeal Life Support Organization includes a predischarge CT or MRI in the discharge checklist described in their recommendations for neonatal-pediatric ECMO patient follow-up.

Post-discharge Imaging

Few studies have addressed the role of neuroimaging in high-risk neonates after discharge from the hospital. Some have suggested performing US 1 month after HIE to detect infants who will develop multicystic encephalopathy. In support of this recommendation, multicystic encephalopathy has been associated with death or poor psychomotor development at 18 months, and the cysts may not be visualized until around 3 weeks of life with resolution of small cysts by 3 months [32]. Other US findings post-discharge include progressive enlargement of the ventricles, inter-hemispheric fissure, and subarachnoid space which can be seen at 5 months of age in infants with severe HIE. Their size has been correlated with poor psychomotor development at the age of 18 months [32].

Measurements of the basal ganglia, thalamus, and cerebellum on MR imaging are still correlated to 1-year outcomes when performed at 3 months of age [135]. When MRI was performed at 8 months of age, structural abnormalities, delayed myelination, or both were present in all infants diagnosed with CP at 18 months [136]. When assessing long-term outcomes, a smaller posterior corpus callosum was demonstrated at 9–10 years of life in children with poorer motor skills determined by the Motor Assessment Battery for Children [137], and an abnormal FA may be seen after moderate HIE well into adolescence. The lower FA in the internal capsule and corpus callosum was demonstrated even in those adolescents considered to have normal development, suggesting that asymptomatic children may still have smaller, fewer, more poorly organized, and less myelinated axons than children who did not have a hypoxic injury at birth [138].

Outside of clinical research protocols, there is currently inadequate data to support routine post-discharge neuroimaging for follow-up of the high-risk infant. The lack of data, however, should not be equated with lack of value. It is clear from studies in infants after HIE that cerebral metabolism remains abnormal for more than 7 months after injury [99, 139] and that infants with more prolonged neuroimaging changes are at higher risk for abnormal neurological outcome than those that resolve earlier in life [31, 32]. Motor deterioration can be seen between 12 and 30 months after HIE [93], and microcephaly may be delayed in HIE more than 12 months [140]. All of this suggests that obtaining neuroimaging after a year of life, when the brain has reached a more adultlike structural state, could potentially provide valuable information regarding a child's neurodevelopmental prognosis. The benefits of obtaining images later in life would need to be weighed against the imaging risks as well as the increasing prognostic value of physical exams as the child ages.

Regardless of their neonatal course, children with global developmental delay may benefit from neuroimaging, as MRI may contribute to etiologic diagnoses in up to 65% of children with global developmental delay. Clinicians should, therefore, consider MRI in children with global delay, especially if they have either had early negative brain imaging or no imaging performed [141].

Conclusion

Currently, the predictive capabilities of neuroimaging for high-risk infants do not adequately allow for either the exclusion of infants from structured follow-up or the selection for additional therapies. Additionally, the most predictive imaging is performed after the acute phase of illness, making it ineffective for counseling parents on limitations of care or withdrawal of support. The main remaining benefit is that of providing parents with a risk-adjusted estimate of developmental outcome, something that has been suggested by families and physicians to be of questionable benefit and potentially emotionally harmful [142, 143]. Therein lies the largest gap in the research surrounding neonatal neuroimaging: we have repeatedly *described* the outcomes of these infants through US, MRI, MRS, CT, and others, but have done a poor job of completing the truly meaningful research of finding ways to utilize the neuroimaging results to *improve* outcomes.

References

1. Volpe, J. J., & Pasternak, J. F. (1977). Parasagittal cerebral injury in neonatal hypoxic-ischemic encephalopathy: Clinical and neuroradiologic features. *The Journal of Pediatrics, 91*(3), 472–476.
2. Volpe, J. J. (1978). Neonatal periventricular hemorrhage: Past, present, and future. *The Journal of Pediatrics, 92*(4), 693–696.
3. Heimburger, R. F., Fry, F. J., Franklin, T. D., Eggleton, R. C., & Gresham, E. (1976). Ultrasound for identification of brain damage in infants and young children. In D. White & R. Barnes (Eds.), *Ultrasound in Medicine: Volume 2 Proceedings of the 20th Annual Meeting of the American Institute of Ultrasound in Medicine* (pp. 239–249). Boston, MA: Springer US.
4. Papile, L. A., Burstein, J., Burstein, R., & Koffler, H. (1978). Incidence and evolution of subependymal and intraventricular hemorrhage: A study of infants with birth weights less than 1,500 gm. *The Journal of Pediatrics, 92*(4), 529–534.
5. Valkama, A. M., Paakko, E. L., Vainionpaa, L. K., Lanning, F. P., Ilkko, E. A., & Koivisto, M. E. (2000). Magnetic resonance imaging at term and neuromotor outcome in preterm infants. *Acta Paediatrica, 89*(3), 348–355.
6. Woodward, L. J., Anderson, P. J., Austin, N. C., Howard, K., & Inder, T. E. (2006). Neonatal MRI to predict neurodevelopmental outcomes in preterm infants. *The New England Journal of Medicine, 355*(7), 685–694. https://doi.org/10.1056/NEJMoa053792.
7. Spittle, A. J., Cheong, J., Doyle, L. W., Roberts, G., Lee, K. J., Lim, J., et al. (2011). Neonatal white matter abnormality predicts childhood motor impairment in very preterm children. *Developmental Medicine and Child Neurology, 53*(11), 1000–1006. https://doi.org/10.1111/j.1469-8749.2011.04095.x.
8. Slaughter, L. A., Bonfante-Mejia, E., Hintz, S. R., Dvorchik, I., & Parikh, N. A. (2016). Early conventional MRI for prediction of neurodevelopmental impairment in extremely-low-birth-weight infants. *Neonatology, 110*(1), 47–54. https://doi.org/10.1159/000444179.
9. Graham, M., Levene, M. I., Trounce, J. Q., & Rutter, N. (1987). Prediction of cerebral palsy in very low birthweight infants: Prospective ultrasound study. *Lancet, 2*(8559), 593–596.
10. Pinto-Martin, J. A., Riolo, S., Cnaan, A., Holzman, C., Susser, M. W., & Paneth, N. (1995). Cranial ultrasound prediction of disabling and nondisabling cerebral palsy at age two in a low birth weight population. *Pediatrics, 95*(2), 249–254.

11. De Vries, L. S., Van Haastert, I. L., Rademaker, K. J., Koopman, C., & Groenendaal, F. (2004). Ultrasound abnormalities preceding cerebral palsy in high-risk preterm infants. *The Journal of Pediatrics, 144*(6), 815–820. https://doi.org/10.1016/j.jpeds.2004.03.034.
12. Nwaesei, C. G., Allen, A. C., Vincer, M. J., Brown, S. J., Stinson, D. A., Evans, J. R., & Byrne, J. M. (1988). Effect of timing of cerebral ultrasonography on the prediction of later neurodevelopmental outcome in high-risk preterm infants. *The Journal of Pediatrics, 112*(6), 970–975.
13. Tudehope, D. I., Masel, J., Mohay, H., O'Callaghan, M., Burns, Y., Rogers, Y., & Williams, G. (1989). Neonatal cranial ultrasonography as predictor of 2 year outcome of very low birth-weight infants. *Australian Paediatric Journal, 25*(2), 66–71.
14. Bassan, H., Limperopoulos, C., Visconti, K., Mayer, D. L., Feldman, H. A., Avery, L., et al. (2007). Neurodevelopmental outcome in survivors of periventricular hemorrhagic infarction. *Pediatrics, 120*(4), 785–792. https://doi.org/10.1542/peds.2007-0211.
15. Maitre, N. L., Marshall, D. D., Price, W. A., Slaughter, J. C., O'Shea, T. M., Maxfield, C., & Goldstein, R. F. (2009). Neurodevelopmental outcome of infants with unilateral or bilateral periventricular hemorrhagic infarction. *Pediatrics, 124*(6), e1153–e1160. https://doi.org/10.1542/peds.2009-0953.
16. Ancel, P. Y., Livinec, F., Larroque, B., Marret, S., Arnaud, C., Pierrat, V., et al. (2006). Cerebral palsy among very preterm children in relation to gestational age and neonatal ultrasound abnormalities: The EPIPAGE cohort study. *Pediatrics, 117*(3), 828–835. https://doi.org/10.1542/peds.2005-0091.
17. Kuban, K. C., Allred, E. N., O'Shea, T. M., Paneth, N., Pagano, M., Dammann, O., et al. (2009). Cranial ultrasound lesions in the NICU predict cerebral palsy at age 2 years in children born at extremely low gestational age. *Journal of Child Neurology, 24*(1), 63–72. https://doi.org/10.1177/0883073808321048.
18. Roze, E., Van Braeckel, K. N., van der Veere, C. N., Maathuis, C. G., Martijn, A., & Bos, A. F. (2009). Functional outcome at school age of preterm infants with periventricular hemorrhagic infarction. *Pediatrics, 123*(6), 1493–1500. https://doi.org/10.1542/peds.2008-1919.
19. Fawer, C. L., Diebold, P., & Calame, A. (1987). Periventricular leucomalacia and neurodevelopmental outcome in preterm infants. *Archives of Disease in Childhood, 62*(1), 30–36.
20. Weiss, H. E., Goldstein, R. B., & Piecuch, R. E. (1999). A critical review of cranial ultrasounds: Is there a closer association between intraventricular blood, white matter abnormalities or cysts, and cerebral palsy? *Clin Pediatr (Phila), 38*(6), 319–323. https://doi.org/10.1177/000992289903800601.
21. Rogers, B., Msall, M., Owens, T., Guernsey, K., Brody, A., Buck, G., & Hudak, M. (1994). Cystic periventricular leukomalacia and type of cerebral palsy in preterm infants. *The Journal of Pediatrics, 125*(1), S1–S8.
22. Whitaker, A. H., Feldman, J. F., Van Rossem, R., Schonfeld, I. S., Pinto-Martin, J. A., Torre, C., et al. (1996). Neonatal cranial ultrasound abnormalities in low birth weight infants: Relation to cognitive outcomes at six years of age. *Pediatrics, 98*(4 Pt 1), 719–729.
23. Whitaker, A. H., Van Rossem, R., Feldman, J. F., Schonfeld, I. S., Pinto-Martin, J. A., Tore, C., et al. (1997). Psychiatric outcomes in low-birth-weight children at age 6 years: Relation to neonatal cranial ultrasound abnormalities. *Archives of General Psychiatry, 54*(9), 847–856.
24. Hintz, S. R., Barnes, P. D., Bulas, D., Slovis, T. L., Finer, N. N., Wrage, L. A., et al. (2015). Neuroimaging and neurodevelopmental outcome in extremely preterm infants. *Pediatrics, 135*(1), e32–e42. https://doi.org/10.1542/peds.2014-0898.
25. Limperopoulos, C., Bassan, H., Gauvreau, K., Robertson, R. L., Jr., Sullivan, N. R., Benson, C. B., et al. (2007). Does cerebellar injury in premature infants contribute to the high prevalence of long-term cognitive, learning, and behavioral disability in survivors? *Pediatrics, 120*(3), 584–593. https://doi.org/10.1542/peds.2007-1041.
26. Broitman, E., Ambalavanan, N., Higgins, R. D., Vohr, B. R., Das, A., Bhaskar, B., et al. (2007). Clinical data predict neurodevelopmental outcome better than head ultrasound in extremely low birth weight infants. *The Journal of Pediatrics, 151*(5), 500–505.e1–2. https://doi.org/10.1016/j.jpeds.2007.04.013.

27. Levene, M. I. (1988). Is neonatal cerebral ultrasound just for the voyeur? *Archives of Disease in Childhood, 63*(1), 1–2.
28. Mann, P. C., Woodrum, D. E., & Wilfond, B. S. (2013). Fuzzy images: Ethical implications of using routine neuroimaging in premature neonates to predict neurologic outcomes. *The Journal of Pediatrics, 163*(2), 587–592. https://doi.org/10.1016/j.jpeds.2013.03.055.
29. Ment, L. R., Bada, H. S., Barnes, P., Grant, P. E., Hirtz, D., Papile, L. A., et al. (2002). Practice parameter: Neuroimaging of the neonate: Report of the Quality Standards Subcommittee of the American Academy of Neurology and the Practice Committee of the Child Neurology Society. *Neurology, 58*(12), 1726–1738.
30. (2001). Routine screening cranial ultrasound examinations for the prediction of long term neurodevelopmental outcomes in preterm infants. *Paediatrics & Child Health, 6*(1), 39–52.
31. Babcock, D. S., & Ball, W., Jr. (1983). Postasphyxial encephalopathy in full-term infants: Ultrasound diagnosis. *Radiology, 148*(2), 417–423. https://doi.org/10.1148/radiology.148.2.6867334.
32. Ilves, P., Lintrop, M., Talvik, I., Muug, K., Maipuu, L., & Metsvaht, T. (2009). Low cerebral blood flow velocity and head circumference in infants with severe hypoxic ischemic encephalopathy and poor outcome. *Acta Paediatrica, 98*(3), 459–465. https://doi.org/10.1111/j.1651-2227.2008.01119.x.
33. Amess, P. N., Penrice, J., Wylezinska, M., Lorek, A., Townsend, J., Wyatt, J. S., et al. (1999). Early brain proton magnetic resonance spectroscopy and neonatal neurology related to neurodevelopmental outcome at 1 year in term infants after presumed hypoxic-ischaemic brain injury. *Developmental Medicine and Child Neurology, 41*(7), 436–445.
34. Thayyil, S., Chandrasekaran, M., Taylor, A., Bainbridge, A., Cady, E. B., Chong, W. K., et al. (2010). Cerebral magnetic resonance biomarkers in neonatal encephalopathy: A meta-analysis. *Pediatrics, 125*(2), e382–e395. https://doi.org/10.1542/peds.2009-1046.
35. Shankaran, S., McDonald, S. A., Laptook, A. R., Hintz, S. R., Barnes, P. D., Das, A., et al. (2015). Neonatal magnetic resonance imaging pattern of brain injury as a biomarker of childhood outcomes following a trial of hypothermia for neonatal hypoxic-ischemic encephalopathy. *The Journal of Pediatrics, 167*(5), 987–993.e983. https://doi.org/10.1016/j.jpeds.2015.08.013.
36. Cheong, J. L., Coleman, L., Hunt, R. W., Lee, K. J., Doyle, L. W., Inder, T. E., & Jacobs, S. E. (2012). Prognostic utility of magnetic resonance imaging in neonatal hypoxic-ischemic encephalopathy: Substudy of a randomized trial. *Archives of Pediatrics & Adolescent Medicine, 166*(7), 634–640. https://doi.org/10.1001/archpediatrics.2012.284.
37. Rutherford, M. A., Pennock, J. M., Counsell, S. J., Mercuri, E., Cowan, F. M., Dubowitz, L. M., & Edwards, A. D. (1998). Abnormal magnetic resonance signal in the internal capsule predicts poor neurodevelopmental outcome in infants with hypoxic-ischemic encephalopathy. *Pediatrics, 102*(2 Pt 1), 323–328.
38. Rutherford, M. A., Pennock, J. M., & Dubowitz, L. M. (1994). Cranial ultrasound and magnetic resonance imaging in hypoxic-ischaemic encephalopathy: A comparison with outcome. *Developmental Medicine and Child Neurology, 36*(9), 813–825.
39. Alderliesten, T., de Vries, L. S., Benders, M. J., Koopman, C., & Groenendaal, F. (2011). MR imaging and outcome of term neonates with perinatal asphyxia: Value of diffusion-weighted MR imaging and (1)H MR spectroscopy. *Radiology, 261*(1), 235–242. https://doi.org/10.1148/radiol.11110213.
40. Hanrahan, J. D., Cox, I. J., Azzopardi, D., Cowan, F. M., Sargentoni, J., Bell, J. D., et al. (1999). Relation between proton magnetic resonance spectroscopy within 18 hours of birth asphyxia and neurodevelopment at 1 year of age. *Developmental Medicine and Child Neurology, 41*(2), 76–82.
41. Roelants-Van Rijn, A. M., van der Grond, J., de Vries, L. S., & Groenendaal, F. (2001). Value of (1)H-MRS using different echo times in neonates with cerebral hypoxia-ischemia. *Pediatric Research, 49*(3), 356–362. https://doi.org/10.1203/00006450-200105000-00015.
42. Archer, L. N., Levene, M. I., & Evans, D. H. (1986). Cerebral artery Doppler ultrasonography for prediction of outcome after perinatal asphyxia. *Lancet, 2*(8516), 1116–1118.

43. Gray, P. H., Tudehope, D. I., Masel, J. P., Burns, Y. R., Mohay, H. A., O'Callaghan, M. J., & Williams, G. M. (1993). Perinatal hypoxic-ischaemic brain injury: Prediction of outcome. *Developmental Medicine and Child Neurology, 35*(11), 965–973.
44. Levene, M. I., Fenton, A. C., Evans, D. H., Archer, L. N., Shortland, D. B., & Gibson, N. A. (1989). Severe birth asphyxia and abnormal cerebral blood-flow velocity. *Developmental Medicine and Child Neurology, 31*(4), 427–434.
45. Eken, P., Toet, M. C., Groenendaal, F., & de Vries, L. S. (1995). Predictive value of early neuroimaging, pulsed Doppler and neurophysiology in full term infants with hypoxic-ischaemic encephalopathy. *Archives of Disease in Childhood. Fetal and Neonatal Edition, 73*(2), F75–F80.
46. Pryds, O., Greisen, G., Lou, H., & Friis-Hansen, B. (1990). Vasoparalysis associated with brain damage in asphyxiated term infants. *The Journal of Pediatrics, 117*(1 Pt 1), 119–125.
47. Seibert, J. J., McCowan, T. C., Chadduck, W. M., Adametz, J. R., Glasier, C. M., Williamson, S. L., et al. (1989). Duplex pulsed Doppler US versus intracranial pressure in the neonate: Clinical and experimental studies. *Radiology, 171*(1), 155–159. https://doi.org/10.1148/radiology.171.1.2648468.
48. Bennhagen, R. G., Weintraub, R. G., Lundstrom, N. R., & Svenningsen, N. W. (1998). Hypoxic-ischaemic encephalopathy is associated with regional changes in cerebral blood flow velocity and alterations in cardiovascular function. *Biology of the Neonate, 73*(5), 275–286.
49. Ilves, P., Talvik, R., & Talvik, T. (1998). Changes in Doppler ultrasonography in asphyxiated term infants with hypoxic-ischaemic encephalopathy. *Acta Paediatrica, 87*(6), 680–684.
50. Ilves, P., Lintrop, M., Metsvaht, T., Vaher, U., & Talvik, T. (2004). Cerebral blood-flow velocities in predicting outcome of asphyxiated newborn infants. *Acta Paediatrica, 93*(4), 523–528.
51. Elstad, M., Whitelaw, A., & Thoresen, M. (2011). Cerebral Resistance Index is less predictive in hypothermic encephalopathic newborns. *Acta Paediatrica, 100*(10), 1344–1349. https://doi.org/10.1111/j.1651-2227.2011.02327.x.
52. Dubowitz, L. M., & Bydder, G. M. (1985). Nuclear magnetic resonance imaging in the diagnosis and follow-up of neonatal cerebral injury. *Clinics in Perinatology, 12*(1), 243–260.
53. Barkovich, A. J., & Truwit, C. L. (1990). Brain damage from perinatal asphyxia: Correlation of MR findings with gestational age. *AJNR. American Journal of Neuroradiology, 11*(6), 1087–1096.
54. McArdle, C. B., Richardson, C. J., Hayden, C. K., Nicholas, D. A., & Amparo, E. G. (1987). Abnormalities of the neonatal brain: MR imaging. Part II. Hypoxic-ischemic brain injury. *Radiology, 163*(2), 395–403. https://doi.org/10.1148/radiology.163.2.3550882.
55. de Bruine, F. T., van den Berg-Huysmans, A. A., Leijser, L. M., Rijken, M., Steggerda, S. J., van der Grond, J., & van Wezel-Meijler, G. (2011). Clinical implications of MR imaging findings in the white matter in very preterm infants: A 2-year follow-up study. *Radiology, 261*(3), 899–906. https://doi.org/10.1148/radiol.11110797.
56. Hart, A., Whitby, E., Wilkinson, S., Alladi, S., Paley, M., & Smith, M. (2011). Neurodevelopmental outcome at 18 months in premature infants with diffuse excessive high signal intensity on MR imaging of the brain. *Pediatric Radiology, 41*(10), 1284–1292. https://doi.org/10.1007/s00247-011-2155-7.
57. Inder, T. E., Wells, S. J., Mogridge, N. B., Spencer, C., & Volpe, J. J. (2003). Defining the nature of the cerebral abnormalities in the premature infant: A qualitative magnetic resonance imaging study. *The Journal of Pediatrics, 143*(2), 171–179. https://doi.org/10.1067/s0022-3476(03)00357-3.
58. Rutherford, M. A., Supramaniam, V., Ederies, A., Chew, A., Bassi, L., Groppo, M., et al. (2010). Magnetic resonance imaging of white matter diseases of prematurity. *Neuroradiology, 52*(6), 505–521. https://doi.org/10.1007/s00234-010-0700-y.
59. Dyet, L. E., Kennea, N., Counsell, S. J., Maalouf, E. F., Ajayi-Obe, M., Duggan, P. J., et al. (2006). Natural history of brain lesions in extremely preterm infants studied with serial magnetic resonance imaging from birth and neurodevelopmental assessment. *Pediatrics, 118*(2), 536–548. https://doi.org/10.1542/peds.2005-1866.

60. Murray, A. L., Scratch, S. E., Thompson, D. K., Inder, T. E., Doyle, L. W., Anderson, J. F., & Anderson, P. J. (2014). Neonatal brain pathology predicts adverse attention and processing speed outcomes in very preterm and/or very low birth weight children. *Neuropsychology, 28*(4), 552–562. https://doi.org/10.1037/neu0000071.
61. Omizzolo, C., Scratch, S. E., Stargatt, R., Kidokoro, H., Thompson, D. K., Lee, K. J., et al. (2014). Neonatal brain abnormalities and memory and learning outcomes at 7 years in children born very preterm. *Memory, 22*(6), 605–615. https://doi.org/10.1080/09658211.2013.809765.
62. Kidokoro, H., Anderson, P. J., Doyle, L. W., Woodward, L. J., Neil, J. J., & Inder, T. E. (2014). Brain injury and altered brain growth in preterm infants: Predictors and prognosis. *Pediatrics, 134*(2), e444–e453. https://doi.org/10.1542/peds.2013-2336.
63. Neubauer, V., Djurdjevic, T., Griesmaier, E., Biermayr, M., Gizewski, E. R., & Kiechl-Kohlendorfer, U. (2017). Routine magnetic resonance imaging at term-equivalent age detects brain injury in 25% of a contemporary cohort of very preterm infants. *PLoS One, 12*(1), e0169442. https://doi.org/10.1371/journal.pone.0169442.
64. Tam, E. W., Rosenbluth, G., Rogers, E. E., Ferriero, D. M., Glidden, D., Goldstein, R. B., et al. (2011). Cerebellar hemorrhage on magnetic resonance imaging in preterm newborns associated with abnormal neurologic outcome. *The Journal of Pediatrics, 158*(2), 245–250. https://doi.org/10.1016/j.jpeds.2010.07.049.
65. Malova, M., Rossi, A., Severino, M., Parodi, A., Morana, G., Sannia, A., et al. (2017). Incidental findings on routine brain MRI scans in preterm infants. *Archives of Disease in Childhood. Fetal and Neonatal Edition, 102*(1), F73–f78. https://doi.org/10.1136/archdischild-2015-310333.
66. Morris, Z., Whiteley, W. N., Longstreth, W. T., Jr., Weber, F., Lee, Y. C., Tsushima, Y., et al. (2009). Incidental findings on brain magnetic resonance imaging: Systematic review and meta-analysis. *BMJ, 339*, b3016. https://doi.org/10.1136/bmj.b3016.
67. Miller, S. P., Ferriero, D. M., Leonard, C., Piecuch, R., Glidden, D. V., Partridge, J. C., et al. (2005). Early brain injury in premature newborns detected with magnetic resonance imaging is associated with adverse early neurodevelopmental outcome. *The Journal of Pediatrics, 147*(5), 609–616. https://doi.org/10.1016/j.jpeds.2005.06.033.
68. Ho, T., Dukhovny, D., Zupancic, J. A., Goldmann, D. A., Horbar, J. D., & Pursley, D. M. (2015). Choosing wisely in newborn medicine: Five opportunities to increase value. *Pediatrics, 136*(2), e482–e489. https://doi.org/10.1542/peds.2015-0737.
69. Miller, S., Ferriero, D., Barkovich, A. J., & Silverstein, F. (2002). Practice parameter: Neuroimaging of the neonate: Report of the Quality Standards Subcommittee of the American Academy of Neurology and the Practice Committee of the Child Neurology Society. *Neurology, 59*(10), 1663; author reply 1663–1664.
70. Mirmiran, M., Barnes, P. D., Keller, K., Constantinou, J. C., Fleisher, B. E., Hintz, S. R., & Ariagno, R. L. (2004). Neonatal brain magnetic resonance imaging before discharge is better than serial cranial ultrasound in predicting cerebral palsy in very low birth weight preterm infants. *Pediatrics, 114*(4), 992–998. https://doi.org/10.1542/peds.2003-0772-L.
71. Smyser, C. D., Kidokoro, H., & Inder, T. E. (2012). Magnetic resonance imaging of the brain at term equivalent age in extremely premature neonates: To scan or not to scan? *Journal of Paediatrics and Child Health, 48*(9), 794–800. https://doi.org/10.1111/j.1440-1754.2012.02535.x.
72. Miller, S. P., Ramaswamy, V., Michelson, D., Barkovich, A. J., Holshouser, B., Wycliffe, N., et al. (2005). Patterns of brain injury in term neonatal encephalopathy. *The Journal of Pediatrics, 146*(4), 453–460. https://doi.org/10.1016/j.jpeds.2004.12.026.
73. Barkovich, A. J., Westmark, K., Partridge, C., Sola, A., & Ferriero, D. M. (1995). Perinatal asphyxia: MR findings in the first 10 days. *AJNR. American Journal of Neuroradiology, 16*(3), 427–438.
74. Li, A. M., Chau, V., Poskitt, K. J., Sargent, M. A., Lupton, B. A., Hill, A., et al. (2009). White matter injury in term newborns with neonatal encephalopathy. *Pediatric Research, 65*(1), 85–89. https://doi.org/10.1203/PDR.0b013e31818912d2.

75. Biagioni, E., Mercuri, E., Rutherford, M., Cowan, F., Azzopardi, D., Frisone, M. F., et al. (2001). Combined use of electroencephalogram and magnetic resonance imaging in full-term neonates with acute encephalopathy. *Pediatrics, 107*(3), 461–468.
76. Kuenzle, C., Baenziger, O., Martin, E., Thun-Hohenstein, L., Steinlin, M., Good, M., et al. (1994). Prognostic value of early MR imaging in term infants with severe perinatal asphyxia. *Neuropediatrics, 25*(4), 191–200. https://doi.org/10.1055/s-2008-1073021.
77. Barkovich, A. J., Hajnal, B. L., Vigneron, D., Sola, A., Partridge, J. C., Allen, F., & Ferriero, D. M. (1998). Prediction of neuromotor outcome in perinatal asphyxia: Evaluation of MR scoring systems. *AJNR. American Journal of Neuroradiology, 19*(1), 143–149.
78. Robertson, R. L., Ben-Sira, L., Barnes, P. D., Mulkern, R. V., Robson, C. D., Maier, S. E., et al. (1999). MR line-scan diffusion-weighted imaging of term neonates with perinatal brain ischemia. *AJNR. American Journal of Neuroradiology, 20*(9), 1658–1670.
79. Rutherford, M., Counsell, S., Allsop, J., Boardman, J., Kapellou, O., Larkman, D., et al. (2004). Diffusion-weighted magnetic resonance imaging in term perinatal brain injury: A comparison with site of lesion and time from birth. *Pediatrics, 114*(4), 1004–1014. https://doi.org/10.1542/peds.2004-0222.
80. Bednarek, N., Mathur, A., Inder, T., Wilkinson, J., Neil, J., & Shimony, J. (2012). Impact of therapeutic hypothermia on MRI diffusion changes in neonatal encephalopathy. *Neurology, 78*(18), 1420–1427. https://doi.org/10.1212/WNL.0b013e318253d589.
81. Boichot, C., Walker, P. M., Durand, C., Grimaldi, M., Chapuis, S., Gouyon, J. B., & Brunotte, F. (2006). Term neonate prognoses after perinatal asphyxia: Contributions of MR imaging, MR spectroscopy, relaxation times, and apparent diffusion coefficients. *Radiology, 239*(3), 839–848. https://doi.org/10.1148/radiol.2393050027.
82. McKinstry, R. C., Miller, J. H., Snyder, A. Z., Mathur, A., Schefft, G. L., Almli, C. R., et al. (2002). A prospective, longitudinal diffusion tensor imaging study of brain injury in newborns. *Neurology, 59*(6), 824–833.
83. Alderliesten, T., de Vries, L. S., Staats, L., van Haastert, I. C., Weeke, L., Benders, M. J., et al. (2017). MRI and spectroscopy in (near) term neonates with perinatal asphyxia and therapeutic hypothermia. *Archives of Disease in Childhood. Fetal and Neonatal Edition, 102*(2), F147–f152. https://doi.org/10.1136/archdischild-2016-310514.
84. Vermeulen, R. J., van Schie, P. E., Hendrikx, L., Barkhof, F., van Weissenbruch, M., Knol, D. L., & Pouwels, P. J. (2008). Diffusion-weighted and conventional MR imaging in neonatal hypoxic ischemia: Two-year follow-up study. *Radiology, 249*(2), 631–639. https://doi.org/10.1148/radiol.2492071581.
85. Barnett, A., Mercuri, E., Rutherford, M., Haataja, L., Frisone, M. F., Henderson, S., et al. (2002). Neurological and perceptual-motor outcome at 5–6 years of age in children with neonatal encephalopathy: Relationship with neonatal brain MRI. *Neuropediatrics, 33*(5), 242–248. https://doi.org/10.1055/s-2002-36737.
86. Cady, E. B., Costello, A. M., Dawson, M. J., Delpy, D. T., Hope, P. L., Reynolds, E. O., et al. (1983). Non-invasive investigation of cerebral metabolism in newborn infants by phosphorus nuclear magnetic resonance spectroscopy. *Lancet, 1*(8333), 1059–1062.
87. Hope, P. L., Costello, A. M., Cady, E. B., Delpy, D. T., Tofts, P. S., Chu, A., et al. (1984). Cerebral energy metabolism studied with phosphorus NMR spectroscopy in normal and birth-asphyxiated infants. *Lancet, 2*(8399), 366–370.
88. Lorek, A., Takei, Y., Cady, E. B., Wyatt, J. S., Penrice, J., Edwards, A. D., et al. (1994). Delayed ("secondary") cerebral energy failure after acute hypoxia-ischemia in the newborn piglet: Continuous 48-hour studies by phosphorus magnetic resonance spectroscopy. *Pediatric Research, 36*(6), 699–706.
89. Wyatt, J. S., Edwards, A. D., Azzopardi, D., & Reynolds, E. O. (1989). Magnetic resonance and near infrared spectroscopy for investigation of perinatal hypoxic-ischaemic brain injury. *Archives of Disease in Childhood, 64*(7 Spec No), 953–963.
90. Azzopardi, D., Wyatt, J. S., Cady, E. B., Delpy, D. T., Baudin, J., Stewart, A. L., et al. (1989). Prognosis of newborn infants with hypoxic-ischemic brain injury assessed by phos-

phorus magnetic resonance spectroscopy. *Pediatric Research, 25*(5), 445–451. https://doi.org/10.1203/00006450-198905000-00004.
91. Roth, S. C., Edwards, A. D., Cady, E. B., Delpy, D. T., Wyatt, J. S., Azzopardi, D., et al. (1992). Relation between cerebral oxidative metabolism following birth asphyxia, and neurodevelopmental outcome and brain growth at one year. *Developmental Medicine and Child Neurology, 34*(4), 285–295.
92. Groenendaal, F., Veenhoven, R. H., van der Grond, J., Jansen, G. H., Witkamp, T. D., & de Vries, L. S. (1994). Cerebral lactate and N-acetyl-aspartate/choline ratios in asphyxiated full-term neonates demonstrated in vivo using proton magnetic resonance spectroscopy. *Pediatric Research, 35*(2), 148–151. https://doi.org/10.1203/00006450-199402000-00004.
93. Miller, S. P., Newton, N., Ferriero, D. M., Partridge, J. C., Glidden, D. V., Barnwell, A., et al. (2002). Predictors of 30-month outcome after perinatal depression: Role of proton MRS and socioeconomic factors. *Pediatric Research, 52*(1), 71–77. https://doi.org/10.1203/00006450-200207000-00014.
94. Peden, C. J., Rutherford, M. A., Sargentoni, J., Cox, I. J., Bryant, D. J., & Dubowitz, L. M. (1993). Proton spectroscopy of the neonatal brain following hypoxic-ischaemic injury. *Developmental Medicine and Child Neurology, 35*(6), 502–510.
95. Shanmugalingam, S., Thornton, J. S., Iwata, O., Bainbridge, A., O'Brien, F. E., Priest, A. N., et al. (2006). Comparative prognostic utilities of early quantitative magnetic resonance imaging spin-spin relaxometry and proton magnetic resonance spectroscopy in neonatal encephalopathy. *Pediatrics, 118*(4), 1467–1477. https://doi.org/10.1542/peds.2005-2976.
96. Barkovich, A. J., Baranski, K., Vigneron, D., Partridge, J. C., Hallam, D. K., Hajnal, B. L., & Ferriero, D. M. (1999). Proton MR spectroscopy for the evaluation of brain injury in asphyxiated, term neonates. *AJNR. American Journal of Neuroradiology, 20*(8), 1399–1405.
97. Brissaud, O., Chateil, J. F., Bordessoules, M., & Brun, M. (2005). Chemical shift imaging and localised magnetic resonance spectroscopy in full-term asphyxiated neonates. *Pediatric Radiology, 35*(10), 998–1005. https://doi.org/10.1007/s00247-005-1524-5.
98. Zarifi, M. K., Astrakas, L. G., Poussaint, T. Y., Plessis Ad, A., Zurakowski, D., & Tzika, A. A. (2002). Prediction of adverse outcome with cerebral lactate level and apparent diffusion coefficient in infants with perinatal asphyxia. *Radiology, 225*(3), 859–870. https://doi.org/10.1148/radiol.2253011797.
99. Hanrahan, J. D., Cox, I. J., Edwards, A. D., Cowan, F. M., Sargentoni, J., Bell, J. D., et al. (1998). Persistent increases in cerebral lactate concentration after birth asphyxia. *Pediatric Research, 44*(3), 304–311. https://doi.org/10.1203/00006450-199809000-00007.
100. Penrice, J., Cady, E. B., Lorek, A., Wylezinska, M., Amess, P. N., Aldridge, R. F., et al. (1996). Proton magnetic resonance spectroscopy of the brain in normal preterm and term infants, and early changes after perinatal hypoxia-ischemia. *Pediatric Research, 40*(1), 6–14. https://doi.org/10.1203/00006450-199607000-00002.
101. Wisnowski, J. L., Wu, T. W., Reitman, A. J., McLean, C., Friedlich, P., Vanderbilt, D., et al. (2016). The effects of therapeutic hypothermia on cerebral metabolism in neonates with hypoxic-ischemic encephalopathy: An in vivo 1H-MR spectroscopy study. *Journal of Cerebral Blood Flow and Metabolism, 36*(6), 1075–1086. https://doi.org/10.1177/0271678x15607881.
102. De Vis, J. B., Hendrikse, J., Petersen, E. T., de Vries, L. S., van Bel, F., Alderliesten, T., et al. (2015). Arterial spin-labelling perfusion MRI and outcome in neonates with hypoxic-ischemic encephalopathy. *European Radiology, 25*(1), 113–121. https://doi.org/10.1007/s00330-014-3352-1.
103. Wu, D., & Zhang, J. (2016). Recent progress in magnetic resonance imaging of the embryonic and neonatal mouse brain. *Frontiers in Neuroanatomy, 10*, 18. https://doi.org/10.3389/fnana.2016.00018.
104. Porter, E. J., Counsell, S. J., Edwards, A. D., Allsop, J., & Azzopardi, D. (2010). Tract-based spatial statistics of magnetic resonance images to assess disease and treatment effects in perinatal asphyxial encephalopathy. *Pediatric Research, 68*(3), 205–209. https://doi.org/10.1203/00006450-201011001-00399, https://doi.org/10.1203/PDR.0b013e3181e9f1ba.

105. Massaro, A. N., Evangelou, I., Fatemi, A., Vezina, G., McCarter, R., Glass, P., & Limperopoulos, C. (2015). White matter tract integrity and developmental outcome in newborn infants with hypoxic-ischemic encephalopathy treated with hypothermia. *Developmental Medicine and Child Neurology, 57*(5), 441–448. https://doi.org/10.1111/dmcn.12646.
106. Roze, E., Benders, M. J., Kersbergen, K. J., van der Aa, N. E., Groenendaal, F., van Haastert, I. C., et al. (2015). Neonatal DTI early after birth predicts motor outcome in preterm infants with periventricular hemorrhagic infarction. *Pediatric Research, 78*(3), 298–303. https://doi.org/10.1038/pr.2015.94.
107. Gao, J., Li, X., Li, Y., Zeng, L., Jin, C., Sun, Q., et al. (2016). Differentiating T2 hyperintensity in neonatal white matter by two-compartment model of diffusional kurtosis imaging. *Scientific Reports, 6*, 24473. https://doi.org/10.1038/srep24473.
108. Barnette, A. R., Horbar, J. D., Soll, R. F., Pfister, R. H., Nelson, K. B., Kenny, M. J., et al. (2014). Neuroimaging in the evaluation of neonatal encephalopathy. *Pediatrics, 133*(6), e1508–e1517. https://doi.org/10.1542/peds.2013-4247.
109. Blennow, M., Ingvar, M., Lagercrantz, H., Stone-Elander, S., Eriksson, L., Forssberg, H., et al. (1995). Early [18F]FDG positron emission tomography in infants with hypoxic-ischaemic encephalopathy shows hypermetabolism during the postasphyctic period. *Acta Paediatrica, 84*(11), 1289–1295.
110. Taboada, D., Alonso, A., Olague, R., Mulas, F., & Andres, V. (1980). Radiological diagnosis of periventricular and subcortical leukomalacia. *Neuroradiology, 20*(1), 33–41.
111. Roland, E. H., Poskitt, K., Rodriguez, E., Lupton, B. A., & Hill, A. (1998). Perinatal hypoxic-ischemic thalamic injury: Clinical features and neuroimaging. *Annals of Neurology, 44*(2), 161–166. https://doi.org/10.1002/ana.410440205.
112. Chau, V., Poskitt, K. J., Sargent, M. A., Lupton, B. A., Hill, A., Roland, E., & Miller, S. P. (2009). Comparison of computer tomography and magnetic resonance imaging scans on the third day of life in term newborns with neonatal encephalopathy. *Pediatrics, 123*(1), 319–326. https://doi.org/10.1542/peds.2008-0283.
113. Flodmark, O., Becker, L. E., Harwood-Nash, D. C., Fitzhardinge, P. M., Fitz, C. R., & Chuang, S. H. (1980). Correlation between computed tomography and autopsy in premature and full-term neonates that have suffered perinatal asphyxia. *Radiology, 137*(1 Pt 1), 93–103. https://doi.org/10.1148/radiology.137.1.7422867.
114. Khalil, A., Suff, N., Thilaganathan, B., Hurrell, A., Cooper, D., & Carvalho, J. S. (2014). Brain abnormalities and neurodevelopmental delay in congenital heart disease: Systematic review and meta-analysis. *Ultrasound in Obstetrics & Gynecology, 43*(1), 14–24. https://doi.org/10.1002/uog.12526.
115. Owen, M., Shevell, M., Majnemer, A., & Limperopoulos, C. (2011). Abnormal brain structure and function in newborns with complex congenital heart defects before open heart surgery: A review of the evidence. *Journal of Child Neurology, 26*(6), 743–755. https://doi.org/10.1177/0883073811402073.
116. Rios, D. R., Welty, S. E., Gunn, J. K., Beca, J., Minard, C. G., Goldsworthy, M., et al. (2013). Usefulness of routine head ultrasound scans before surgery for congenital heart disease. *Pediatrics, 131*(6), e1765–e1770.
117. Kinney, H. C., Panigrahy, A., Newburger, J. W., Jonas, R. A., & Sleeper, L. A. (2005). Hypoxic-ischemic brain injury in infants with congenital heart disease dying after cardiac surgery. *Acta Neuropathologica, 110*(6), 563–578. https://doi.org/10.1007/s00401-005-1077-6.
118. Clouchoux, C., du Plessis, A. J., Bouyssi-Kobar, M., Tworetzky, W., McElhinney, D. B., Brown, D. W., et al. (2013). Delayed cortical development in fetuses with complex congenital heart disease. *Cerebral Cortex, 23*(12), 2932–2943. https://doi.org/10.1093/cercor/bhs281.
119. Limperopoulos, C., Tworetzky, W., McElhinney, D. B., Newburger, J. W., Brown, D. W., Robertson, R. L., Jr., et al. (2010). Brain volume and metabolism in fetuses with congenital heart disease: Evaluation with quantitative magnetic resonance imaging and spectroscopy. *Circulation, 121*(1), 26–33. https://doi.org/10.1161/circulationaha.109.865568.
120. Licht, D. J., Shera, D. M., Clancy, R. R., Wernovsky, G., Montenegro, L. M., Nicolson, S. C., et al. (2009). Brain maturation is delayed in infants with complex congenital heart defects.

The Journal of Thoracic and Cardiovascular Surgery, 137(3), 529–536.; ; discussion 536–527. https://doi.org/10.1016/j.jtcvs.2008.10.025.

121. Andropoulos, D. B., Hunter, J. V., Nelson, D. P., Stayer, S. A., Stark, A. R., McKenzie, E. D., et al. (2010). Brain immaturity is associated with brain injury before and after neonatal cardiac surgery with high-flow bypass and cerebral oxygenation monitoring. *The Journal of Thoracic and Cardiovascular Surgery, 139*(3), 543–556. https://doi.org/10.1016/j.jtcvs.2009.08.022.
122. Latal, B., Kellenberger, C., Dimitropoulos, A., Hagmann, C., Balmer, C., Beck, I., & Bernet, V. (2015). Can preoperative cranial ultrasound predict early neurodevelopmental outcome in infants with congenital heart disease? *Developmental Medicine and Child Neurology*. https://doi.org/10.1111/dmcn.12701.
123. Cheng, H. H., Wypij, D., Laussen, P. C., Bellinger, D. C., Stopp, C. D., Soul, J. S., et al. (2014). Cerebral blood flow velocity and neurodevelopmental outcome in infants undergoing surgery for congenital heart disease. *The Annals of Thoracic Surgery, 98*(1), 125–132. https://doi.org/10.1016/j.athoracsur.2014.03.035.
124. Williams, I. A., Tarullo, A. R., Grieve, P. G., Wilpers, A., Vignola, E. F., Myers, M. M., & Fifer, W. P. (2012). Fetal cerebrovascular resistance and neonatal EEG predict 18-month neurodevelopmental outcome in infants with congenital heart disease. *Ultrasound in Obstetrics & Gynecology, 40*(3), 304–309. https://doi.org/10.1002/uog.11144.
125. Mulkey, S. B., Swearingen, C. J., Melguizo, M. S., Schmitz, M. L., Ou, X., Ramakrishnaiah, R. H., et al. (2013). Multi-tiered analysis of brain injury in neonates with congenital heart disease. *Pediatric Cardiology, 34*(8), 1772–1784. https://doi.org/10.1007/s00246-013-0712-6.
126. Bellinger, D. C., Watson, C. G., Rivkin, M. J., Robertson, R. L., Roberts, A. E., Stopp, C., et al. (2015). Neuropsychological status and structural brain imaging in adolescents with single ventricle who underwent the Fontan procedure. *Journal of the American Heart Association, 4*(12). https://doi.org/10.1161/jaha.115.002302.
127. Marino, B. S., Lipkin, P. H., Newburger, J. W., Peacock, G., Gerdes, M., Gaynor, J. W., et al. (2012). Neurodevelopmental outcomes in children with congenital heart disease: Evaluation and management: A scientific statement from the American Heart Association. *Circulation, 126*(9), 1143–1172. https://doi.org/10.1161/CIR.0b013e318265ee8a.
128. Bulas, D. I., Taylor, G. A., O'Donnell, R. M., Short, B. L., Fitz, C. R., & Vezina, G. (1996). Intracranial abnormalities in infants treated with extracorporeal membrane oxygenation: Update on sonographic and CT findings. *AJNR. American Journal of Neuroradiology, 17*(2), 287–294.
129. Griffin, M. P., Minifee, P. K., Landry, S. H., Allison, P. L., Swischuk, L. E., & Zwischenberger, J. B. (1992). Neurodevelopmental outcome in neonates after extracorporeal membrane oxygenation: Cranial magnetic resonance imaging and ultrasonography correlation. *Journal of Pediatric Surgery, 27*(1), 33–35.
130. Rollins, M. D., Yoder, B. A., Moore, K. R., Barnhart, D. C., Jones, C., Null, D. M., & DiGeronimo, R. J. (2012). Utility of neuroradiographic imaging in predicting outcomes after neonatal extracorporeal membrane oxygenation. *Journal of Pediatric Surgery, 47*(1), 76–80. https://doi.org/10.1016/j.jpedsurg.2011.10.016.
131. Lago, P., Rebsamen, S., Clancy, R. R., Pinto-Martin, J., Kessler, A., Zimmerman, R., et al. (1995). MRI, MRA, and neurodevelopmental outcome following neonatal ECMO. *Pediatric Neurology, 12*(4), 294–304.
132. Glass, P., Bulas, D. I., Wagner, A. E., Rajasingham, S. R., Civitello, L. A., Papero, P. H., et al. (1997). Severity of brain injury following neonatal extracorporeal membrane oxygenation and outcome at age 5 years. *Developmental Medicine and Child Neurology, 39*(7), 441–448.
133. Lazar, E. L., Abramson, S. J., Weinstein, S., & Stolar, C. J. (1994). Neuroimaging of brain injury in neonates treated with extracorporeal membrane oxygenation: Lessons learned from serial examinations. *Journal of Pediatric Surgery, 29*(2), 186–190; discussion 190–181.
134. Vaucher, Y. E., Dudell, G. G., Bejar, R., & Gist, K. (1996). Predictors of early childhood outcome in candidates for extracorporeal membrane oxygenation. *The Journal of Pediatrics, 128*(1), 109–117.

135. Spring In "t Veld, L. G., de Vries, L. S., Alderliesten, T., Benders, M. J., & Groenendaal, F. (2016). Serial 1- and 2-dimensional cerebral MRI measurements in full-term infants after perinatal asphyxia. *Neonatology, 110*(1), 27–32. https://doi.org/10.1159/000444121.
136. Byrne, P., Welch, R., Johnson, M. A., Darrah, J., & Piper, M. (1990). Serial magnetic resonance imaging in neonatal hypoxic-ischemic encephalopathy. *The Journal of Pediatrics, 117*(5), 694–700.
137. Van Kooij, B., Van Handel, M., Uiterwaal, C. S., Groenendaal, F., Nievelstein, R. A., Rademaker, K. J., et al. (2008). Corpus callosum size in relation to motor performance in 9- to 10-year-old children with neonatal encephalopathy. *Pediatric Research, 63*(1), 103–108. https://doi.org/10.1203/PDR.0b013e31815b4435.
138. Nagy, Z., Lindstrom, K., Westerberg, H., Skare, S., Andersson, J., Hallberg, B., et al. (2005). Diffusion tensor imaging on teenagers, born at term with moderate hypoxic-ischemic encephalopathy. *Pediatric Research, 58*(5), 936–940. https://doi.org/10.1203/01.pdr.0000186516.85702.61.
139. Robertson, N. J., Cox, I. J., Cowan, F. M., Counsell, S. J., Azzopardi, D., & Edwards, A. D. (1999). Cerebral intracellular lactic alkalosis persisting months after neonatal encephalopathy measured by magnetic resonance spectroscopy. *Pediatric Research, 46*(3), 287–296. https://doi.org/10.1203/00006450-199909000-00007.
140. Cordes, I., Roland, E. H., Lupton, B. A., & Hill, A. (1994). Early prediction of the development of microcephaly after hypoxic-ischemic encephalopathy in the full-term newborn. *Pediatrics, 93*(5), 703–707.
141. Shevell, M., Ashwal, S., Donley, D., Flint, J., Gingold, M., Hirtz, D., et al. (2003). Practice parameter: Evaluation of the child with global developmental delay: Report of the Quality Standards Subcommittee of the American Academy of Neurology and The Practice Committee of the Child Neurology Society. *Neurology, 60*(3), 367–380.
142. Janvier, A., & Barrington, K. (2012). Trying to predict the future of ex-preterm infants: Who benefits from a brain MRI at term? *Acta Paediatrica, 101*(10), 1016–1017. https://doi.org/10.1111/j.1651-2227.2012.02788.x.
143. Pearce, R., & Baardsnes, J. (2012). Term MRI for small preterm babies: Do parents really want to know and why has nobody asked them? *Acta Paediatrica, 101*(10), 1013–1015. https://doi.org/10.1111/j.1651-2227.2012.02767.x.

Social-Emotional Development in Early Childhood: Normative, NICU Considerations, and Application in NICU Follow-Up Programs for At-Risk Infants and Their Families

Robin H. Adair

Abstract Experienced NICU clinicians are familiar with the dramatic change in social interactions seen in their newborn patients as premature infants progress toward their due dates or near- or full-term infants progress from acute illness to recovery. The interactions gradually increase in frequency and become more sophisticated as the infant matures and/or stabilizes. Talking to the infant evolves from minimal to encouraging to fun, while the infant's ability to respond evolves from minimal to aware to engaging. These changes—both infant and caregiver—are both cause and effect of the infant's social-emotional development.

Social-emotional development is a relatively new term for what has historically been referred to as infant mental health. In 2001, *Zero to Three*, an international organization dedicated to science and policy in support of early childhood development, convened a multi-professional steering committee charged with defining "infant mental health." The resulting definition became widely accepted:

> Infant Mental Health is the developing capacity of the child from birth to three to: experience, regulate, and express emotions; form close interpersonal relationships; and explore the environment and learn - all in the context of family, community and cultural expectations for young children. Infant mental health is synonymous with healthy social and emotional development. (*Zero to Three*, 2001; from Zeanah PD, Stafford BS, Nagle GA, Rice T. *Addressing Social-Emotional Development and Infant Mental Health in Early Childhood Systems, National Center for Infant and Early Childhood Health Policy,* January 2005; http://files.eric.ed.gov/fulltext/ED496853.pdf)

More recently, in its web-based parenting resource, "Tips for Promoting Social-Emotional Development," *Zero to Three* states:

R. H. Adair (✉)
Division, Developmental and Behavioral Pediatrics, University of Rochester School of Medicine and Dentistry, Rochester, NY, USA
e-mail: robin_adair@urmc.rochester.edu

H. Needelman, B. J. Jackson (eds.), *Follow-Up for NICU Graduates*,
https://doi.org/10.1007/978-3-319-73275-6_4

Social-emotional wellness is often known as infant mental health by early childhood professionals. In a nutshell, it is developing the capacity to experience and regulate emotions, form secure relationships, and explore and learn. (https://www.zerotothree.org/resources/225-tips-for-promoting-social-emotional-development)

Social-Emotional Development in the First Year of Life

Foundations of an infant's social-emotional development have been well described by many well-known child development scholars. Key tenets of these theorists are the roles of temperament and sensorimotor interactions of the infant and their interplay while important trusting relationships are developing, listed below and illustrated online (web addresses provided in Appendix 1).

- According to Erik Erikson's theory of psychosocial development (Appendix 1, Reference 1), a primary goal of the first year of life is to establish trust that the world, in the form of the infant's caregivers, will safely and adequately meet the infant's needs. According to Sigmund Freud (Appendix 1, Reference 1), that specifically includes satisfying the infant's oral needs, nutritive and nonnutritive.
- According to Jean Piaget's theory of cognitive development (Appendix 1, Reference 1), the first year of life is steeped in making sense of sensations, internal and environmental, and bodily movements, both random (nondirected) and intentional.
- According to John Bowlby who developed the theory of attachment, and Mary Ainsworth, who built on his work (Appendix 1, Reference 1), an emerging goal at the end of the first year of life is infant attachment to one or more caregivers. The theory was confirmed using Ainsworth's research paradigm, the "Strange Situation," in which a child's reactions to the coming and going of both a friendly stranger and their parent is classified. Secure attachment frees the infant to comfortably explore his/her environment while feeling confident he/she can return to a safe home-base (caregiver) when reassurance is needed.
- According to Stella Chess and Alexander Thomas (with Herbert Birch; Appendix 1, Reference 2), each newborn arrives with nine innate temperamental traits that influence how he/she behaves, particularly to novel stimulation. These traits are activity, distractibility, persistence, sensory threshold (to trigger a response), intensity (of response), approach/withdrawal, mood (general, not "at the moment"), adaptability, and regularity. Each trait ranges from low to high. While three temperament profiles are recognized (easy, difficult, and slow to warm-up) and certain trait directions can be more challenging for others (e.g., a highly persistent child), no trait direction is regarded as consistently good or bad. "Goodness of fit" between a child and his/her caregivers was also conceptualized by Chess and Thomas and is regarded as critical to the emotional development of the child, regardless of the child's temperamental profile.

A child's capacity to develop is highly dependent on one-on-one interactions with caregivers. The child-caregiver pair is referred to as a "dyad" in early childhood development literature, and interactions between them are called dyadic. Dyadic interactions are viewed as crucial to the infant's social-emotional development. "Responsive parenting" training was shown to improve parent responsiveness as well as early social-emotional development measures, particularly for very low birth weight infants [1]. Breakdowns in significant dyadic relationships in the first year of life contribute to distrust according to Erikson and problematic attachment (ambivalent insecure, avoidant/resistant insecure, and disorganized insecure) according to Ainsworth.

"Mutual regulation," a type of interaction found within the normative dyad in the first year of life, was characterized by the research of Edward Tronick using the "Still Face" paradigm (Appendix 1, reference 3). In this research setting, the caregiver (typically the mother) is asked to interact face-to-face with their infant, abruptly stop interacting and continue to look at their infant expressionlessly, and then resume interacting. The parent's change in behavior elicits a range of behaviors in the infant, demonstrating the dynamic exchange of verbal cues and facial expressions, as well as other nonverbal communication, between a child and caregiver used to modify one another's behaviors. This exchange is a cornerstone to healthy social-emotional development and clearly depends on contributions from both the infant and caregiver.

A powerful, positive caregiver trait for promoting a young child's social-emotional development is "mentalization," promulgated by Peter Fonagy and Anthony Bateman (Appendix 1, Reference 4). When using mentalization, the caregiver is attuned to the mind, or perspective, of the infant/child and aware of and responsive to the child's interests and emotions in the moment. A related concept, "reflective functioning" of the caregiver, calls for the caregiver to remain separate from the child. An example of a parent using mentalization and reflective functioning would be a parent who calmly says to their mildly-injured, crying child, "I can tell it hurts! But it will be better soon." A parental response of "Calm down, what's the big deal?" would be under-mentalized. A parental response of "OH MY, this is awful! Come here! I can't believe this happened!" would likely be inadequately separated.

A long-term outcome of an individual's social-emotional development in early childhood is self-regulation, the internal ability to stay emotionally and behaviorally in control of oneself, in culturally appropriate ways, despite distractions, impulses, distress, and other strong influences, both positive and negative [2]. In early childhood, self-regulation is a preferred synonym for self-control. Attributes in older children and adults that would fall into this skill domain are attention, impulse control, and mood stability. Mental health disorders, such as attention deficit hyperactivity disorder, internalizing disorders (e.g., depression, anxiety), and learning disorders have been associated with both high-risk newborn medical status (e.g., very preterm birth; [3]) and adverse early childhood social risk factors and are discussed elsewhere in this book.

Development and Social-Emotional Support in the NICU

As NICU medical interventions advanced, it became increasingly evident that developmental supports should be part of infant care. An understanding of the developmental status of the infant in the NICU is a prerequisite of the crafting of both developmental and social-emotional interventions in the nursery.

The prematurely born infant's successful coordination of sucking, swallowing, and breathing while feeding emerges around 34 weeks corrected gestational age [4], soon followed by successful weight gain and temperature control without increased ambient temperature. Such physiological stability, or lack thereof, is fairly easy to measure, directly as well as by observing the diminishing need for support from nurses and parents.

A prematurely born infant's abilities to process sound, sight, touch, and other sensations also become more reliable after 34 weeks corrected gestational age. Startling to sound becomes listening; winces and foggy gazes become visual fixation and then tracking. The infant's developing ability to "organize" him-/herself in response to stimulation results in less crying and quicker recovery from distress.

One example of developmental support in the NICU is kangaroo mother care (skin-to-skin contact between mother and infant with frequent breastfeeding) for low birth weight infants. It emerged in the 1970s when standard care involved isolating the infants from infectious contaminants using isolettes and limited family visiting policies. A 2016 Cochrane review examined 21 studies that included 3042 infants and concluded that, for stabilized low birth weight infants, kangaroo mother care was preferable to conventional neonatal care, mainly in low-resource settings [5]. Research has shown that briefer periods of skin-to-skin holding of an infant by either the mother or father conveys benefits in the NICU such as promoting parental attachment, decreasing stress, and increasing breast milk volumes [6].

"Developmental care," a structured approach to NICU nursing care, has become a routine part of many NICU nursing protocols and forms the foundation for social-emotional well-being of the infant. It was introduced by Heidelise Als in the 1990s as the Neonatal Individualized Developmental Care and Assessment Program (NIDCAP). It calls for nursing observation of infant cues and developing a care plan that includes "decreasing noxious stimuli, containing the infant in a snug and secure position, performing care activities according to a schedule and allowing the infant time for uninterrupted rest and sleep." Research on NIDCAP and other forms of developmental care has shown variable results [7].

Applying Knowledge of Social-Emotional Development in the NICU Follow-Up Setting

Recognizing that "emotional, behavioral, and relationship problems can develop in very young children, especially those living in high-risk families and communities," the American Academy of Pediatrics recently released a Policy Statement, "Addressing

Early Childhood Emotional and Behavioral Problems" [8]. It summarizes empirically supported treatments for children of all ages. Regarding infants and toddlers, the statement advocates for treatments such as dyadic (child-caregiver) interventions to promote attachment and/or child self-regulation. This comprehensive statement also addresses systematic barriers to interventions and offers recommendations.

While NICU follow-up programs have not historically been identified sites for early identification of these concerns, applying what is now known about social-emotional development is feasible, and fundamental aspects of established interventions can be integrated in the follow-up clinic. The following specifics are suggested.

Role Model

- Utilize mentalization and reflective functioning when speaking with parents and other caregivers.
  - This means keeping in mind the parent/caregiver(s) may be anxious, defensive, depressed, or otherwise unable to report or listen accurately, or not. Your job is to "get a read" on the state of the caregiver and adjust your interactions to meet their state.
  - At the same time, maintain a separation from the parent/caregiver(s), making it possible to objectively make observations and conduct assessments.

"Give the Child a Voice"

- Utilize mentalization and reflective functioning when interacting with the child, regardless of his/her age, to help parents understand what the child is probably experiencing. Periodically throughout the visit:
  - Speak for the child (e.g., "I don't know, Mom, I'm not sure I like this person.")
  - State what you think the child could be thinking or feeling (e.g., "You don't want me to listen to your heart. I'm sorry, I will be quick.")
  - Point out something you think will interest the child (e.g. "What do you think of this toy?")

Narrate

- Comment on what you are seeing for the benefit of the parent/caregiver(s):
  - Point out (preferably with a smile) "good problems" that parallel likely experiences the caregiver has at home (e.g., "I am glad he wants to check out my stethoscope, even though it makes my job more difficult!").
  - Point out something appropriate the infant is doing in which you think the parent will take pride (e.g., "I like how persistently she is working on that puzzle.")

Conclude with a Connection

- During the visit wrap-up, include relatable comments, whether clinical (e.g., "Your little one is a fighter.") or mundane (e.g., about the weather), injected into the specifics of your assessment and recommendations. Tell parents "what a pleasure it is to see your baby." Ask if the toddler gives high-5 (most do!). Make eye contact and smile, maybe even shake hands.

See Appendix 2 for resources that can help caregivers as well as NICU follow-up practitioners acquire a working knowledge of social-emotional development.

NICU follow-up clinicians interested in establishing more formal social-emotional developmental care in their programs will find the "DC:0-5™: Diagnostic Classification of Mental Health and Developmental Disorders of Infancy and Early Childhood," published in 2016, a useful reference. It provides evidence-based guidance on systematically identifying clinical disorders (Axis I), relational context (Axis II), physical health conditions and considerations (Axis III), psychosocial stressors (Axis IV), and developmental competence (Axis V) [9].

Summary

Social-emotional development in early childhood, also referred to as infant mental health, is now viewed as a developmental track, much in the way that language, motor, cognitive (problem-solving), and other skill areas are viewed. Infants are recognized as being born with innate traits that influence how they behave. For high-risk infants hospitalized in the NICU, these innate traits, as well as their abilities to engage with their environment, are generally not robust before 34 weeks corrected gestational age or, in the case of the near- or full-term infant, until the acute illness has stabilized. Developmental care nursing protocols exist for providing support to general and social-emotional development in the NICU environment. NICU follow-up clinicians can apply concepts of social-emotional development, such as psychosocial development, attachment, mutual regulation, and temperament, to inform their interactions with infants, young children, and caregivers in the NICU follow-up setting.

Appendix 1: Online References

1. Background information can be found at https://www.verywell.com/child-development-theories-2795068.
2. Their 1970 publication can be found at http://www.acamedia.info/sciences/sci-literature/origin_of_personality.htm

3. Background information and the classic mutual regulation video can be found at http://scienceblogs.com/thoughtfulanimal/2010/10/18/ed-tronick-and-the-still-face/
4. Background information and application in psychopathology can be found at https://www.mentalhelp.net/articles/attachment-theory-expanded-mentalization.

Appendix 2: Parent Resources

Zero To Three, Developing social-emotional skills (https://www.zerotothree.org/resources/series/developing-social-emotional-skills).

American Academy of Pediatrics, Healthychildren.org, Ages & Stages (https://healthychildren.org/english/ages-stages/baby/Pages/default.aspx).

Your Baby is Speaking to You: A Visual Guide to the Amazing Behaviors of Your Newborn and Growing Baby. A photograph-enriched book by Kevin Nugent (New York: Houghton Mifflin Harcourt, 2011; www.hmhbooks.com).

You Are My World: How a Parent's Love Shapes a Baby's Mind. A small photo book with short captions for parents, by Amy Hatkoff (New York: Stewart, Tabori & Chang, 2007; www.stcbooks.com).

Handbook of Infant Mental Health, Third Edition. Edited by Charles Zeanah (New York: Guilford, 2011; www.guilford.com), an in-depth guide for motivated clinicians.

References

1. Landry, S. H., Smith, K. E., & Swank, P. R. (2006). Responsive parenting: Establishing early foundations for social, communication, and independent problem-solving skills. *Developmental Psychology, 42*(4), 627–642.
2. Beeghly M, Perry B. D., & Tronick E. (2016). Self-regulatory processes in early development. In *The Oxford handbooks of treatment processes and outcomes in psychology: A multidisciplinary, biopsychosocial approach.* https://doi.org/10.1093/oxfordhb/9780199739134.013.3.
3. Montagna, A., & Nosarti, C. (2016). Socio-emotional development following very preterm birth: Pathways to psychopathology. *Frontiers in Psychology, 7*(Article 80), 1–23.
4. McGuire, W., Henderson, G., & Fowlie, P. W. (2004). Feeding the preterm infant. *BMJ, 329*(7476), 1227–1230.
5. Conde-Agudelo, A., & Diaz-Rossello, J. L. (2016). Kangaroo mother care to reduce morbidity and mortality in low birthweight infants. *Cochrane Database of Systematic Reviews, 8*(8), CD002771.
6. Furman, L. (2017). Kangaroo mother care 20 years later: Connecting infants and families. *Pediatrics, 139*(1), e20163332.
7. Goldstein, R. F. (2012). Developmental care for premature infants: A state of mind. *Pediatrics, 129*(5), e1332–e1323.
8. AAP Council on Early Childhood, AAP Committee on Psychosocial Aspects of Child and Family Health, AAP Section on Developmental and Behavioral Pediatrics. (2016). Addressing early childhood emotional and behavioral problems. *Pediatrics, 138*(6), e20163023.
9. ZERO TO THREE. (2016). *DC:0-5™: Diagnostic classification of mental health and developmental disorders of infancy and early childhood.* Washington, DC: ZERO TO THREE.

Part III
Populations at Risk

Premature Infants: Issues Associated with Prematurity

Marie A. Clark and Nina Sand-Loud

Abstract This is the population that has been discussed most extensively in the literature. This chapter will discuss the changing neurodevelopmental outcomes at varying gestational age/birth weight over time as it relates to new interventions in the NICU. Risk factors such as intraventricular hemorrhage and bronchopulmonary dysplasia within this population are addressed.

Introduction

A wealth of literature exists elucidating the neurodevelopmental outcomes of premature infants. Continuous advances in care supported by this research have contributed to the growing and changing nature of this population. This body of work has also influenced our understanding of risk factors, such as intraventricular hemorrhage and bronchopulmonary dysplasia, and their impact on development.

Definitions

Prematurity has classically been divided into categories of risk based upon gestational age at birth. Infants born below 25 weeks are defined as extremely preterm, <32 weeks as very preterm, between 32 and 34 weeks as moderately preterm, and between 34 and 37 weeks as late preterm. The largest category, late preterm birth, will be explored in detail in another chapter. Birth weight is another measure of risk,

M. A. Clark (✉)
UH Rainbow Babies and Children's Hospital, Cleveland, OH, USA

Case Western Reserve University, Cleveland, OH, USA
e-mail: Marie.clark@uhhospitals.org

N. Sand-Loud
Dartmouth Geisel School of Medicine, Hanover, NH, USA

Children's Hospital at Dartmouth-Hitchcock, Lebanon, NH, USA
e-mail: Nina.sand-loud@hitchcock.org

H. Needelman, B. J. Jackson (eds.), *Follow-Up for NICU Graduates*,
https://doi.org/10.1007/978-3-319-73275-6_5

and categories have been developed based upon infants' weight at birth. A birth weight below <1000 g is specified as extremely low birth weight, <1500 g as very low birth weight, and <2500 g as low birth weight. The use of these cut points in research has proven useful in providing more specific information about risk factors and outcomes for premature infants and about the differential impact of birth at earlier gestational ages and lower birth weights.

Epidemiology

According to a systematic review conducted by the World Health Organization, the approximate percentage of worldwide preterm birth was 9.6%, with most of these births (85%) occurring in Africa and Asia [1]. The rate in the United States is comparable to the worldwide figure, as the National Center for Health Statistics reports that the preterm birth rate in 2015 was 9.62%, a slight increase from 9.57% in 2014 [2]. Overall, the trend in preterm birth has fallen since a peak in 2006. Rates within the United States vary by state and also by race and ethnicity, with the highest rates in non-Hispanic black infants (13.39%) [2]. Preterm birth is the most frequent cause of infant death in the United States. One third of infant deaths can be attributed to prematurity [3]. Despite this association, the survival rate of preterm infants has overall increased with time, and it is their particular profile of risk for adverse neurodevelopmental outcomes that makes follow-up so important for this population [4].

Premature infants are at risk for multiple adverse neurodevelopmental outcomes, including cognitive, language, visual-perceptual, sensory, language, attention, and learning problems. These infants account for a large proportion of the affected children in each disability category: approximately 45% of children with cerebral palsy, 35% of children with vision impairment, and 25% of children with cognitive or hearing impairment [5].

Extremely Preterm Infants

Several longitudinal studies have evaluated survivors of extremely preterm birth over long periods of time to assess neurodevelopmental progress. In the EPICure (Extremely Preterm Infant Cure) study, a cohort of 308 extremely preterm infants born in the United Kingdom in 1995 were followed with developmental assessments conducted at 30 months, 6 years, and 11 years of age [6–9]. At 30 months, 30% of the infants demonstrated delays (scores below 2 standard deviations from the mean) on the Bayley Scales of Infant and Toddler Development (BSID). At that time, 2% had severe vision impairment, 3% had significant hearing impairment, and 10% had severe motor impairment. At 6 years of age, 22% of the children continued to have severe disabilities, 24% demonstrated a moderate level of disability, and 34% were characterized as having mild disabilities. At 11 years of age, survivors were compared with controls born at term and were noted to have lower test scores for cognitive ability, reading, and mathematics, as well as lower performance ratings by their teachers [5, 8]. Overall, about 45% of extremely premature infants in the study were characterized as having serious functional disability, defined as the most severe impairment in measures of motor, cognitive, and neurosensory impairments, compared with only 1% of matched controls [8]. Another EPICure cohort was recruited in 2006, in order to compare outcomes with the original cohort. Data thus far indicates that while survival rates have improved, the pattern of major morbidity in the neonatal period and the proportion of survivors affected remain unchanged [10].

A cohort of extremely low-birth-weight infants (<28 weeks or <1000 g) born in 1997 in Victoria, Australia, were followed through 18 years of age [11]. Medical findings at 18 years of age include increasing small airway obstruction demonstrated on spirometry compared to controls born at term and relatively mild increase in systolic and diastolic blood pressures [12, 13]. Advances in imaging have provided information about alterations in neurobiology that are associated with adverse outcomes in this group of adolescents. For example, investigators utilized diffusion and structural magnetic resonance imaging to elucidate the corticospinal tract microstructure and found that the extremely preterm group had alterations in microstructure that were associated with a diagnosis of cerebral palsy [14]. Other deficits noted in adolescence in comparison with peers born at term include poorer executive functioning skills, significantly worse visual acuity, and more problems with visual perception [15, 16]. Despite these limitations, it is reassuring that the survivors who reached age 18 reported a similar quality of life, self-esteem, and social and risk-taking behaviors as controls born at term [17].

A longitudinal study from a large urban center in Cleveland, Ohio, followed a cohort of extremely low-birth-weight infants born between born 1992 and 1995 [18]. The Ohio study followed these infants to 14 years of age, at which point they continued to show lower scores on cognitive tests and higher rates of learning disabilities in math and need for special education compared with control children [19]. Behavior problems also persisted within this cohort, with higher parent ratings compared to controls for symptoms of inattentive attention deficit hyperactivity disorder (ADHD), anxiety, and social problems [20]. In addition, the extremely preterm boys in this group were more likely to be bullied compared to peers born at term [21].

Other longitudinal studies bear out similar findings. There have been similar results seen in studies of extremely low-birth-weight infants through the US National Institute of Child Health and Human Development (NICHD) Neonatal Research Network, which followed over 5000 extremely low-birth-weight infants born between 1998 and 2001 [22]. Increasing rates of impairment were seen with decreasing birth weight.

Very Preterm Infants

Although the risk for neurodevelopmental disability decreases in the very preterm population, survivors of very preterm birth have also shown evidence of neurodevelopmental deficits. The French EPIPAGE (Epidemiologique des Petites Ages Gestationnels) study followed infants born in 1997 at gestational ages between 27 and 32 weeks. Of this cohort, 77% were examined at 5 years of age [23]. Among members of the cohort, cerebral palsy was present in 9%. Survivors also demonstrated higher rates of impaired cognitive function and severe bilateral vision impairment compared with controls.

A meta-analysis of literature published between 1998 and 2008 reviewed the developmental outcomes of very preterm (<33 weeks) and very-low-birth-weight infants. Findings indicated that very preterm infants had more behavioral problems (inattention), poorer executive function (verbal fluency, working memory, and cognitive flexibility), and lower test scores in mathematics, reading, and spelling compared with peers born at term [24]. These differences in academic achievement, behavioral difficulties, and neurocognitive problems persisted through adulthood.

Moderately Preterm Infants

There has been increasing evidence of neurodevelopmental impairments even in moderate to late preterm infants. In a retrospective study in Holland, 307 infants who were born between 32 and 37 weeks were more likely to need special education (7.7 vs. 2.8%), had a three-point lower IQ score, and were more likely to have internalizing behavior and ADHD at 7–9 years of age [25]. In the Lollipop (Longitudinal Preterm Outcome Project) study, preterm children born between 2002 and 2003 were evaluated. Preschool children born moderately preterm were twice as likely to be developmentally delayed, compared with children born full term [26]. At 7 years, those children born moderately preterm were found to have worse performance on cognitive testing and neuropsychological functioning [27].

Emotional and Behavioral Impairment

All groups of lower-birth-weight infants have been found to have an increased frequency of specific behavioral problems and psychological problems compared with normal birth weight peers. These findings become more apparent as the cohorts are followed over time. Problems include ADHD, general anxiety, depression, and poor social interaction [28–31]. In the aforementioned EPICure study, children with extreme prematurity were more likely to have behavior problems at 6 years of age compared to those born full term based on reports by teachers and parents [32]. There also is an increased risk of autistic spectrum disorders in those infants born with extremely low birth weights. In a follow-up assessment as part of the EPICure study at 11 years of age, 7% of the children were diagnosed with an autism spectrum disorder compared with none of the term controls [33]. Similar behavior and psychological changes have been found in children and adults born with very low birth weight [34, 35].

In summary, outcome studies at school age and adolescence have documented an increased risk of neurodevelopmental disability with decreasing gestational age and birth weight, although infants of any gestational age may have neurodevelopmental deficits. For this reason, early assessment of neurodevelopmental function of survivors of prematurity is essential.

Neonatal Complications

With advances in neonatal care, there has been a reduction in the likelihood of significant complications of prematurity such as air leak and severe intraventricular hemorrhage (IVH). However, preterm infants are at risk for multiple complications, which may impact their developmental trajectory. It remains clear that infants with lower gestational age, infants with white matter findings such as cystic periventricular leukomalacia (PVL), and those with severe intraventricular hemorrhage (IVH) are at highest risk for severe disability [5]. Chronic lung disease and necrotizing enterocolitis (NEC) are also predictors of poor developmental outcome [36, 37]. Other severe perinatal infections such as sepsis and meningitis are related to poorer outcomes [38, 39]. There has also been impaired motor and cognitive performance seen in those very preterm infants with poor postnatal growth [40]. We examine some of these complications in more detail below.

Necrotizing Enterocolitis: Necrotizing enterocolitis (NEC) is a gastrointestinal process which most commonly occurs in premature infants and involves infection and necrosis of the intestinal mucosa. NEC can often be managed medically with bowel rest and antibiotic therapy. Medically managed NEC does not seem to confer additional risk to the child [41]. However, when NEC is more severe and surgery is indicated, outcomes become much more grave for infants. Follow-up data indicate that up to a third of patients die, 64% of patients experience late-onset sepsis, and 11% suffer white matter injury. Strictures and short bowel syndrome are also common in survivors. Surgical NEC is associated with an increased odds ratio of impaired neurodevelopment at 18–22 months [41–43]. Systematic reviews have also indicated that worse disease is more highly associated with impairment [44, 45].

Bronchopulmonary Dysplasia: Bronchopulmonary dysplasia (BPD) is a chronic lung disease which has been associated with increased morbidity and mortality in preterm infants. BPD is currently associated with the process of disrupted lung development leading to decreased surface area for gas exchange in the alveoli and increased pulmonary resistance. The definition has evolved over the years with the availability of surfactant and modification of ventilation strategies. Risk factors for

BPD include birth at earlier gestational ages, antenatal infection such as chorioamnionitis, genetic susceptibility, and use of mechanical ventilation [46]. Despite improvements in respiratory care over the years, infants with severe BPD remain at an increased mortality risk, and those who survive are more likely suffer complications such as high rates of rehospitalization for respiratory disease, asthma-like disease, and pulmonary artery hypertension [47, 48]. Infants with BPD also are more likely to have poorer neurodevelopmental outcomes. Rates of cerebral palsy are higher in infants with BPD [49]. In addition, BPD is associated with lower scores in intelligence testing, academic difficulties, delayed speech and language development, impairment in visual-motor functioning, as well as behavioral problems [50]. Infants with the most severe lung disease may require placement of a tracheostomy. Follow-up of this uniquely vulnerable population indicates that they are at an increased risk of death or neurodevelopmental impairment, although this risk was lower in infants who received their tracheostomies earlier [51].

Brain Injury: Intraventricular hemorrhage (IVH) and periventricular leukomalacia (PVL) are important causes of brain injury in the preterm infant. IVH in preterm infants results from hemorrhage originating in the germinal matrix, and severity is defined by the extent of the hemorrhage. These bleeds are denoted by the Papile Classification system: Grade 1 bleeds do not extend beyond the germinal matrix, Grade II bleeds extend into the lateral ventricles without ventricular dilatation, Grade III bleeds include ventricular dilatation, and Grade IV include infarction of the surrounding periventricular white matter [52]. While it is well-understand that severe bleeds (Grades III and IV) are associated with neurodevelopmental impairment, recent findings have elucidated that even preterm infants with isolated Grade I and II bleeds have increased rates of moderate-severe neurosensory impairment [53, 54]. White matter injury is another important mechanism of brain injury in the preterm infant. Immature white matter in the preterm brain is damaged through hypoxia, ischemia, and inflammation. The most severe version of white matter injury is the cystic form, also known as cystic periventricular leukomalacia (PVL) [55]. Cranial ultrasound of the preterm infant brain is able to detect cystic PVL injury, but brain MRI is necessary for the more subtle findings of non-cystic white matter injury [56]. White matter injury is correlated with impairments in motor development and cognition. More severe white matter injury is associated with a greater risk of more severe deficits. Behavioral deficits such as inattention and executive dysfunction have also been noted [57].

Sensory Impairments: Impairments in both hearing and vision are also risks of premature birth. Preterm infants are more frequently associated with sensorineural hearing loss. Although sensorineural hearing loss is multifactorial and may have a genetic or infectious etiology, risk factors specific to the preterm infant include perinatal complications and high and prolonged doses of ototoxic drugs such as aminoglycoside antibiotics and furosemide [58]. Additional factors that impart risk are need for ventilation, use of oxygen supplementation, respiratory failure, low Apgar scores, acidosis, treatment for hypotension, patent ductus arteriosus ligation, hyponatremia, and high levels of environmental noise [59]. One 30-year longitudinal study found a permanent hearing loss rate of 3% (compared to the general population rate of 0.01%), with a significant percentage of those presenting with delayed onset and progressive hearing loss [59]. Because of the incidence of hearing loss, the Joint Committee on Infant Hearing issued a policy statement recommending that neonates with an ICU stay >5 days or those with risk factors should receive auditory brainstem response (ABR) testing [60].

Retinopathy of Prematurity: Retinopathy of prematurity (ROP) is a disease that affects the immature vasculature in the eye of the preterm infant. ROP is defined by anatomic location of disease, designated as Zones I, II, and III extending out in concentric circles from the optic disk. The stages of ROP describe the abnormal vascular response where the vascularized and avascular portions of the retina meet. Immature vasculature is present at Stage 0. Stage 1 denotes the presence of a demarcation line separating the vascularized and unvascularized portions of the retina. Stage 2 indicates a ridge at the demarcation line, and Stage 3 contains neovascularization from the ridge to the vitreous matter. Progression to Stage 4 indicates partial retinal detachment, and Stage 5 is total retinal detachment [61]. Plus disease is a separate constellation of findings, which consists of increased venous dilatation and arteriolar tortuosity of the posterior retinal vessels, iris vascular engorgement, rigid pupil, and vitreous haze [61]. The rates of ROP, as well as its severity, increase in preterm infants born at earlier gestational ages and lower birth weight. Other risk factors include longer NICU stay, respiratory conditions, fetal hemorrhage, IVH, and blood transfusion [62].

As a result of high rates of ROP, American Academy of Pediatrics (AAP) policy recommends that infants born less than or equal to 1500 g or at 30 weeks gestational age or lower or those with an unstable clinical course have retinal screening examinations to detect ROP [63]. Treatment is initiated when infants have any of the following: any stage ROP with plus disease in Zone 1, Stage 3 ROP without plus disease in Zone 1, Stage 2 or 3 ROP with plus disease in Zone II. Treatments for ROP include retinal ablative therapy, intravitreal injection of anti-vascular endothelial growth factor, or, more rarely, cryotherapy. Newer therapies such as retinal ablative therapy have resulted in improved outcomes, but more advanced ROP continues to have poor outcomes in visual acuity [64]. Long-term visual impairment in children born prematurely may also be related to cerebral damage.

Improvement in Outcomes

Interventions to improve outcome for premature infants have included changes in both obstetrical and pediatric practices. Among the obstetrical factors has been the use of prenatal steroids and recommendations regarding the need to attempt to prevent late preterm delivery. In caring for the newborn, there has been improved neonatal resuscitation, the use of artificial surfactant, and changing recommendations in the use of postnatal steroids, particularly the relatively recent admonition against the use of dexamethasone [65]. Coincident with changes in ventilation strategies has been a decrease in the risk-adjusted mortality and bronchopulmonary dysplasia [66]. Improved nutritional strategies have led to both better early head growth and improved scores on developmental testing [67].

Infants who have experienced hypoxic-ischemic encephalopathy and those with congenital anomalies are also at risk for neurodevelopmental challenges. These groups will be discussed in detail in the following chapters.

In summary, there is an increased risk for neurodevelopmental disorders in all groups of infants requiring neonatal intensive care. The risk of impairment increases with decreasing gestational age and decreased weight and increasing medical complexity. Disabilities include cognitive impairments, motor deficits including delays in fine and gross motor skills, as well as cerebral palsy and hearing and vision impairments. There is also an increased risk for specific psychological and behavioral impairments, including ADHD, autism spectrum disorders, anxiety, and depression.

References

1. Beck, S., Wojdyla, D., Say, L., et al. (2010). The worldwide incidence of preterm birth: A systematic review of maternal mortality and morbidity. *Bulletin of the World Health Organization, 88*(1), 31–38. https://doi.org/10.1590/S0042-96862010000100012.
2. Hamilton, B., Martin, J., & Osterman, M. *National vital statistics system. Births: Preliminary Data for 2015*. Retrieved November 13, 2016, from http://www.cdc.gov/nchs/data/nvsr/nvsr65/nvsr65_03.pdf. Published June 2, 2016.
3. Callaghan, W. M., MacDorman, M. F., Rasmussen, S. A., Qin, C., & Lackritz, E. M. (2006). The contribution of preterm birth to infant mortality rates in the United States. *Pediatrics, 118*(4), 1566–1573. https://doi.org/10.1542/peds.2006-0860.
4. Stoll, B. J., Hansen, N. I., Bell, E. F., et al. (2015). Trends in care practices, morbidity, and mortality of extremely preterm neonates, 1993–2012. *Journal of the American Medical Association, 314*(10), 1039. https://doi.org/10.1001/jama.2015.10244.
5. Allen, M. C. (2008). Neurodevelopmental outcomes of preterm infants. *Current Opinion in Neurology, 21*(2), 123–128. https://doi.org/10.1097/WCO.0b013e3282f88bb4.
6. Wood, N. S., Marlow, N., Costeloe, K., Gibson, A. T., & Wilkinson, A. R. (2000). Neurologic and developmental disability after extremely preterm birth. EPICure Study Group. *The New England Journal of Medicine, 343*(6), 378–384. https://doi.org/10.1056/NEJM200008103430601.
7. Marlow, N., Wolke, D., Bracewell, M. A., Samara, M., & EPICure Study Group. (2005). Neurologic and developmental disability at six years of age after extremely preterm birth. *The New England Journal of Medicine, 352*(1), 9–19. https://doi.org/10.1056/NEJMoa041367.
8. Johnson, S., Fawke, J., Hennessy, E., et al. (2009). Neurodevelopmental disability through 11 years of age in children born before 26 weeks of gestation. *Pediatrics, 124*(2), e249–e257. https://doi.org/10.1542/peds.2008-3743.
9. Johnson, S., Hennessy, E., Smith, R., Trikic, R., Wolke, D., & Marlow, N. (2009). Academic attainment and special educational needs in extremely preterm children at 11 years of age: The EPICure study. *Archives of Disease in Childhood. Fetal and Neonatal Edition, 94*(4), F283–F289. https://doi.org/10.1136/adc.2008.152793.
10. Costeloe, K. L., Hennessy, E. M., Haider, S., Stacey, F., Marlow, N., & Draper, E. S. (2012). Short term outcomes after extreme preterm birth in England: Comparison of two birth cohorts in 1995 and 2006 (the EPICure studies). *BMJ, 345*, e7976. https://doi.org/10.1136/bmj.e7976.
11. Hutchinson, E. A., De Luca, C. R., Doyle, L. W., Roberts, G., Anderson, P. J., & Victorian Infant Collaborative Study Group. (2013). School-age outcomes of extremely preterm or extremely low birth weight children. *Pediatrics, 131*(4), e1053–e1061. https://doi.org/10.1542/peds.2012-2311.

12. Doyle, L. W., Adams, A.-M., Robertson, C., et al. (2016). Increasing airway obstruction from 8 to 18 years in extremely preterm/low-birthweight survivors born in the surfactant era. *Thorax, 72*(8), 712–719. https://doi.org/10.1136/thoraxjnl-2016-208524.
13. Kowalski, R. R., Beare, R., Doyle, L. W., Smolich, J. J., Cheung, M. M. H., & Victorian Infant Collaborative Study Group. (2016). Elevated blood pressure with reduced left ventricular and aortic dimensions in adolescents born extremely preterm. *The Journal of Pediatrics, 172*, 75–80.e2. https://doi.org/10.1016/j.jpeds.2016.01.020.
14. Kelly, C. E., Chan, L., Burnett, A. C., et al. (2015). Brain structural and microstructural alterations associated with cerebral palsy and motor impairments in adolescents born extremely preterm and/or extremely low birthweight. *Developmental Medicine and Child Neurology, 57*(12), 1168–1175. https://doi.org/10.1111/dmcn.12854.
15. Burnett, A. C., Scratch, S. E., Lee, K. J., et al. (2015). Executive function in adolescents born <1000 g or <28 weeks: A prospective cohort study. *Pediatrics, 135*(4), e826–e834. https://doi.org/10.1542/peds.2014-3188.
16. Molloy, C. S., Wilson-Ching, M., Anderson, V. A., et al. (2013). Visual processing in adolescents born extremely low birth weight and/or extremely preterm. *Pediatrics, 132*(3), e704–e712. https://doi.org/10.1542/peds.2013-0040.
17. Roberts, G., Burnett, A. C., Lee, K. J., et al. (2013). Quality of life at age 18 years after extremely preterm birth in the post-surfactant era. *The Journal of Pediatrics, 163*(4), 1008–1013.e1. https://doi.org/10.1016/j.jpeds.2013.05.048.
18. Hack, M., & Klein, N. (2006). Young adult attainments of preterm infants. *Journal of the American Medical Association, 295*(6), 695–696. https://doi.org/10.1001/jama.295.6.695.
19. Litt, J. S., Gerry Taylor, H., Margevicius, S., Schluchter, M., Andreias, L., & Hack, M. (2012). Academic achievement of adolescents born with extremely low birth weight. *Acta Paediatrica, 101*(12), 1240–1245. https://doi.org/10.1111/j.1651-2227.2012.02790.x.
20. Taylor, H. G., Margevicius, S., Schluchter, M., Andreias, L., & Hack, M. (2015). Persisting behavior problems in extremely low birth weight adolescents. *Journal of Developmental and Behavioral Pediatrics, 36*(3), 178–187. https://doi.org/10.1097/DBP.0000000000000139.
21. Yau, G., Schluchter, M., Taylor, H. G., et al. (2013). Bullying of extremely low birth weight children: Associated risk factors during adolescence. *Early Human Development, 89*(5), 333–338. https://doi.org/10.1016/j.earlhumdev.2012.11.004.
22. Gargus, R. A., Vohr, B. R., Tyson, J. E., et al. (2009). Unimpaired outcomes for extremely low birth weight infants at 18 to 22 months. *Pediatrics, 124*(1), 112–121. https://doi.org/10.1542/peds.2008-2742.
23. Larroque, B., Ancel, P.-Y., Marret, S., et al. (2008). Neurodevelopmental disabilities and special care of 5-year-old children born before 33 weeks of gestation (the EPIPAGE study): A longitudinal cohort study. *Lancet, 371*(9615), 813–820. https://doi.org/10.1016/S0140-6736(08)60380-3.
24. Aarnoudse-Moens, C. S. H., Weisglas-Kuperus, N., van Goudoever, J. B., & Oosterlaan, J. (2009). Meta-analysis of neurobehavioral outcomes in very preterm and/or very low birth weight children. *Pediatrics, 124*(2), 717–728. https://doi.org/10.1542/peds.2008-2816.
25. Van Baar, A. L., Vermaas, J., Knots, E., de Kleine, M. J. K., & Soons, P. (2009). Functioning at school age of moderately preterm children born at 32 to 36 weeks' gestational age. *Pediatrics, 124*(1), 251–257. https://doi.org/10.1542/peds.2008-2315.
26. Kerstjens, J. M., de Winter, A. F., Bocca-Tjeertes, I. F., ten Vergert, E. M. J., Reijneveld, S. A., & Bos, A. F. (2011). Developmental delay in moderately preterm-born children at school entry. *The Journal of Pediatrics, 159*(1), 92–98. https://doi.org/10.1016/j.jpeds.2010.12.041.
27. Cserjesi, R., Van Braeckel, K. N. J. A., Butcher, P. R., et al. (2012). Functioning of 7-year-old children born at 32 to 35 weeks' gestational age. *Pediatrics, 130*(4), e838–e846. https://doi.org/10.1542/peds.2011-2079.
28. Hack, M., Youngstrom, E. A., Cartar, L., et al. (2004). Behavioral outcomes and evidence of psychopathology among very low birth weight infants at age 20 years. *Pediatrics, 114*(4), 932–940. https://doi.org/10.1542/peds.2003-1017-L.

29. Hille, E. T. M., Dorrepaal, C., Perenboom, R., et al. (2008). Social lifestyle, risk-taking behavior, and psychopathology in young adults born very preterm or with a very low birthweight. *The Journal of Pediatrics, 152*(6), 793–800, 800.e1-e4.. https://doi.org/10.1016/j.jpeds.2007.11.041.
30. Conrad, A. L., Richman, L., Lindgren, S., & Nopoulos, P. (2010). Biological and environmental predictors of behavioral sequelae in children born preterm. *Pediatrics, 125*(1), e83–e89. https://doi.org/10.1542/peds.2009-0634.
31. Lindström, K., Lindblad, F., & Hjern, A. (2011). Preterm birth and attention-deficit/hyperactivity disorder in schoolchildren. *Pediatrics, 127*(5), 858–865. https://doi.org/10.1542/peds.2010-1279.
32. Samara, M., Marlow, N., Wolke, D., & EPICure Study Group. (2008). Pervasive behavior problems at 6 years of age in a total-population sample of children born at </= 25 weeks of gestation. *Pediatrics, 122*(3), 562–573. https://doi.org/10.1542/peds.2007-3231.
33. Johnson, S., Hollis, C., Kochhar, P., Hennessy, E., Wolke, D., & Marlow, N. (2010). Autism spectrum disorders in extremely preterm children. *The Journal of Pediatrics, 156*(4), 525–531.e2. https://doi.org/10.1016/j.jpeds.2009.10.041.
34. Delobel-Ayoub, M., Arnaud, C., White-Koning, M., et al. (2009). Behavioral problems and cognitive performance at 5 years of age after very preterm birth: The EPIPAGE Study. *Pediatrics, 123*(6), 1485–1492. https://doi.org/10.1542/peds.2008-1216.
35. Hack, M., Taylor, H. G., Schluchter, M., Andreias, L., Drotar, D., & Klein, N. (2009). Behavioral outcomes of extremely low birth weight children at age 8 years. *Journal of Developmental and Behavioral Pediatrics, 30*(2), 122–130. https://doi.org/10.1097/DBP.0b013e31819e6a16.
36. Eichenwald, E. C., & Stark, A. R. (2008). Management and outcomes of very low birth weight. *The New England Journal of Medicine, 358*(16), 1700–1711. https://doi.org/10.1056/NEJMra0707601.
37. Neubauer, A.-P., Voss, W., & Kattner, E. (2008). Outcome of extremely low birth weight survivors at school age: The influence of perinatal parameters on neurodevelopment. *European Journal of Pediatrics, 167*(1), 87–95. https://doi.org/10.1007/s00431-007-0435-x.
38. Van Vliet, E. O. G., de Kieviet, J. F., Oosterlaan, J., & van Elburg, R. M. (2013). Perinatal infections and neurodevelopmental outcome in very preterm and very low-birth-weight infants: A meta-analysis. *JAMA Pediatrics, 167*(7), 662–668. https://doi.org/10.1001/jamapediatrics.2013.1199.
39. Mitha, A., Foix-L'Hélias, L., Arnaud, C., et al. (2013). Neonatal infection and 5-year neurodevelopmental outcome of very preterm infants. *Pediatrics, 132*(2), e372–e380. https://doi.org/10.1542/peds.2012-3979.
40. Cooke, R. W. I., & Foulder-Hughes, L. (2003). Growth impairment in the very preterm and cognitive and motor performance at 7 years. *Archives of Disease in Childhood, 88*(6), 482–487.
41. Hintz, S. R., Kendrick, D. E., Stoll, B. J., et al. (2005). Neurodevelopmental and growth outcomes of extremely low birth weight infants after necrotizing enterocolitis. *Pediatrics, 115*(3), 696–703. https://doi.org/10.1542/peds.2004-0569.
42. Wadhawan, R., Oh, W., Hintz, S. R., et al. (2014). Neurodevelopmental outcomes of extremely low birth weight infants with spontaneous intestinal perforation or surgical necrotizing enterocolitis. *Journal of Perinatology, 34*(1), 64–70. https://doi.org/10.1038/jp.2013.128.
43. Murthy, K., Yanowitz, T. D., DiGeronimo, R., et al. (2014). Short-term outcomes for preterm infants with surgical necrotizing enterocolitis. *Journal of Perinatology, 34*(10), 736–740. https://doi.org/10.1038/jp.2014.153.
44. Schulzke, S. M., Deshpande, G. C., & Patole, S. K. (2007). Neurodevelopmental outcomes of very low-birth-weight infants with necrotizing enterocolitis: A systematic review of observational studies. *Archives of Pediatrics & Adolescent Medicine, 161*(6), 583–590. https://doi.org/10.1001/archpedi.161.6.583.
45. Rees, C. M., Pierro, A., & Eaton, S. (2007). Neurodevelopmental outcomes of neonates with medically and surgically treated necrotizing enterocolitis. *Archives of Disease in Childhood. Fetal and Neonatal Edition, 92*(3), F193–F198. https://doi.org/10.1136/adc.2006.099929.

46. Baraldi, E., & Filippone, M. (2007). Chronic lung disease after premature birth. *The New England Journal of Medicine, 357*(19), 1946–1955. https://doi.org/10.1056/NEJMra067279.
47. Davidson, L. M., & Berkelhamer, S. K. (2017). Bronchopulmonary dysplasia: Chronic lung disease of infancy and long-term pulmonary outcomes. *Journal of Clinical Medicine, 6*(1), 4. https://doi.org/10.3390/jcm6010004.
48. Smith, V. C., Zupancic, J. A. F., McCormick, M. C., et al. (2004). Rehospitalization in the first year of life among infants with bronchopulmonary dysplasia. *The Journal of Pediatrics, 144*(6), 799–803. https://doi.org/10.1016/j.jpeds.2004.03.026.
49. Van Marter, L. J., Kuban, K. C. K., Allred, E., et al. (2011). Does bronchopulmonary dysplasia contribute to the occurrence of cerebral palsy among infants born before 28 weeks of gestation? *Archives of Disease in Childhood. Fetal and Neonatal Edition, 96*(1), F20–F29. https://doi.org/10.1136/adc.2010.183012.
50. Anderson, P. J., & Doyle, L. W. (2006). Neurodevelopmental outcome of bronchopulmonary dysplasia. *Seminars in Perinatology, 30*(4), 227–232. https://doi.org/10.1053/j.semperi.2006.05.010.
51. DeMauro, S. B., D'Agostino, J. A., Bann, C., et al. (2014). Developmental outcomes of very preterm infants with tracheostomies. *The Journal of Pediatrics, 164*(6), 1303–1310.e2. https://doi.org/10.1016/j.jpeds.2013.12.014.
52. Papile, L.-A., Burstein, J., Burstein, R., & Koffler, H. (1978). Incidence and evolution of subependymal and intraventricular hemorrhage: A study of infants with birth weights less than 1,500 gm. *The Journal of Pediatrics, 92*(4), 529–534. https://doi.org/10.1016/S0022-3476(78)80282-0.
53. O'Shea, T. M., Allred, E. N., Kuban, K. C. K., et al. (2012). Intraventricular hemorrhage and developmental outcomes at 24 months of age in extremely preterm infants. *Journal of Child Neurology, 27*(1), 22–29. https://doi.org/10.1177/0883073811424462.
54. Bolisetty, S., Dhawan, A., Abdel-Latif, M., et al. (2014). Intraventricular hemorrhage and neurodevelopmental outcomes in extreme preterm infants. *Pediatrics, 133*(1), 55–62. https://doi.org/10.1542/peds.2013-0372.
55. Gano, D. (2016). White matter injury in premature newborns. *Neonatal Network, 35*(2), 73–77. https://doi.org/10.1891/0730-0832.35.2.73.
56. Miller, S. P., Cozzio, C. C., Goldstein, R. B., et al. (2003). Comparing the diagnosis of white matter injury in premature newborns with serial MR imaging and transfontanel ultrasonography findings. *AJNR. American Journal of Neuroradiology, 24*(8), 1661–1669.
57. Woodward, L. J., Clark, C. A. C., Bora, S., & Inder, T. E. (2012). Neonatal white matter abnormalities an important predictor of neurocognitive outcome for very preterm children. *PLoS One, 7*(12), e51879. https://doi.org/10.1371/journal.pone.0051879.
58. Borradori, C., Fawer, C. L., Buclin, T., & Calame, A. (1997). Risk factors of sensorineural hearing loss in preterm infants. *Biology of the Neonate, 71*(1), 1–10.
59. Robertson, C. M. T., Howarth, T. M., Bork, D. L. R., & Dinu, I. A. (2009). Permanent bilateral sensory and neural hearing loss of children after neonatal intensive care because of extreme prematurity: A thirty-year study. *Pediatrics, 123*(5), e797–e807. https://doi.org/10.1542/peds.2008-2531.
60. American Academy of Pediatrics, Joint Committee on Infant Hearing. (2007). Year 2007 position statement: Principles and guidelines for early hearing detection and intervention programs. *Pediatrics, 120*(4), 898–921. https://doi.org/10.1542/peds.2007-2333.
61. International Committee for the Classification of Retinopathy of Prematurity. (2005). The International classification of retinopathy of prematurity revisited. *Archives of Ophthalmology, 123*(7), 991–999. https://doi.org/10.1001/archopht.123.7.991.
62. Lad, E. M., Hernandez-Boussard, T., Morton, J. M., & Moshfeghi, D. M. (2009). Incidence of retinopathy of prematurity in the United States: 1997 through 2005. *American Journal of Ophthalmology, 148*(3), 451–458. https://doi.org/10.1016/j.ajo.2009.04.018.
63. Fierson, W. M., & American Academy of Pediatrics Section on Ophthalmology; American Academy of Ophthalmology; American Association for Pediatric Ophthalmology and

Strabismus; American Association of Certified Orthoptist. (2013). Screening examination of premature infants for retinopathy of prematurity. *Pediatrics, 131*(1), 189–195. https://doi.org/10.1542/peds.2012-2996.
64. Holmström, G., & Larsson, E. (2013). Outcome of retinopathy of prematurity. *Clinics in Perinatology, 40*(2), 311–321. https://doi.org/10.1016/j.clp.2013.02.008.
65. Tam, E. W. Y., Chau, V., Ferriero, D. M., et al. (2011). Preterm cerebellar growth impairment after postnatal exposure to glucocorticoids. *Science Translational Medicine, 3*(105), 105ra105. https://doi.org/10.1126/scitranslmed.3002884.
66. Vendettuoli, V., Bellù, R., Zanini, R., Mosca, F., Gagliardi, L., & Network, I. N. (2014). Changes in ventilator strategies and outcomes in preterm infants. *Archives of Disease in Childhood. Fetal and Neonatal Edition, 99*(4), F321–F324. https://doi.org/10.1136/archdischild-2013-305165.
67. Hay, W. W. (2008). Strategies for feeding the preterm infant. *Neonatology, 94*(4), 245–254. https://doi.org/10.1159/000151643.

Premature Infants: The Behavioral Phenotype of the Preterm Survivor

Paige Terrien Church, Rudaina Banihani, Maureen Luther, Patricia Maddalena, and Elizabeth Asztalos

Abstract Approximately one of ten children was born preterm based on current global estimates. While the outcomes of the preterm survivor are promising, with a small fraction experiencing significant neurodevelopmental disability, many children born preterm will experience subtle, discreet, and additive challenges that can contribute to school challenges. This chapter describes the pattern of neurodevelopmental outcomes for the child born preterm, now increasingly referred to as the behavioral phenotype of prematurity. Awareness of this pattern is critical for the practitioner following these children.

Introduction

A behavioral phenotype describes a cluster of behavioral, cognitive, motor, and social strengths and difficulties seen in a population with a common biological disorder [1]. This descriptor emphasizes the organic etiology of patterns of behavior and neurodevelopment. The child surviving a preterm birth has a behavioral phenotype with the common biological disorder being the atypical brain development or dysmaturation (see etiology) [2, 3]. Given that the trend of preterm birth continues to be one of increasing incidence with current global estimates at 11.1% [4], understanding this cohort of children within this context is critical for identification and support.

P. T. Church (✉) · R. Banihani · E. Asztalos
Department of Paediatrics, University of Toronto, Toronto, ON, Canada

Department of Newborn and Developmental Paediatrics, Sunnybrook Health Sciences Centre, Toronto, ON, Canada
e-mail: paige.church@sunnybrook.ca; rudaina.banihani@sunnybrook.ca; elizabeth.asztalos@sunnybrook.ca

M. Luther · P. Maddalena
Sunnybrook Health Sciences Centre, Neonatal Follow Up Clinic, Toronto, ON, Canada
e-mail: maureen.luther@sunnybrook.ca

H. Needelman, B. J. Jackson (eds.), *Follow-Up for NICU Graduates*,
https://doi.org/10.1007/978-3-319-73275-6_6

In this review, we will describe the behavioral phenotype of prematurity, using literature on the outcomes of prematurity and applying fundamental neurodevelopmental concepts. We will also review the biological underpinnings of the behavioral phenotype of prematurity and the neurological dysmaturation that occurs as a result of preterm delivery. We will present the spectrum of possible clinical presentations of the behavioral phenotype of prematurity, with an emphasis on those that are more prevalent. We will discuss the challenges with identification and management of the features of the behavioral phenotype of prematurity, particularly as it relates to school functioning.

Key Developmental Concepts

The World Health Organization (WHO) defines preterm birth as any infant that is delivered prior to 37 weeks of gestation [5]. Traditional neonatal follow-up has focused on those infants born less than 28 weeks, referred to as the extremely low gestational age neonates (ELGANs) or, for earlier literature, those infants born with very (less than 1500 g, VLBW) or extremely low (less than 1000 g, ELBWs) birth weight. There is increasing literature that the features of the behavioral phenotype of prematurity are present in any child born preterm, with milder presentations for those born closer to term and more significant presentations for those born increasingly preterm [3, 6–8].

Normal behavior and development represent a spectrum ranging from expected or normal to atypical or abnormal. This has been outlined by the Diagnostic and Statistical Manual for Primary Care (DSM-PC) [9] and provides the context for which one can consider the outcomes of the preterm survivor. On one end of the spectrum is normal behavior or development, meaning that the child is following the expected course and behaving in the expected manner for the child's age. Shifting away from completely expected but still within the normal range is *variant* behavior or development, which describes those behavioral or developmental patterns that parents may identify as concerning but remain within that which is developmentally appropriate. *Problem* behavior or development describes a pattern that results in a degree of dysfunction but is only observed in specific environments. It is not pervasive and does respond to environmental modifications. Lastly, *disordered* behavior describes that which is pervasive and interferes with skill acquisition, causing considerable dysfunction. These behaviors are observed in multiple settings and are not as responsive to environmental modifications. The behavioral phenotype of prematurity describes patterns of behavior and development that can be classified as normal, variant, problem, or disordered [3].

In addition, behavior and development are the outward reflection of brain integrity [10]. More significant injuries or greater degrees of dysmaturation within the developing brain reflect overall a greater degree of insult; as a result, these tend to be less survivable and less prevalent. The clinical manifestation of these more significant injuries is consistent with a greater degree of disability. Milder degrees of injury/dysmaturation are more common as they are more likely to be survived and the associated presentation is characterized by less challenge. Furthermore, with greater degrees of injury, the likelihood of an isolated injury only affecting one domain of development becomes far less [10].

Another core feature of developmental medicine is that of plasticity. The developing brain is highly adaptive and responsive to environmental stimulation, with the concept "nature through nurture" replacing the age old debate of nature versus nurture [11]. This responsiveness is both a blessing and a curse. It allows for the developing brain to be stimulated and nurtured and healthy growth to occur. It also, however, leaves the brain vulnerable to negative inputs, including stress, inflammation, pain, toxins, and malnutrition [12–14]. Injurious events can have multiple developmental trajectories, depending on the timing of the event, the nature of the injury, and the recovery. Severe injury may be so widespread that there is no room for plasticity or recovery. Mild and moderate injuries, however, may result in minimal impact whatsoever, observable recovery, and/or the impact of injury manifesting itself only after a period of apparent recovery. This has been called the "sleeper effect," with the injury becoming apparent as brain development proceeds and the injured area, always there, becomes developmentally relevant [15].

While there may not currently be a cure for this neurological injury and its patterns, an understanding of the principles of developmental medicine, the spectrum of neuropathology and neuroplasticity, as well as the spectrum of neurodevelopmental outcomes is critical to truly comprehending these survivors and providing them with support following discharge.

Traditional Outcomes of Prematurity

Traditionally, the focus of neonatal follow-up has been on those outcomes associated with significant morbidity, including cerebral palsy (CP), vision/hearing impairment, and cognitive impairment. Cerebral palsy, vision/hearing impairment, and cognitive impairment are often analyzed in the neonatal follow-up literature as a composite outcome, described as neurodevelopmental impairment (NDI) [16]. The presence of any one or more of these outcomes results in the child being subsequently classified as having a NDI. While these outcomes represent highly variable presentations and impacts on function, they are grouped according to the overarching premise that these individual outcomes represent a significant, or major, morbidity or disability.

In keeping with the concepts described above, however, these outcomes occur with low prevalence and represent significant underlying brain injury/dysmaturation. The incidence of these major morbidities ranges from 6 to 45% depending on the gestational age studied, the era studied, and the specific outcome examined [16].

Etiology of the Behavioral Phenotype of Prematurity

The common biological disorder underlying the behavioral phenotype of prematurity appears to be cerebral dysmaturation [2, 17], with or without additional injury [17–19], resulting in overall smaller brain volume [20]. The development of the brain for a preterm infant varies from that of a term child in that there is an interruption of normal neural development with the preterm birth. In addition, there is developmental susceptibility to stressors at a cell functional/structural level as well as at a macroscopic level with a high potential for injury. Lastly, there are aberrant recovery attempts [2, 3, 6, 17]. In addition to the microscopic and macroscopic changes in the developing preterm brain, there is also aberrant sensory system development [21]. Normal sensory development is one characterized by an inter-redundant quality to sensory development, with one stimulus experienced across multiple sensory systems simultaneously. This inter-redundant quality is theorized to be fundamental in the development of selective attention and perceptual organization and is lost with preterm birth [21]. Further to this, parenting is altered. The stress of a preterm birth is significant for parents, with prolonged separation and chronic stress experienced [22]. These exposures have lingering impacts on how parents parent and attachment [23].

While certainly the preterm survivor demonstrates remarkable neuroplasticity, imaging of former preterms in adolescence continues to demonstrate qualitative differences with altered connectivity appreciated and less white matter added, when compared to term controls [24]. The end result is a complex and diffuse pattern of dysmaturation [2]. The variability observed in outcomes relates to the varying degree that neurodevelopment is impacted and the severity of the injury. This pattern of the preterm brain development is now increasingly being referred to as "encephalopathy of prematurity" or primary cerebral dysmaturation [2, 17] and is the common biological thread for the preterm behavioral phenotype.

Behavioral Phenotype of Prematurity

The behavioral phenotype of prematurity is characterized by a spectrum of individual expression, ranging from normal behavior, to variant behavior, to problem, and, at times, disordered behavior [1, 3]. Much of the current data on outcomes of prematurity is based on the use of assessment tools geared to identification of disordered behavior only and as a result may reflect an underestimation of the true impact of challenge. The behavioral phenotype of prematurity is characterized by clusters of challenges, with each challenge becoming an additive risk factor for school success [3, 6, 25]; isolated challenge in only one domain is rare. The presentation tends to follow an inverse relationship with gestational age, being more obvious in those at the more immature gestations and less obvious in the late preterm [6–8].

The use of the concept of a behavioral phenotype for discussion of children born preterm has the advantage of encapsulating the myriad of outcomes in the context of a spectrum and thereby better alerting providers, parents, and educators to the potential for challenges. Caution, however, is warranted as the disadvantage is that describing the behavioral phenotype of prematurity may be misinterpreted as an inevitable end result [26]. Given how responsive the developing brain is to input, awareness and intervention can hopefully lead to mitigation of the difficulties and bolstering of the strengths [11, 12].

Cognitive Development and Learning

At the core of the behavioral phenotype of prematurity are splintered challenges within cognition and learning. Cognition is an overarching term that describes the child's inherent tools that are available for learning and knowledge building but not equivalent to intellect. Intellect is the body of knowledge and organic learning capacity built with these tools [6]. Assessment of cognition reflects a trend but is not specifically linked to future intellectual abilities [27]. The lack of prediction of future intelligence with early developmental assessments is partly due to the plasticity of the brain and the potential for recovery [12] as well as the potential that the injury will manifest with time and maturation [15]. It also is due to the limitations with developmental testing in the infant and toddler years, which measures emerging motor and then sensorimotor skills as a marker of neurological integrity. They are not, however, specific to the higher-order processes associated with later learning [6, 27]. With maturation, and increasingly discrete developmental domains, delineation of the cognitive delay can be better explored with a series of discrete assessments rather than a broad overview assessment [28].

The most significant degree of impairment for learning associated with premature birth is that of intellectual disability (ID). ID generally refers to significant limitations in one's ability for mental capacity, problem-solving, adaptive skills, and learning [29]. The measurement of intellect is reflected in the intellectual quotient (IQ), which is a calculated score usually comparing the studied child to sampled peers. While the definition of ID is variable, in general, for most neonatal follow-up literature, ID reflects an IQ of less than 70. The majority of preterm survivors have average or borderline average IQs [6, 30, 31]. The likelihood of average to low average IQ diminishes with decreasing gestational age; 75% at 25 weeks are likely to have borderline to average IQ versus less than 50% of those at 23 weeks [31]. This trend is also apparent in a comparison of the late preterm to term controls with the late preterm having a greater likelihood of cognitive delay [7].

While the majority of preterm survivors do have average intellectual abilities, they also have an increased risk for learning disabilities (traditionally defined as a discrepancy between one's intellectual abilities and school-based achievement)

[32, 33]. Studies have documented that children born preterm demonstrate lower scores on measures of academic achievement and are more likely to be identified as having additional educational resources allocated. Mathematics, reading, processing speed, and working memory in particular appear to be areas of significant difficulty, independent of cognitive abilities [33–35].

Behavioral Regulation

Hand in hand with learning capacity is one's learning, social, and executive functioning skills. For the preterm survivor, this includes challenges with self-regulation. The definitions for self-regulation are variable, but it represents a pattern of adjusting cognitive, social, and behavioral responses to best fit the social context [36]. Initially, regulation is provided externally through parents and, over time, becomes an internalized skill [36].

Self-regulation is a challenge with an early presentation for the preterm survivor. Initially, difficulties with excessive crying, sleep dysregulation, and feeding are common presentations and appear to indicate future challenge with behavior [37, 38]. Early dysregulation also appears related to the deterioration of developmental outcome measurements over time, with more dysregulated children demonstrating less progress in the acquisition of milestones [39–41].

Executive Function and Attention

As the child matures, integral aspects of self-regulation include executive function and attention, which, for the preterm infant, are challenges [42]. Executive function describes one's inhibitory control, problem-solving, and ability to do goal-directed activities [43]. Executive function reflects the developing complex neural circuitry between the prefrontal cortex, basal ganglia and thalamus, and cerebellum [44]. This circuitry develops later in gestation, with the prefrontal cortex developing latest [44]; as a result, this network is highly vulnerable to aberrant development associated with preterm birth. Clinically, the end result is one of executive dysfunction for the child born preterm [45–48]. The dysfunction has been described as "global" rather than limited to a specific aspect of executive function [45]. Executive dysfunction has also been found to persist into adolescence [48].

Similar neural circuitry and vulnerability exist for attentional regulation, which is an aspect of executive function [41, 49]. Attention deficit hyperactive disorder (ADHD) is classified by the Diagnostic Statistical Manual of Mental Disorders-5 (DSM-5) as a cluster of inattentive, hyperactive, or impulsive behaviors observed in more than one setting for over 6 months [50]. There is extensive literature documenting behaviors consistent with ADHD among the preterm. The preterm has been found to be two to three times as likely to demonstrate disordered behavior as it relates to hyperactivity, conduct, and peer relationships at school age [42, 47, 51, 52]. Attention and cognition appear to be linked with those demonstrating weaker attention also demonstrating greater challenges achieving cognitive tasks [52]. Clinically, these children can be driven to move and have difficulty sustaining attention and persistence to task, which becomes an additive deficit as they are already facing challenges in the mastery of various developmental tasks. They are easily distracted and may be impulsive and hyperactive [3]. The deficits in attention, however, may not relate to the traditional pathophysiology of traditional ADHD in the term child. Studies on the preterm have documented specific deficits in response time and visuospatial working memory, and these weaknesses appear to mediate the observed inattention [53].

Critical to this discussion is the concept previously discussed of atypical behaviors presenting within a spectrum ranging from variants to problems to disorders [9]. While many studies demonstrate behaviors consistent with ADHD, there may be many more preterm survivors who present with problems maintaining attention. These deficits, while not as pervasive as ADHD, can be equally challenging when combined with other deficits, such as low average intellectual abilities, learning problems, and executive dysfunction [3]. As with executive dysfunction, the challenges around weak attention appear to persist into adulthood [54].

Socio-emotional

The socio-emotional aspects of the behavioral phenotype of prematurity are generally characterized by social interest but difficulty with social inclusion/participation. These children have been referred to as demonstrating nomadic behavior, tending to circle the periphery of the playground, preferring solitary play or adults rather than peers [3, 55–57]. The etiology of the challenge for the preterm with social navigation is unclear but may reflect a combination of factors including language delays/difficulties with auditory processing, ADHD behaviors and poor impulse control, nonverbal learning disability, and parental overprotection related to vulnerable child syndrome [55–57].

In addition, the child born preterm has been reported to struggle with a greater degree of anxiety and depression. They are noted to struggle with adaptive functioning compared with their term colleagues. Multiple risk factors have been identified, including medical comorbidities, overall cognitive status with those demonstrating delays struggling to a greater degree, degree of prematurity, and social risk [40, 55].

Autism Spectrum Disorder

Associated with the behavioral phenotype of prematurity are reports of an increased possibility of autism spectrum disorder (ASD). ASD is a neurodevelopmental disability characterized by restricted social skills, impaired communication, and the presence of atypical, repetitive behaviors [50]. Historically, the prevalence of ASD among preterm children has been difficult to establish due to small populations studied, varying diagnostic tools, and confounding comorbidities such as visual impairment and intellectual disability [58–61]. Despite these challenges, there have been several studies that have established a prevalence between 7 and 8% [62, 63]. The prevalence appears to be greater among those most preterm, occurring in 15% for those 23–24 weeks gestation age (GA) and decreasing to 3.4% for those at 27 weeks GA. The children with ASD were more likely to have comorbid diagnoses of ID [63].

While there is increasing evidence that children born preterm have a higher prevalence of ASD, there are features of the behavioral phenotype of prematurity that can be erroneously attributed to ASD, including increased risk for vision impairment, motor impairment, anxiety (particularly separation anxiety), delayed auditory processing, and language delays [3]. Distinguishing these behavioral characteristics from ASD can be challenging and underscores the benefit of a full neurodevelopmental assessment with Autism Diagnostic Observation Schedule, 2nd Edition (ADOS-2) for confirmation.

Language

Language difficulties have been found to impact significantly on any child's ability to self-regulate [36], to integrate socially, and to perform optimally in the school setting [64]. Early in childhood, the preterm survivor may demonstrate average expressive language and average receptive language skills. As these children are followed into the school years, however, there appears to be a discrepancy in language abilities that develops between term children and preterm survivors [64, 65]. As with most other aspects of the behavioral phenotype of prematurity, splintered skills are appreciated. Specific deficits appreciated include social pragmatics (how language is used socially), complexity of language used and comprehended, phonological processing, and syntax [64, 66, 67]. The more preterm the infant, the more severe the deficits appreciated [68].

Clinically, this presents as language that tends to be simplified, with less elaborate vocabulary and sentence structure. The child may take longer to process language, thereby appearing disengaged or inattentive. Children born preterm may demonstrate delays with social pragmatics of speech, not understanding social nuances. On the playground, this can translate into significant social isolation, despite apparent social interest.

Auditory Processing

Linked to the use of language is the processing of it. For the preterm survivor, this is an area of potential challenge. The incidence of hearing impairment (generally defined as the need for amplification) is extremely low, less than 5% [16, 69, 70], and most ascribed to aberrant auditory nerve maturation, or auditory dys-synchrony. In addition, preterm children appear to struggle to a greater degree with auditory processing [71]. As with all other aspects of the behavioral phenotype of prematurity, most outcomes captured in the literature reflect the most significant degree of dysfunction/disorder, but there may be many more that struggle with hearing or processing that do not meet criteria for a formal identification [3].

Clinically, what this looks like is the child born preterm taking longer to process verbal instructions and demonstrating a greater degree of use of visual cues to complete instructions. The children may withdraw socially as they are inefficient at processing conversation and struggle to keep up with the pace of the conversation. The children may demonstrate behaviors suggestive of ADHD as the hearing loss or delayed processing may present with inconsistent and patchy hearing, creating distress or discomfort for the child. Quiet environments may result in apparent hearing, and noisier environments lead to greater dysregulation due to perceived cacophony. Careful consideration for hearing loss is warranted for any diagnosis of ADHD. Identification and diagnosis require awareness and focused assessment.

Motor

There is a spectrum of outcomes possible within motor development of the preterm, but the most significant is that of the neuromotor disability, cerebral palsy (CP). CP is an "umbrella term covering a group of nonprogressive, but often changing, motor impairment syndromes secondary to lesions or anomalies of the brain arising in the early stages of its development" [72]. Generally, this outcome has remained stable across the years of study, with 10–15% of survivors affected, depending on the degree of prematurity [16, 69]. Of these affected children with CP, the majority are ambulatory and demonstrate associated mild to moderate cognitive and behavioral impacts [3, 73–75]. It is these comorbid cognitive, learning, and behavioral conditions that disproportionately contribute to the child's impaired function rather than the motor impairment [3, 75].

More prevalent for the child born preterm is atypical and delayed motor development. Early on, this is in the form of what has historically been referred to as transient dystonia of the preterm, affecting those under 2 years of age [76]. This is characterized by abnormal muscle tone presenting progressively from the cephalad to caudal trunk and symmetric distribution. On exam, there is proximal hypotonia, extensor posturing, exaggerated primitive reflexes, and scapular retraction [77, 78]. Clinically, parents will report "irritability," "jitteriness," being "easy to startle," "stiff in handling," and "very good at standing up" [3, 76]. The result of this presentation is that of transient delays in motor skill acquisition, such as delayed independent sitting, trunk rotation, reaching, manipulating objects, and potentially delayed independent ambulation [79–81]. Dystonia is not predictive of CP but does signal potential future neuromotor, cognitive, and/or behavioral difficulties [82, 83].

Even those preterm children whose neuromotor exam appears "normal" early on in infancy have a greater probability of presenting with developmental coordination disorder (DCD) later in early childhood [84]. DCD is defined by the DSM-5 as a neurodevelopmental disorder characterized by significant difficulty in the acquisition of functional motor skills which interferes with daily function and impacts on school success and play [50]. The child born preterm faces six times greater possibility of developing DCD, using the more stringent diagnostic criteria of performing less successfully than 95% of the child's peers on functional tasks [84]. Studies demonstrate significant discrepancy in their manual dexterity, ball skills, and balance among preterm survivors free of cognitive or motor disability when compared to term controls [85]. Clinically, this presents with greater difficulty mastering patterns of motor activity such as mastering a bicycle, and the difficulty is resistant to repetitive practice.

In addition, the preterm children demonstrate increased difficulty with visual processing and visuomotor skills. Coordination and execution of simple tasks such as writing letters is characterized by greater difficulty, in both how the letter is perceived and written. Pencil pressure is often diminished resulting in barely perceptible output or significantly exaggerated with the answer carved into the paper. Regardless, the result is more effortful written output resulting in eventual oversimplification of the written work, which often is not in keeping with the capacity of the child [3, 6].

Early Intervention, School Functioning, and Impact of Environment

The preterm infant does face challenges but the developing brain is highly malleable [11, 12]. In the spirit of nature through nurture [11], optimizing protective factors such as minimizing neonatal morbidity [13] by improving primary caregiver education level appear to be linked to promotion of healthier outcomes [14]. Furthermore, while the literature on early intervention for the preterm population has been limited by variable study designs and outcomes measured, it has been documented to improve cognitive outcomes in the preschool years [11].

The effect of early intervention into the school years, however, seems not sustained. The shift to school is a challenging time for the preterm child. Roberts et al. evaluated the preterm child's school readiness, including motor, cognitive, language, social, and self-regulatory abilities, and demonstrated a consistent gap between the preterm school child and the term classmate [25]. Prediction, however, of these challenges is imprecise for the following reasons [28]. Composite assessments of early development, such as those used in traditional neonatal follow-up programs including the Bayley Scales of Infant Development, may identify a concern but are inadequate to determine a specific future academic challenge [27]. More specific and discreet assessments have been identified as better at prediction but may still fail to identify the complete profile due to the spectrum of presentation from problem behaviors to disorders [3, 6]. Furthermore, these difficulties may not have been measureable prior to school and not appreciated by traditional surveillance programs [3, 25, 86].

The impact of early intervention may be compromised by the lack of sustainability in the educational system. For a positive intervention to be sustained, one must be aware of the challenge and sustain the strategies that have been successful. Awareness of prematurity may provide the necessary stimulus for further education on the premature behavioral phenotype within the educational sector. Current literature, however, is limited and indicates inconsistent disclosure by parents of their child's medical history [87] and a lack of knowledge of the potential academic challenges for a preterm child among educators [88].

Conclusion

The behavioral phenotype of prematurity describes a spectrum of splintered skills in the domains of cognition, motor, socio-emotional, behavioral, regulatory, and language. It is not synonymous with describing every preterm child but encapsulates the patterns that are often encountered in children born preterm, with multiple areas of challenge potentially acting synergistically common. Underlying this phenotype is a common etiology of dysmaturation, altered sensory development, as well as altered parenting. Nonetheless, the environment offers opportunities for input and enrichment.

The challenge is that these outcomes are not perceived by traditional neonatal follow-up or society as disabling but do contribute to significant challenges in the school years. Furthermore, most aspects of the behavioral phenotype of prematurity arise in early school years, following discharge from neonatal follow-up and the team that is most likely to be aware of the dysmaturation, impact on parenting, and outcome literature. Compounding this is the lack of knowledge among educators to identify and optimize outcomes. Thus, we are essential to providing care for these children and their families. While the challenges represent problems, many are highly responsive to intervention when offered early [3].

References

1. Hodapp, R. M., & Fidler, D. J. (1999). Special education and genetics: Connections for the 21st century. *The Journal of Special Education, 33*, 130–137.
2. Bach, S. A., & Miller, S. P. (2014). Brain injury in premature neonates: A primary cerebral dysmaturation disorder? *Annals of Neurology, 75*, 469–486.
3. Church, P. T., Luther, M., & Asztalos, E. (2012). The perfect storm: Minor morbidities in the preterm survivor. *Current Pediatric Reviews, 8*, 142–151.
4. Blencowe, H., Cousens, S., Oestergaard, M. Z., et al. (2012). National, regional, and worldwide estimates of preterm birth rates in the year 2010 with time trends since 1990 for selected countries: A systematic analysis and implications. *Lancet, 379*, 2162–2172.
5. March of Dimes, PMNCH, Save the Children, WHO. (2012). In C. P. Howson, M. V. Kinney, & J. E. Lawn (Eds.), *Born too soon: The global action report on preterm birth*. Geneva: World Health Organization.

6. Alyward, G. P. (2005). Neurodevelopmental Outcomes of Infants Born Prematurely. *Journal of Developmental and Behavioral Pediatrics, 26*, 427–440.
7. Chyi, L., Lee, H. C., Hintz, S. R., Gould, J. B., & Sutcliffe, T. L. (2008). School outcomes of late preterm infants: Special needs and challenges for infants born at 32 to 36 weeks gestation. *The Journal of Pediatrics, 153*, 25–31.
8. Talge, N. M., Holzman, C., Wang, J., Lucia, V., Gardner, J., & Breslau, N. (2010). Late-preterm birth and its association with cognitive and social outcomes at 6 years of Age. *Pediatrics, 126*, 1124–1131.
9. Wolraich, M. L., Felice, M. E., & Drotar, D. (1996). *The classification of child and adolescent mental diagnoses in primary care. Diagnostic and statistical manual for primary care (DSM-PC) child and adolescent version*. Elk Grove Village: American Academy of Pediatrics.
10. Accardo, P. J., Accardo, J. A., & Capute, A. J. (2008). A neurodevelopmental perspective on the continuum of developmental disabilities. In P. J. Accardo (Ed.), *Capute & Accardo's neurodevelopmental disabilities in infancy and childhood* (3rd ed.). Baltimore: Brooks Publishing.
11. National Research Council and Institute of Medicine. (2000). From neurons to neighborhoods: The science of early childhood development. In J. P. Shonkoff & D. A. Phillips (Eds.), *Board on children, youth, and families, commission on behavioral and social sciences and education*. Washington, DC: National Academy Press.
12. Kolb, B., & Gibb, R. (2011). Brain plasticity and behaviour in the developing brain. *Journal of the Canadian Academy of Child and Adolescent Psychiatry, 20*, 265–276.
13. Asztalos, E. V., Church, P. T., Riley, P., Shah, P., & Canadian Neonatal Network and Canadian Neonatal Follow Up Network Investigators. (2017). Neonatal factors associated with a good neurodevelopmental outcome in very preterm infants. *American Journal of Perinatology, 34*, 388–396.
14. Asztalos, E. V., Church, P. T., Riley, P., Shah, P. S., & Canadian Neonatal Network and Canadian Neonatal Follow Up Network Investigators. (2017). Association between primary caregiver education and cognitive and language development of preterm neonates. *American Journal of Perinatology, 34*, 364–371.
15. Luciana, M. (2003). Cogntive development in chidren born preterm: Implications for theories of brain plasticity following early injury. *Development and Psychopathology, 15*, 1017–1047.
16. Synnes, A., Luu, T. M., Moddemann, D., et al. (2017). Determinants of developmental outcomes in a very preterm canadian cohort. *Archives of Disease in Childhood. Fetal and Neonatal Edition, 102*, F235–F234.
17. Volpe, J. J. (2009). Brain injury in preterm infants: A complex amalgam of destructive and developmental disturbances. *The Journal-Lancet, 8*, 110–124.
18. Volpe, J. J. (2009). Cerebellum of the premature infant: Rapidly developing, vulnerable, clinically important. *Journal of Child Neurology, 24*, 1085–1104.
19. Volpe, J. J. (1997). Brain Injury in the preterm infant—from pathogenesis to prevention. *Brain and Development, 19*, 519–534.
20. Inder, T. E., Wells, S. J., Mogridge, N. B., Spencer, C., & Volpe, J. J. (2003). Defining the nature of the cerebral abnormalities in the premature infant: A qualitative magnetic resonance imaging study. *The Journal of Pediatrics, 143*, 171–179.
21. Lickliter, R. (2011). The integrated development of sensory development. *Clinics in Perinatology, 38*, 591–603.
22. Karatzias, A., Chouliara, Z., Maxton, F., Freer, Y., & Power, K. (2007). Post-traumatic symptomatology with premature infants: A systematic review of the literature. *Journal of Prenatal and Perinatal Psychology and Health, 21*, 249–260.
23. Shah, P. E., Clements, M., & Poehlmann, J. (2011). Maternal resolution of grief after preterm birth: Implications for infant attachment security. *Pediatrics, 127*, 284–292.
24. Ment, L. R., Kesler, S., Vohr, B., et al. (2009). Longitudinal brain volume changes in preterm and term control subjects during late childhood and adolescence. *Pediatrics, 123*, 503–511.
25. Roberts, G., Lim, J., Doyle, L. W., & Anderson, P. J. (2011). High rates of school readiness difficulties at 5 years of age in very preterm infants compared with term controls. *Journal of Developmental and Behavioral Pediatrics, 32*, 117–124.

26. O'Brien, G. (2006). Behavioural phenotypes: Causes and clinical implications. *Advances in Psychiatric Treatment, 12*, 338–348.
27. Hack, M., Taylor, G., Drotar, D., et al. (2005). Poor predictive validity of the bayley scales of infant development for cognitive function of extremely low birth weight children at school age. *Pediatrics, 116*, 333–340.
28. Alyward, G. P. (2005). The Conundrum of Prediction. *Pediatrics, 116*, 491–492.
29. Elias, E. R., & Crocker, A. C. (2009). Intellectual disability. In W. B. Carey, A. C. Crocker, W. L. Coleman, E. R. Elias, & H. M. Feldman (Eds.), *Developmental-behavioral pediatrics* (4th ed.). Philadelphia: Saunders Elsevier.
30. Bos, A. F., & Roze, E. (2011). Neurodevelopmental outcome in preterm infants. *Developmental Medicine and Child Neurology, 53*, 35–39.
31. Marlowe, N. M., Wolke, D. M., Bracewell, M. A., et al. (2005). Neurologic and developmental disability at six years of age after extremely preterm birth. *NEJM, 353*, 9–19.
32. Johnson, S., Hennessy, E., Smith, R., Trikic, R., Wolke, D., & Marlow, N. (2009). Academic attainment and special education needs in extremely preterm children at 11 years: The EPICure study. *Archives of Disease in Childhood. Fetal and Neonatal Edition, 94*, F283–F289.
33. Simms, V., Gilmore, C., Cragg, L., Marlow, N., Wolke, D., & Johnson, S. (2013). Mathematics difficulties in extremely preterm children: Evidence of a specific deficit in mathematics processing. *Pediatric Research, 73*, 236–244.
34. Mulder, H., Pitchford, N. J., & Marlow, N. (2010). Processing speed and working memory underlie academic attainment in very preterm children. *Archives of Disease in Childhood. Fetal and Neonatal Edition, 95*(4), F267–F272.
35. Whitfield, M., Grunau, R. V. E., & Holsti, L. (1997). Extremely premature ($\leq$800 g) schoolchildren: Multiple areas of hidden disability. *Archives of Disease in Childhood, 77*, F85–F90.
36. Lynn, L. N., Cuskelly, M., O'Callaghan, M. J., & Gray, P. H. (2011). Self-regulation: A new perspective on learning problems experienced by children born extremely preterm. *Australian Journal of Educational and Developmental Psychology, 11*, 1–10.
37. Hemmi, M. H., Wolke, D., & Schneider, S. (2011). Associations between problems with crying, sleeping, and/or feeding in infancy and long-term behavioural outcomes in childhood: A meta-analysis. *Archives of Disease in Childhood, 96*, 622–629.
38. Bilgin, A., & Wolke, D. (2016). Regulatory problems in very preterm and full-term infants over the first 18 months. *Journal of Developmental and Behavioral Pediatrics, 37*(4), 298–305.
39. Wolke, D., Schmid, G., Schreier, A., & Meyer, R. (2009). Crying and feeding problems in infancy and cognitive outcome in preschool children born at risk: A prospective population study. *Journal of Developmental and Behavioral Pediatrics, 30*, 226–238.
40. Arpi, E., & Ferrari, F. (2013). Preterm birth and behavior problems in infants and preschool-age children: A review of the recent literature. *Developmental Medicine and Child Neurology, 55*, 788–769.
41. Lowe, J., Woodward, B., & Papile, L. A. (2005). Emotional regulation and its impact on development in extremely low birth weight infants. *Journal of Developmental and Behavioral Pediatrics, 26*, 209–213.
42. Scott, M. N., Taylor, H. G., Fristad, M. A., et al. (2012). Behavior disorders in extremely preterm/extremely low birth weight children in kindergarten. *Journal of Developmental and Behavioral Pediatrics, 33*, 1–12.
43. Miller, K. J. (2005). Executive functions. *Pediatric Annals, 34*, 311–317.
44. Powell, K. B., & Voeller, K. K. S. (2004). Prefrontal executive function syndromes in children. *Journal of Child Neurology, 19*, 785–797.
45. Anderson, P. J., Doyle, L. W., & Victorian Infant Collaborative Study Group. (2004). Executive functioning in school-aged children who were born very preterm or with extremely low birth weight in the 1990's. *Pediatrics, 114*, 50–57.
46. Marlow, N., Hennessy, E. M., Bracewell, M. A., & Wolke, D. (2007). Motor and executive function at 6 years of age after extremely preterm birth. *Pediatrics, 120*, 793–803.

47. Aarnoudse-Moens, C. S. H., Weisglas-Kuperus, N., van Goudoever, B., & Oosterlaan, J. (2009). Meta-analysis of neurobehavioral outcomes in very preterm and/or very low birth weight children. *Pediatrics, 124*, 717–728.
48. Burnett, A. C., Scratch, S. E., Lee, K. J., et al. (2015). Executive function in adolescents born < 1000 g or < 28 weeks: A prospective cohort study. *Pediatrics, 135*, e826–e834.
49. Prock, L. A., & Rappaport, L. (2009). Attention and deficits of attention. In W. B. Carey, A. C. Crocker, W. L. Coleman, E. R. Elias, & H. M. Feldman (Eds.), *Developmental-behavioral pediatrics* (4th ed.). Philadelphia: Saunders Elsevier.
50. American Psychiatric Association. (2013). *Diagnostic and statistical manual of mental disorders* (5th ed.). Arlington: American Psychiatric Association.
51. Farooqi, A., Hagglof, B., Sedin, G., Gothefors, L., & Serenius, F. (2007). Mental health and social competencies of 10-to 12-year-old children born at 23 to 25 weeks of gestation in the 1990s: A Swedish national prospective follow-up study. *Pediatrics, 120*, 118–133.
52. Delobel-Ayoub, M., Arnaud, C., White-Koning, M., et al. (2009). Behavioral problems and cognitive performance at 5 years of age after very preterm birth: The EPIPAGE study. *Pediatrics, 123*, 1485–1492.
53. de Kieviet, J. F., van Elburg, R. M., Lafeber, H. N., & Oosterlaan, J. (2012). Attention problems of very preterm children compared with age-matched term controls at school-age. *The Journal of Pediatrics, 161*, 824–829.
54. Taylor, H. G., Margevicius, S., Schluchter, M., Andreias, L., & Hack, M. (2015). Persisting behavior problems in extremely low birth weight adolescents. *Journal of Developmental and Behavioral Pediatrics, 36*(3), 178–187.
55. Montagna, A., & Nosarti, C. (2016). Socio-emotional development following very preterm birth: Pathways to psychopathology. *Frontiers in Psychology, 7*, 23.
56. Msall, M. E., & Park, J. J. (2008). The spectrum of behavioral outcomes after extreme prematurity: Regulatory, attention, social, and adaptive dimensions. *Seminars in Perinatology, 32*, 42–50.
57. Williamson, K. E., & Jakobson, L. S. (2014). Social perception in children born at very low birth weight and its relationship with social/behavioral outcomes. *Journal of Child Psychology and Psychiatry, 55*, 990–998.
58. Msall, M. E., Phelps, D. L., Hardy, R. J., Dobson, V., Quinn, G. E., Summers, C. G., & Tremont, M. R. (2004). Educational and social competencies at 8 years in children with threshold retinopathy of prematurity in the CRYO-ROP multicenter study. *Pediatrics, 133*, 790–799.
59. Indredavik, M. S., Vik, T., Heyerdahl, S., Kulseng, S., Fayers, P., & Brubakk, A.-M. (2004). Psychiatric symptoms and disorders in adolescents with low birth weight. *Archives of Disease in Childhood. Fetal and Neonatal Edition, 89*, F445–F450.
60. Ek, U., Fernell, E., Jacobsen, L., & Gillberg, C. (1998). Relationship between blindness due to retinopathy of prematurity and autistic spectrum disorders: A population based study. *Developmental Medicine and Child Neurology, 40*, 297–301.
61. Limperopoulos, C., Bassan, H., Sullivan, N. R., et al. (2008). Positive screening for autism in ex-preterm infants: Prevalence and risk factors. *Pediatrics, 121*, 758–765.
62. Johnson, S., Hollis, C., Kochhar, P., Hennessey, E., Wolke, D., & Marlow, N. (2010). Autism spectrum disorders in extremely preterm children. *The Journal of Pediatrics, 156*, 525–531.
63. Joseph, R. M., O'Shea, T. M., Allred, E. N., et al. (2017). Prevalence and associated features of autism spectrum disorder in extremely low gestational age newborns at age 10 years. *Autism Research, 10*(2), 224–232.
64. Barre, N., Morgan, A., Doyle, L. W., & Anderson, P. J. (2011). Language abilities in children who were very preterm and/or very low birth weight: A meta-analysis. *The Journal of Pediatrics, 158*, 766–774.
65. Putnick, D. L., Bornstein, M. H., Eryigit-Madzwamuse, S., & Wolke, D. (2017). Long-term stability of language performance in very preterm, moderate-late preterm, and term children. *The Journal of Pediatrics, 181*, 74–79.

66. Guarini, A., Marini, A., Savini, S., Alessandroni, R., Faldella, G., & Sansavini, A. (2016). Linguistic features in children born very preterm at preschool age. *Developmental Medicine and Child Neurology, 58*, 949–956.
67. Foster-Cohen, S. H., Friesen, M. D., Champion, P. R., & Woodward, L. J. (2010). High prevalence/low severity language delay in preschool children born very preterm. *Journal of Developmental and Behavioral Pediatrics, 31*, 658–667.
68. Foster-Cohen, S., Edgin, J. O., Champion, P. R., & Woodward, L. J. (2007). Early delayed language in very preterm infants: Evidence from the MacArthur-Bates CDI. *Journal of Child Language, 34*, 655–675.
69. Saigal, S., & Doyle, L. W. (2008). An overview of mortality and sequelae of preterm birth from infancy to adulthood. *Lancet, 371*, 261–269.
70. Robertson, C. M. T., Howarth, T. M., Bork, D. L. R., & Dinu, I. A. (2009). permanent bilateral sensory and neural hearing loss of children after neonatal intensive care because of extreme prematurity: A thirty year study. *Pediatrics, 123*, e797–e807.
71. Fellman, V., Kushnerenko, E., Mikkola, K., Ceponiene, R., Leipala, J., & Naatanen, R. (2004). Atypical auditory event-related potentials in preterm infants during the first year of life: A possible sign of cognitive dysfunction? *Pediatric Research, 56*, 291–297.
72. Mutch, L., Alberman, E., Hagberg, G., Kodama, K., & Perat, M. V. (1992). Cerebral palsy epidemiology: Where are we now and where are we going? *Developmental Medicine and Child Neurology, 34*, 547–551.
73. Ancel, P. Y., Livinec, F., Larroque, B., et al. (2006). Cerebral palsy among very preterm children in relation to gestational age and neonatal ultrasound abnormalities: The EPIPAGE cohort study. *Pediatrics, 117*, 828–836.
74. Roze, E., Van Braeckel, K. N. J. A., van der Veere, C. N., et al. (2009). Functional outcome at school age of preterm infants with periventricular hemorrhagic infarction. *Pediatrics, 123*, 1493–1500.
75. Sigurdardottir, S., Indredavik, M. S., Eiriksodottir, A., Einarsdottir, K., Gudmundsson, H. S., & Vik, T. (2010). Behavioural and emotional symptoms of preschool children with cerebral palsy: A population based study. *Developmental Medicine and Child Neurology, 52*, 1056–1061.
76. Drillien, C. M. (1972). Abnormal neurologic signs in the first year of life in low-birthweight infants: Possible prognostic significance. *Developmental Medicine and Child Neurology, 14*, 575–584.
77. Georgieff, M. K., & Bernbaum, J. C. (1986). Abnormal Shoulder girdle muscle tone in premature infants during their first 18 months of life. *Pediatrics, 77*, 664–669.
78. deGroot, L., Hopkins, B., & Touwen, B. (1997). Motor asymmetries in preterm infants at 18 weeks corrected age and outcomes at 1 year. *Early Human Development, 48*, 35–46.
79. Samson, J. F., & deGroot, L. (2001). Study of a group of extremely preterm infants (25-27 Weeks): How do they function at 1 year of age? *Journal of Child Neurology, 16*, 832–837.
80. deGroot, L., Hopkins, B., & Touwen, B. (1995). Muscle power, sitting unsupported and trunk rotation in pre-term infants. *Early Human Development, 43*, 37–46.
81. deGroot, L., deGroot, C. J., & Hopkins, B. (1997). An instrument to measure independent walking: Are there differences between preterm and fullterm infants? *Journal of Child Neurology, 12*, 37–41.
82. Breslau, N., Chilcoat, E. O., Johnson, E. O., Andreski, P., & Lucia, V. C. (2000). Neurologic soft signs and low birthweight: Their association and neuropsychiatric implications. *Biological Psychiatry, 47*, 71–79.
83. Harmon, H. M., Taylor, G. H., Minich, N., Wilson-Costello, D., & Hack, M. (2015). Early school outcomes for extremely preterm infants with transient neurological abnormalities. *Developmental Medicine and Child Neurology, 57*, 865–871.
84. Edwards, J., Berube, M., Erlandson, K., et al. (2011). Developmental coordination disorder in school-aged children born very preterm and/or at very low birth weight: A systematic review. *Journal of Developmental and Behavioral Pediatrics, 32*, 678–687.

85. Goyen, T.-A., & Lui, K. (2009). Developmental coordination disorder in "apparently normal" schoolchildren born extremely preterm. *Archives of Disease in Childhood, 94*, 298–302.
86. Kuppala, V. S., Tabangin, M., Haberman, B., Steichen, J., & Yolton, K. (2012). Current state of high-risk infant follow-up care in the United States: results of a national survey of academic follow up programs. *Journal of Perinatology, 32*, 293–298.
87. Church, P. T., Cavanagh, A., Lee, S. K., Shah, V. *Academic challenges for preterm infants: Parent and educators' perspectives*. Manuscript in draft.
88. Johnson, S., Gilmore, C., Gallimore, I., Jaekel, J., & Wolke, D. (2015). The long-term consequences of preterm birth: What do teachers know? *Developmental Medicine and Child Neurology, 57*(6), 571–577.

The Late Preterm Infant

Katherine Steingass, Lindsay Bartram, and Anita Narayanan

Abstract Late preterm refers to infants born between 34 0/7 weeks and 36 6/7 weeks gestational age. In 2005, the National Institute of Child Health and Human Development introduced this nomenclature to replace the previous descriptor "near term" to more accurately categorize this group as preterm and at risk for requiring specialized care and monitoring. Late preterm infants are often comparable in birth weight and appearance to term infants and may be in the well-baby nursery or with their mother after birth, but they are physiologically immature in a number of domains rendering them medically vulnerable in comparison to term infants. While late preterm birth does not carry nearly as high a risk of severe neonatal morbidities and neurodevelopmental impairment as very preterm birth, this is a population at risk for early medical complications and later developmental and behavioral difficulties.

An infant girl was born at 35 weeks gestation weighing 2600 g. Labor was induced preterm due to maternal preeclampsia. The infant did not require admission to the neonatal intensive care unit (NICU) but remained hospitalized in the nursery for 4 days due to hyperbilirubinemia requiring phototherapy. The infant is breastfed but did receive a supplemental formula feed due to hypoglycemia. During preparations for discharge, the parents ask if their infant is considered a "preemie" and if any special monitoring or follow-up is needed.

In 2005, the National Institute of Child Health and Human Development introduced the nomenclature "late preterm" to replace the previous descriptor "near term" to more accurately categorize this group as preterm and at risk for requiring specialized care and monitoring [1]. They defined late preterm as births occurring between 34 0/7 weeks and 36 6/7 weeks gestational age (GA). The American Academy of Pediatrics (AAP) and the American College of Obstetricians and Gynecologists also use late preterm to describe infants born between 34 0/7 and 36 6/7 weeks GA [2], while the World Health Organization categorizes infants born

K. Steingass (✉) · L. Bartram · A. Narayanan
Nationwide Children's Hospital, Columbus, OH, USA
e-mail: Katherine.steingass@nationwidechildrens.org

H. Needelman, B. J. Jackson (eds.), *Follow-Up for NICU Graduates*,
https://doi.org/10.1007/978-3-319-73275-6_7

from 32 to <37 weeks GA as moderate to late preterm [3]. In this chapter, late preterm (LPT) is used to refer to information and studies pertaining to infants born between 34 0/7 and 36 6/7 weeks gestation. Moderate to late preterm (MLPT) is used when studies included infants born between 32 0/7 and 36 6/7 weeks gestation.

The final weeks of gestation are a critical period of growth and development of the brain and other organ systems. At 34 weeks gestation, the brain's volume is only 65% of its volume at term, and cortical volume is approximately half the volume it will be at 40 weeks [4]. In addition to increasing in volume, multiple maturational processes in the brain such as synaptogenesis and myelination are ongoing during this time. Late preterm birth may interrupt these neurologic processes with the transition from the intrauterine to extrauterine environment [5]. Late preterm infants are often comparable in birth weight and appearance to term infants and may be in the well-baby nursery or with their mother after birth, but they are physiologically immature in a number of domains rendering them medically vulnerable in comparison to term infants [1, 6, 7].

Late preterm infants have increased neonatal morbidity and mortality compared to term infants [1, 6, 7]. They are more likely to be admitted to the NICU and have longer hospital stays [6]. While almost 80% of LPT infants have no significant neonatal complications, LPT infants are seven times more likely to experience morbidities than term infants [8]. Late preterm birth is also a significant risk factor for hospital readmission in the neonatal period, especially if infants are discharged early from the birth hospitalization (<48 h) [6, 9]. Failure to account for the increased risk for complications such as hyperbilirubinemia and dehydration and treating LPT infants similar to term infants may be a factor in the risk for readmission. The risk for medical complications in LPT infants might best be considered as "intermediate risk" between very preterm and term infants.

Multiple studies suggest increased rates of adverse developmental and behavioral outcomes for children born LPT although the literature is quite variable. The disorders seen in this population tend not to be severe disabilities (e.g., intellectual disability), but children may have more mild deficits in multiple areas such as academic difficulties and behavior problems [7, 10]. Similar to neonatal medical complications, children born LPT are likely best considered as at "intermediate risk" for adverse developmental outcomes. Indeed, studies assessing outcomes for individuals born across the full range of GA indicate a continuum of increasing risk with decreasing GA rather than a threshold effect [11–13].

While LPT infants are increasingly being recognized as a population at risk, their outcomes have not been studied as extensively or as systematically as very preterm infants. Many of the studies on outcomes past the neonatal period are retrospective in nature [10]. A note of caution in considering reports on developmental and behavioral outcomes for children born late preterm is that many studies have not accounted for neonatal morbidities or need for NICU care making it difficult to determine the risk for healthy LPT children who did not require specialized care [7, 10]. Another complicating factor is that many LPT births occur due to medical indications for early delivery such as maternal diabetes, preeclampsia, intrauterine growth restriction (IUGR), and chorioamnionitis which may adversely affect neurodevelopment apart from the early delivery [6, 7]. Additionally, some studies have included both moderate and late preterm infants, and results may not be generalizable to all LPT infants.

Although the absolute risk for serious complications and severe adverse long-term outcomes in LPT infants is low, a much higher number of infants are born LPT than very preterm making this a significant public health issue [7]. Late preterm birth is associated with greater healthcare costs than term delivery [9]. In 2013, LPT births accounted for 8% of all births in the United States and 70% of all preterm births [14]. Thus, LPT birth likely accounts for a substantial portion of the total costs associated with prematurity [6]. Additionally, in a large, population-based study, almost three-quarters of the total long-term disability associated with prematurity occurred in individuals born MLPT [13].

The 2012 AAP policy statement Hospital Discharge of the High-risk Neonate recommends periodic developmental evaluation to identify children with developmental delays and to facilitate referrals to early intervention services [15]. Many NICUs, especially those in tertiary care centers, have specialized follow-up programs. However, no standardized guidelines exist for follow-up, and there is significant variability between programs (data presented at 2015 Society for Developmental and Behavioral Pediatrics Annual Meeting). A 2004 report described evidence supporting the importance of monitoring and studying the neurodevelopmental outcomes of high-risk infants and provided objectives for follow-up programs and research studies. This report focused on very preterm infants and those with specific perinatal problems, but it does provide some guidance on collaboration with community physicians which is applicable to follow-up care of LPT infants [15].

Late preterm infants are often not followed in NICU follow-up programs. In a 2014 survey of North American NICU follow-up program directors, the mean maximum GA accepted in NICU follow-up programs was 33.1 weeks, and the mean maximum birth weight was 1782 g (data presented at 2015 Society for Developmental and Behavioral Pediatrics Annual Meeting). Given the numbers of infants born LPT, it is not feasible for most NICU follow-up programs to follow all LPT infants. Thus, much of the follow-up care for this population will occur within the medical home. Primary care physicians should perform routine developmental surveillance and screening in LPT infants as in all children while keeping in mind that these children are at increased risk compared to those born at term.

Specialized NICU follow-up programs can play an important role in educating community medical and early intervention providers regarding the needs of this population as well as providing consultation to primary care providers regarding individual infants in their practices. Some LPT infants who are identified as being at particularly high risk for adverse neurodevelopmental outcomes will benefit from referral to a specialized NICU follow-up program. Follow-up programs can also work with NICU and nursery discharge planners to ensure that parents of LPT infants receive appropriate education regarding their infant's risks and follow-up needs.

Medical Issues in the Late Preterm Infant

Feeding and Growth

Challenges of Feeding the Late Preterm Infant

The risks of feeding problems are significantly higher in LPT compared to term infants [16]. The sucking ability that allows for sufficient intake for growth begins to develop at 34 weeks. However, LPT infants have immature oro-buccal coordination and swallowing mechanisms that interfere with the establishment and maintenance of adequate oral feeds. In addition to the oral-motor difficulties, cardiorespiratory instability and poor suck-swallow-breathe coordination can place LPT infants at risk for fatiguing easily during feeding leading to inadequate caloric intake [17, 18]. The oral-motor skills needed for feeding improve between 35 and 38 weeks as the nervous system matures [19].

Feeding difficulties can contribute to problems with dehydration, hypoglycemia, and hyperbilirubinemia in the neonatal period. Late preterm infants are more likely to develop hypoglycemia compared to term infants due to reduced glycogen stores and deficient activity of enzymes involved in glucose regulation [16, 20]. Decreasing GA and birth weight are associated with an increasing risk of needing nutritional support (e.g., total parental nutrition, intravenous fluids, oral-gastric feeding) in the neonatal period [21]. Late preterm infants should demonstrate at least 24 h of successful feeding prior to discharge to reduce the risk of readmission [6].

Breastfeeding and the LPT Infant

Breast milk is important to the LPT infant as the best form of nutrition to support brain development and growth [22]. However, LPT infants often have difficulties initiating and maintaining successful breastfeeding. Breastfeeding initiation rates among LPT mother/infant dyads are less than that of term infants, as are rates of exclusive breastfeeding at several months after birth [18].

There are multiple barriers to successful breastfeeding in the LPT population including both infant and maternal factors. Immaturities in the infant's oral-motor skills for feeding can cause difficulties with latching, sucking, and transferring milk [23]. Even LPT infants with no associated medical problems may fatigue easily and have hypotonia, poor head control, and difficulties establishing a good latch [24]. Fatigue may be mistaken for satiety [17, 18]. Late preterm infants also tend to have longer sleep intervals leading to decreased opportunities to nurse. In addition, LPT infants are more likely to have metabolic disturbances such as hypoglycemia that necessitate formula supplementation [18].

Maternal barriers to successful breastfeeding at any GA include type 1 diabetes, obesity, cesarean section delivery, infections, multiple births, pregnancy-induced hypertension, and medications used during the ante- and intrapartum periods [18, 23].

Some of these issues may be more common in mothers of LPT infants as they can contribute to preterm delivery. Maternal anxiety stemming from an early or traumatic birth and the fragility of the preterm infant has also been cited as a contributor to breastfeeding failure. Anxiety has been implicated in delaying lactogenesis, although the mechanism for this is unclear [18, 23].

Lactogenesis may also be negatively affected due to decreased ability of the infant to remove milk from the breast resulting in feedback inhibition, delayed lactogenesis, and a downregulation of milk volume [19]. Maternal milk supply is best established by the proximity of the mother and infant for frequent breastfeeding; however LPT infants and their mothers may be separated immediately after birth for medical interventions [18].

Neonatal intensive care unit admission in and of itself does not negatively affect breastfeeding initiation in LPT infants. Admission to the NICU is associated with higher rates of initiation and continuation of breastfeeding compared with care outside of a NICU. The positive effect of NICU admission may be partially due to increased exposure to messages and educational interventions promoting breastfeeding. Many NICUs have lactation consultants available which has been shown to be effective in increasing breastfeeding rates [25].

If infants are not able to breastfeed well enough to establish the milk supply, use of an electric grade pump may be needed to stimulate and increase milk production [23]. Breastfeeding positions that provide head support, such as the football or cross-cradle holds, will also help the LPT infant feed effectively. The head of the LPT infant is heavy in relation to the weak neck musculature, and many traditional breastfeeding positions allow the head to fall forward or backward interfering with the infant's ability to maintain an adequate latch on the breast [26].

The Academy of Breastfeeding Medicine in 2011 created a clinical protocol on breastfeeding specific to LPT infants. The protocol recommends that mothers use a combination of frequent skin-to-skin care, direct breastfeeding, and breast milk expression with a hospital grade pump. Close observation of breastfeeding in the hospital, specific discharge instructions, and early and frequent post-discharge follow-up with a healthcare provider or lactation consultant are also recommended [27].

Nursery and NICU discharge planning should include education of caregivers regarding feeding and signs of dehydration and hyperbilirubinemia [6]. Late preterm infants who are exclusively breastfed at hospital discharge are at risk for rehospitalization secondary to dehydration and hyperbilirubinemia [18]. Follow-up care providers, whether in primary care or NICU follow-up clinics, need to be aware of the special breastfeeding needs of LPT infants. The first visit should occur within 2 days of discharge from the nursery, during which time a full history should be obtained regarding breastfeeding and pumping, frequency and duration of feeds, infant behaviors, urine and stool output, and weight. Late preterm infants should have at least weekly follow-ups until they reach 40 weeks GA or until full breastfeeding is established. The current recommendation for weight gain in LPT infants as they approach 40 weeks is approximately 30 g per day [19].

Later Feeding Difficulties

Little research has been done regarding feeding in MLPT children past infancy, but one study suggests that these children continue to have increased rates of feeding difficulties. In a study of 2-year-olds, those born LPT were at a 1.6 times increased risk of feeding difficulties compared to term. This included problems with both oral-motor skills and refusal/picky eating. Appropriately addressing early feeding difficulties and providing anticipatory guidance to families may decrease the risk of ongoing feeding problems [28].

Nutritional Requirements

Knowledge of nutritional requirements specific to MLPT infants is minimal and is calculated based on extrapolating data from very preterm and term infants. Moderate-to-late preterm infants have higher energy and protein requirements for growth compared to term infants. In addition, stores of essential nutrients such as iron and other minerals are acquired via transplacental transfer, a significant proportion of which occurs during the third trimester of gestation. Moderate-to-late preterm infants are born before this is complete and thus may require higher amounts of minerals including calcium and phosphorus for bone mineralization postnatally [29].

Moderate and late preterm infants may also be at risk for iron deficiency [30]. Some authors have recommended that all MLPT infants receive multivitamins with iron [23, 31]. A multicenter randomized controlled trial suggested benefit from iron supplementation in MLPT infants up to age 1 year based on higher ferritin levels compared to placebo [31].

Long-chain polyunsaturated fatty acids, especially docosahexaenoic acid, play a key role in the development of the central nervous and visual systems. Stores of these also primarily accumulate during the third trimester. Formulas containing docosahexaenoic acid may improve developmental outcomes for MLPT infants who are formula-fed [29].

Given that MLPT infants have nutritional requirements different from those of term infants, breast milk or formula designed for term infants may not be sufficient to meet the nutritional needs of some MLPT infants especially if they have low birth weight (<2500 g), IUGR, or poor postnatal growth. These infants may benefit from supplementation with specialized preterm post-discharge formula which provides increased calories and protein per ounce as well as higher amounts of specific vitamins and minerals (e.g., vitamin E, calcium, phosphorus, magnesium) [29]. Further research is needed to determine whether all MLPT infants need supplementation and for how long it is required.

Growth

Intrauterine growth restriction is common in MLPT infants which further increases their risk for morbidity including poor postnatal growth. They also have higher rates of breastfeeding failure and feeding difficulties which may contribute to later growth problems [32]. Late preterm infants are almost twice as likely term infants to have gastroesophageal reflux and be prescribed medications for reflux by 2 years corrected age [28]. Reflux may be associated with poor nutritional intake and growth potential in this group.

It is widely known that growth problems are common in very preterm infants, but growth has been less well researched in MLPT infants. These infants are born at the time of peak in utero growth velocity, and some authors hypothesize that missing this peak results in growth restraint at least during the first year of life [33]. Some studies suggest an increased risk for postnatal growth problems in MLPT children, but the evidence is inconclusive. Multiple investigations report that MLPT children are smaller than term peers in early childhood (up to at least 18–24 months) [32–34]. However, studies have differed in regard to how quickly they catch up with some showing ongoing growth restriction at age 3–4 years [33, 35] and others finding growth comparable to term counterparts by this age [34]. Potential reasons for the conflicting findings are the different GA ranges for the study groups (MLPT vs. LPT) and the exclusion of infants born small for gestational age (SGA) in the study that found no growth differences by age 4 years [34]. In the other studies, a history of SGA correlated with later short stature or underweight status although children with birth weights appropriate for GA were still more likely to exhibit growth restraint than term peers in early childhood across studies [32, 33, 35].

Monitoring growth parameters and intervening to promote appropriate growth when needed are important due to associations between physical growth and cognitive development [36]. However, excessive catch-up growth should be avoided as rapid weight gain as fat puts LPT infants at increased risk for central adiposity, hyperinsulinemia, and metabolic syndrome as young adults [21].

Recommendations for Follow-Up Care

- Arrange for medical follow-up within 24–48 h of nursery discharge. In most cases, follow-up should be with a primary care provider within a medical home. Close follow-up is recommended until the infant demonstrates consistent, appropriate weight gain.
- Refer to a lactation specialist if the mother/infant dyad is having difficulty initiating or maintaining breastfeeding.
- Refer infants who have persistent feeding difficulties to a speech or occupational therapist for a feeding evaluation and therapy.
- Provide anticipatory guidance to families regarding infant feeding.
- Consider prescribing or recommending iron supplementation, especially for infants who are breastfed.
- Consider recommending preterm discharge formula, especially for infants with low birth weight (<2500 g), a history of IUGR, or poor postnatal growth.
- Monitor growth within primary care.
- Consider referral of LPT infants with low birth weight or a history of IUGR or with postnatal difficulties with weight gain to a NICU follow-up program that includes a dietician.

Hyperbilirubinemia

Hyperbilirubinemia is more common and more severe in LPT infants and has a longer duration than in term infants [16, 20]. Late preterm infants are approximately five times more likely to have jaundice requiring phototherapy compared to term, and the risk increases with decreasing GA [20]. Risk of hyperbilirubinemia is even further increased in moderate preterm infants [37]. Late preterm infants are more also likely than term infants to have bilirubin-induced neurotoxicity including kernicterus from severe hyperbilirubinemia [38, 39].

Similar processes are involved in the development of hyperbilirubinemia in term and preterm infants. However, MLPT infants have increased bilirubin production due to a shortened red blood cell life span and decreased bilirubin conjugation due to immaturity of liver enzymes [39]. Breastfeeding and other feeding difficulties in these infants can lead to decreased intake resulting in increased enterohepatic circulation and decreased elimination of bilirubin [39, 40]. Factors contributing to the increased risk of kernicterus in preterm infants are thought to include low serum albumin with decreased bilirubin-binding capacity and possibly increased permeability of the central nervous system to bilirubin due to immaturity of the blood-brain barrier. Medical complications such as hypoxemia and infection can further decrease bilirubin-binding capacity and increase CNS permeability [39].

In 2009, the AAP published an update and clarification to the 2004 guidelines for the management of hyperbilirubinemia in infants born at ≥35 weeks gestation. The update recommends assessing both a prehospital discharge bilirubin level (serum or transcutaneous) and clinical risk factors to predict the likelihood of subsequent hyperbilirubinemia. The clinical risk factors that are most predictive of hyperbilirubinemia are lower GA and exclusive breastfeeding. Nomograms are available to guide initiation of treatment with phototherapy and exchange transfusion based on an infant's age in hours, total bilirubin level, and level of risk [41].

Treatment of hyperbilirubinemia in the LPT infant (≥35 weeks) is similar to treatment of term infants although the threshold for initiating treatment is lower based on the nomogram [41]. Treatment may include phototherapy, hydration, and, in extreme cases, exchange transfusion. The Academy of Breastfeeding Medicine (ABM) clinical protocol guidelines for the management of jaundice in the breastfeeding infant ≥35 weeks gestation endorses that all modes of treatment of hyperbilirubinemia are compatible with breastfeeding and breastfeeding should be encouraged. Supplementation of breastfeeding with cow's milk-based or hydrolyzed protein formulas has been shown to inhibit the intestinal absorption of bilirubin; therefore supplementation of breastfeeding with small amounts of infant formula can be used to lower bilirubin levels in breastfed infants [40].

Limited evidence is available regarding management of hyperbilirubinemia in infants <35 weeks gestation, but a consensus-based report published in 2012 provides suggested thresholds for initiation of phototherapy and exchange transfusion based on GA [42].

Hyperbilirubinemia accounts for a significant proportion of early hospital readmission (<14 days after birth) in LPT infants [43]. Measures that may reduce the risk of readmission include avoidance of early discharge <48 h after birth, appropriate discharge education for caregivers, and close outpatient follow-up [6, 41].

Recommendations for Follow-Up Care

- Consider MLPT infants at increased risk for hyperbilirubinemia and secondary neurologic sequelae.
- Arrange for outpatient follow-up within 2 days if discharged at less than 72 h after birth.
- Consider follow-up measurement of total serum or transcutaneous bilirubin level depending on bilirubin risk zone at discharge and other risk factors for hyperbilirubinemia.

Respiratory

Moderate-to-late preterm infants have increased rates of respiratory morbidity in the neonatal period including respiratory distress syndrome (RDS), transient tachypnea of the newborn (TTN), persistent pulmonary hypertension, and pneumonia [16, 20, 44]. These may result in need for respiratory support and NICU admission. A systematic review found that LPT infants were approximately 17 times more likely to experience RDS than term infants and were also significantly more likely to require mechanical ventilation [20].

Immaturities of the respiratory system in LPT infants lead to vulnerability to short- and long-term respiratory morbidities [7]. Moderate-to-late preterm infants are born during the saccular stage of lung development (28–36 weeks) [45]. During the third trimester, the number of bronchi and alveoli in the lungs increases, capillaries develop, and surfactant is produced [44]. At 34 weeks gestation, the total lung volume is less than half its volume when mature, and mature alveoli may not be present until 36 weeks [7, 45]. After LPT birth, the immature lung may have inadequate surfactant, poor gas exchange, and decreased intrapulmonary fluid absorption leading to RDS and TTN [44]. Pregnancy and perinatal complications such as maternal diabetes can also contribute to respiratory morbidities [7].

Late preterm infants are at increased risk of apnea of prematurity due to immaturity of the respiratory control centers in the brainstem [6]. In a systematic review, the relative risk for apnea of prematurity was 15.7 in LPT infants compared to term [20]. While in most cases, infants with apnea of prematurity will remain hospitalized until they demonstrate respiratory maturity, home apnea monitoring may be a reason an infant is referred to a specialized NICU follow-up program [15].

Late preterm infants may be at increased risk for cardiorespiratory compromise when placed in the semi-reclined position in car seats due to their relative hypotonia and small size as most car safety seats are designated for infants weighing more than 4 or 5 lb. The AAP recommends that all infants born at <37 weeks gestation be monitored in their own car safety seat for a period of 90–120 min or the duration of the car ride home, whichever is longer, prior to discharge from hospital to observe for apnea, bradycardia, or oxygen desaturation. Infants who do not pass this car seat challenge may require a car bed and will need a follow-up evaluation to determine when it is safe to transition to a traditional car safety seat [46].

Some reports suggest that LPT infants are at risk for more severe respiratory illness with viruses such as respiratory syncytial virus (RSV) with higher rates of hospitalization and greater resource utilization than term infants [44, 45, 47]. However, more recent studies have not confirmed this risk [48, 49]. While RSV prophylaxis with palivizumab should be considered for preterm infants <29 weeks, the AAP does not recommend this for MLPT infants without chronic lung disease or congenital heart disease as there is minimal health benefit and it is not considered cost-effective. Healthcare providers can recommend that caregivers follow general measures to prevent respiratory infections including careful hand hygiene, avoidance of tobacco exposure, and influenza immunization. For infants at high risk, avoidance of crowds and restricted participation in group childcare may also be recommended [49].

Although bronchopulmonary dysplasia is uncommon in MLPT infants, LPT birth may still have long-term effects on the pulmonary system [45]. Several reports suggest that MLPT birth is a risk factor for recurrent wheezing [50] or asthma [32, 51, 52] although not all investigators have found an association [53]. Like other short- and long-term morbidities, risk of asthma seems to increase with decreasing GA [52]. In a prospective cohort study, MLPT children experienced more respiratory symptoms than term children through age 5 years and were described as being more similar to the extremely preterm comparison group than the term group [54]. In this study, the most important risk factors for continuing respiratory problems in MLPT children were eczema, respiratory problems and passive smoke exposure in the first year of life, higher social class, and family history of asthma. Neonatal factors associated with later asthma in other studies of MLPT infants include apnea, hypoxia, and mechanical ventilation [51, 52]. Interruption of lung development at a critical stage with subsequent effects on physiology is proposed to be a mechanism by which MLPT birth may affect long-term lung function [7, 45].

Recommendations for Follow-Up Care

- Arrange a follow-up car seat challenge for infants discharged home in a car bed.
- RSV prophylaxis with palivizumab is not recommended for MLPT infants without chronic lung disease or congenital heart disease.

Vision

Retinopathy of prematurity (ROP) is one the most common causes of visual impairment in the United States. It occurs due to the abnormal vascularization of the retina in premature infants. The risk of ROP increases with decreasing GA. Infants born <27 weeks gestation are at greatest risk [55]. ROP in infants >32 weeks is rare [37].

While MLPT infants do not fall under the GA recommendation for ROP screening (<30 weeks), MLPT infants with a birth weight <1500 g or a birth weight 1500–2000 g with significant neonatal morbidity (e.g., requirement for cardiorespiratory support) should be screened. This may require ophthalmologic follow-up after NICU discharge to ensure the retina has matured and that ROP, if present, has resolved [56].

Although MLPT infants are at low risk for ROP, some research suggests increased rates of other ocular problems in later childhood. One study found a 2.4-fold increased prevalence of refractive errors compared to term children although there was no increased risk of strabismus [57]. Another study found a slightly higher rate of visual impairment at age 2 years, but the prevalence was too low to assess significance [58].

Recommendations for Follow-Up Care

- Screening and ophthalmological follow-up for ROP is not necessary for most MLPT infants. Good communication between NICU staff and primary care providers is important in ensuring that individual MLPT infants who have additional risk factor for ROP receive appropriate follow-up evaluations.
- Screen for visual problems within primary care following the AAP recommendations for all children.

Hearing

Infants requiring hospitalization in the NICU have elevated rates of hearing loss compared to the general population [59, 60]. The NICU population is especially at increased risk of sensorineural hearing loss (SNHL) and auditory neuropathy [59, 60]. This has not been well studied specifically in LPT infants, but studies of very preterm infants suggest that the risk for sensorineural hearing loss is mediated by higher rates of neonatal morbidities that may affect hearing (i.e., hypoxia, hyperbilirubinemia) and ototoxic exposures (i.e., aminoglycoside antibiotics, loop diuretics, noise from life support equipment) [60]. Auditory neuropathy results from dysfunction of the inner hair cells, neurons of the spiral ganglion, or the auditory nerve. The cochlea appears to receive sounds normally, but processing of the signal from the cochlea to the auditory nerve is abnormal. Risk factors associated with the development of auditory neuropathy in neonates include hypoxia, hyperbilirubinemia, ototoxic medications, and neonatal infections [61, 62]. Late preterm infants who experience these morbidities and exposures are also at risk.

The Joint Committee on Infant Hearing (JCIH) recommends universal screening by 1 month of age as unidentified hearing loss at birth adversely affects speech and language development and later school academic achievement. Screening in newborn nurseries is done by evoked otoacoustic emissions (OAE), automated auditory brainstem response (ABR), or a combination of the two. Otoacoustic emissions are quicker but only assess the function of the peripheral auditory system. Therefore it will not detect auditory neuropathy. Automated ABR assesses the function of the peripheral auditory system as well as the auditory nerve and the auditory pathway of the brainstem [59].

Due to the increased risk of auditory neuropathy for infants in the NICU, LPT infants admitted to the NICU should be screened using ABR, not OAE. Late preterm infants in the well-baby nursery may be screened with either OAE or ABR or a combination of the two, similar to term infants [59]. However caution should be exercised in regard to the timing of OAE. One study found that LPT infants were more likely than term infants to have failure on initial OAE if tested at <42 h of age [63].

Some MLPT infants with specific neonatal risk factors are at increased risk for delayed-onset hearing loss and should have a follow-up audiologic assessment at age 24–30 months. Risk factors include [59]:

- NICU stay more than 5 days
- Mechanical ventilation
- Extracorporeal membrane oxygenation
- Ototoxic medication
- Hyperbilirubinemia

Recommendations for Follow-Up Care

- Refer children with neonatal risk factors for delayed-onset hearing loss for follow-up audiologic assessment by age 24–30 months.

Neurodevelopmental Outcomes of Late Preterm Infants

Neonatal Brain Injury

The preterm brain is at risk for injury due to a combination of intrinsic and extrinsic factors that may interact. Intrinsically, the preterm brain is immature. Incomplete development of the cerebral vasculature and poor vascular autoregulation create a susceptibility to decreased perfusion and hemorrhage with changes in cerebral blood flow. Immature brain cells are also more susceptible to free radical toxicity. Extrinsically, perinatal morbidities such as RDS can threaten the delivery of oxygen to these vulnerable tissues, and inflammatory mediators secondary to infection can create a cascade of cytokine and free radical production which can directly injure pre-oligodendrocytes [5, 64].

While the brain of LPT infants is more mature and resilient than that of very preterm infants, it is still more vulnerable to injury than at term. Multiple developmental processes are ongoing within the brain in the final 6 weeks of gestation including gyral and sulcal development, axonal elongation, synaptogenesis, and maturation of oligodendrocytes and myelination [5]. Disruption of these processes can affect neurodevelopmental outcomes.

Intraventricular hemorrhage (IVH) and periventricular leukomalacia (PVL) are feared neurologic complications in very preterm infants and are associated with increased rates of neurodevelopmental disabilities. Data on incidence of IVH and PVL in the LPT population is limited, but available evidence suggests that the frequency is low [20, 65]. Reported rates may be underestimates though as LPT infants do not routinely undergo screening cranial ultrasonography.

A systematic review of morbidity in LPT infants reported an absolute risk of 0.41% for any IVH and 0.01% for IVH grades III–IV [20]. While low, this still represents a relative risk of 4.9 for any IVH compared to term infants. Greater than 90% of IVH in preterm infants originates in the germinal matrix which is an active site of proliferation of neuronal and glial precursors until 32–34 weeks gestation. The immature vessels providing a rich blood supply to the germinal matrix are highly susceptible to hemorrhage during hypertension or increased cerebral blood flow. The risk of IVH decreases with each increasing week of gestation as the germinal matrix decreases in size and involutes almost completely by term [66].

The risk of PVL is greatest prior to 32 weeks gestation. This is related to the stage of development of pre-oligodendrocytes which are proposed to be the key target of injury in PVL. As pre-oligodendrocytes differentiate into mature oligodendrocytes, the cerebral white matter becomes less vulnerable to injury from hypoxia and ischemia. This process continues through the LPT period and into early infancy explaining the decreasing (but not absent) risk of PVL with increasing GA [5, 64]. While LPT infants are at significantly lower risk for PVL than very preterm infants, they are still at greater risk than those born at term.

Cranial ultrasonography can often detect cystic PVL which is associated with motor deficits such as cerebral palsy (CP), but magnetic resonance imaging (MRI) is much more sensitive in detecting the diffuse PVL that correlates more with adverse cognitive and behavioral outcomes [5, 64]. As MRI is infrequently performed in LPT infants, data on the incidence of PVL in this population is very limited. In one population-based study, the rate of cystic PVL at 34 weeks was 0.1% [37].

The Practice Parameter on Neuroimaging in the Neonate from the American Academy of Neurology and the Practice Committee of the Child Neurology Society recommends cranial ultrasounds for all infants born at <30 weeks gestation [67], but screening ultrasounds are not routinely done in MLPT infants. Several studies have examined cranial ultrasound abnormalities and associated factors in MLPT infants since this practice parameter was published. Some authors advocate for a later GA cutoff for routine screening based on findings that 27.1% of infants born at 33 weeks had cranial ultrasound abnormalities compared to 3.7% at 36 weeks and 1.4% at term [68]. However, others report severe abnormalities in only 1–1.5% of MLPT infants and note that some of the other abnormalities detected in their study populations were transient or of unclear clinical significance [69, 70].

The use of screening cranial ultrasound in MLPT infants is likely to be most cost-effective if based on the presence of additional risk factors and abnormalities. Factors associated with an increased likelihood of cranial ultrasound abnormalities in MLPT infants include lower GA, abnormalities on neurologic exam, microcephaly, RDS, need for assisted ventilation or surfactant, and low 5-min Apgar score (<7) [68, 69].

Although rare, neurologic injury secondary to IVH or PVL can occur in MLPT infants. This population may be best considered at intermediate risk for these morbidities compared to their very preterm and term counterparts. Medical providers need to maintain an index of suspicion to obtain neuroimaging for this population when there are concerns for neurologic injury. Identification of children at highest risk for adverse neurodevelopmental outcomes is important to guide developmental monitoring and referrals to early intervention.

Recommendations for Follow-Up Care

- Consider neuroimaging in MLPT infants with additional risk factors and/or abnormalities on neurologic exam.
- Consider referral to a specialized NICU follow-up program for MLPT infants with abnormalities on neuroimaging, particularly IVH and PVL, which may require a higher level of monitoring for adverse sequelae including CP and other developmental delays.

Cerebral Palsy

The overall prevalence of CP is 2–3 per 1000 live births [71, 72]. Similar to many neonatal morbidities and other developmental outcomes, the risk of CP is inversely related to GA at birth [73]. Cerebral palsy occurs in approximately 15% of extremely preterm infants and 6% of very preterm infants [73]. This decreases to approximately 0.5–1% of MLPT infants [11, 20, 74–76]. However, the rate of CP in MLPT infants is still approximately two to three times higher than in term infants [11, 20, 74–76], and children born MLPT account for 15–20% of all CP cases [71, 72]. Children born LPT who do not have CP are more likely to have motor coordination difficulties than their full-term counterparts [74].

Factors associated with an increased risk of CP within the MLPT population include need for resuscitation at birth, being SGA, need for antibiotics during initial hospitalization, and intracranial hemorrhage, whereas maternal receipt of antenatal steroids is associated with a decreased risk [74, 75]. Structured neurological assessments such as the Hammersmith Neonatal Neurological Examination and the General Movements Assessment can be useful in identifying MLPT infants at high risk for CP to facilitate referrals to early intervention [77, 78].

Recommendations for Follow-Up Care

- Consider neuroimaging and/or use of structured neurological assessments in MLPT infants with perinatal complications known to be associated with higher rates of CP to better define their risk.
- Monitor motor development through developmental surveillance and screening in primary care according to the AAP recommendations for all children.
- Consider referral to a specialized NICU follow-up program for closer monitoring for MLPT infants identified as at high risk for CP.
- Refer infants with motor delays and those identified as at high risk for CP to early intervention programs.

Epilepsy

Very preterm birth is associated with an increased risk for epilepsy that is inversely proportional to gestational age [79, 80], but it is less clear whether this risk also pertains to individuals born LPT. A Swedish study reported an increased risk of epilepsy in adulthood in individuals born at 35–36 weeks gestation [79], but others have not found a significant difference in rates of neonatal seizures and epilepsy in LPT children compared to term [20, 76].

Recommendations for Follow-Up Care

- No specialized monitoring indicated

Developmental Outcomes

Late preterm infants can have different developmental trajectories than their term counterparts. For all infants, an inverse relationship exists between GA and risk of developmental delays [81]. Other factors also affect developmental outcomes, such as level of medical complexity in the neonatal period [82]. Since MLPT infants often appear healthy at birth, there is a tendency to assume that there is no additional risk compared to term infants. However, these children may have differences in general and cognitive development that could lead to suboptimal school performance if left unrecognized [10, 83, 84]. As a group, LPT infants should be considered intermediate risk for developmental delays. Early identification of LPT infants with developmental delays can facilitate access to early intervention services and improve outcomes.

Early Childhood Outcomes

Many LPT infants will follow a typical developmental trajectory, but some will meet developmental milestones at a slower pace than term peers. Using the corrected GA, which is determined from the expected date of birth rather than from the actual birth date, allows providers to more accurately assess developmental milestones by taking into consideration the level of neurological maturity. When corrected GA is used instead of chronological age in standardized developmental testing, the classification of developmental delay decreased from >18% to 13% in one study [85]. Corrected GA is typically used until at least age 24 months when development is expected to be "caught up" to chronological age [2].

Studies comparing early developmental outcomes for LPT infants to those of term infants have reported variable findings. When healthy LPT infants are evaluated as a group, cognitive scores on developmental testing have been shown to be similar to term infants after correcting for GA and once catch-up growth has occurred at

24 months [86, 87]. Broad developmental screening at 48 months typically results in cognitive, motor, and language scores similar to term peers [10].

In contrast, there is also evidence that from birth to age 3 years, LPT infants are 36% more likely to have lower scores on standardized developmental testing than term infants and are more likely to be enrolled in early intervention services [10, 88]. The variability of these findings may represent differences in the populations under investigation. For example, risk factors for poorer developmental outcomes include lower GA, complications during pregnancy and delivery, and social factors such as lower socioeconomic status and parental education level [83]. Late preterm birth is just one of the many factors that affect developmental outcomes. It is promising that even the more medically complex LPT infants who required higher levels of care in the neonatal period will often catch up to full-term peers by 36 months [82].

Recommendations for Follow-Up Care

- Explain to families that the LPT brain is immature and places infants at intermediate risk for developmental delays.
- Explain to families the difference in chronological age and corrected GA and that acquisition of milestones may follow the corrected GA until approximately 24 months.
- Monitor development through developmental surveillance and screening in primary care according to the AAP recommendations for all children.
- Consider referral to a NICU follow-up program for more formal developmental monitoring for infants with a complicated or prolonged NICU course or who are exhibiting significant delays based on developmental surveillance or screening.
- Refer infants with identified delays to early intervention services.

Cognitive Outcomes

While study outcomes have been variable in regard to specific effects of LPT birth on cognition, cognitive scores are inversely proportional to GA at birth across the full span of GA [89]. Late preterm birth appears to be a risk factor for below-average full-scale IQ and performance IQ at 6 years on the Wechsler Intelligence Scale for Children-Revised [90]. A study of children born MLPT showed an increased incidence of below-average IQ and intellectual disability as compared to their term peers at 5, 6, and 10 years [91]. However, LPT birth alone does not seem to be a good predictor of cognitive delays. The cognitive deficits are sometimes so small that they are unlikely to be clinically relevant; for example, the average IQ deficit in one study was only 2.3 points [58, 92].

Late preterm infants who are more medically complex, such as those who experienced more neonatal morbidities or required a higher level of care at birth, demonstrate an increased risk for cognitive delays [58]. In contrast, 90% of LPT infants who had an uncomplicated neonatal course have cognitive scores that fall within normal limits by preschool age [93]. Even though overall cognition appears to be intact, subtle deficits may be revealed in nonverbal learning skills and visuospatial reasoning, which is the ability to identify and imagine visual and spatial relationships among objects [16, 94, 95]. An example of visuospatial reasoning would be the capacity to form an image in your mind of a neighborhood and understand directions when given a map of the area. Children who were born very preterm are also more likely to exhibit deficits in these specific areas [15].

Children born LPT perform worse in math, language, literacy, verbal inhibitory control, and short-term verbal memory at age 4 years [96, 97]. Though these differences are fairly subtle and difficult to evaluate on routine screening, they may affect academic progress if left unidentified.

Late preterm infants may have an increased risk of learning problems once they reach school age. They demonstrate less school readiness and are more likely to require individualized education programs, special education placement, and one-on-one assistants in the classroom compared to age-matched peers who were born full term [10, 84]. It is estimated that 10% of all children who require special education placement have a history of MLPT birth [98]. Children born MLPT are approximately twice as likely to repeat a grade in elementary school, and kindergarten retention rates are as high as 19% [99, 100]. Risk of school failure increases with decreasing GA [101].

Similar to risk factors for early developmental delays in the LPT population, risk factors for poorer cognition include birth weight <2 kg, lower socioeconomic status, and lower maternal education [83, 92, 102]. A recently published prospective cohort study also revealed that children born MLPT with shorter height at 1 and 4 years have a higher incidence of low IQ compared to those with normal linear growth. Contrary to studies in children born very preterm, head circumference did not seem to affect IQ scores in this group [36]. Male gender has also been associated with decreased cognitive performance in young children, but by preschool, males and females perform similarly [93].

A proposed contributing factor for school problems is age at school entrance. Some LPT infants will meet the age requirement for kindergarten entry earlier than if they had been born at term, which may place them at a disadvantage if their neurocognitive development and social-emotional maturity are less developed than their classmates [89].

Recommendations for Follow-Up Care

- Provide anticipatory guidance regarding increased risk of learning problems in children born LPT.
- Provide close surveillance of early learning skills at health supervision visits.
- Refer for psychoeducational testing if learning difficulties are present.
- Consider MLPT birth along with other risk factors for school difficulties when targeting populations for prevention and intervention programs such as Head Start.

Adult Outcomes

Two long-term cohort studies, one from Norway and one from Finland, provide adult outcome measures in the LPT population. In the Finnish cohort, adults who were born LPT were more likely to earn lower incomes, be manual workers, have memory impairment, and have lower scores on general cognitive tests [103, 104]. These findings were especially pronounced in individuals with lower levels of lifetime educational attainment [104]. In contrast, the Norwegian cohort found that adults who were born LPT were just as likely to complete high school, obtain college and postgraduate degrees, and have high job-related income as their term-born peers [11]. The results placed the LPT population in the intermediate-risk category for intellectual disability, psychological and emotional disorders, and mental or physical disorders affecting ability to work with risk increasing as GA decreases.

Although these long-term studies provide some understanding of developmental trajectories in individuals born LPT, they are imperfect. The two studies are difficult to compare due to the difference in methods and participant selection. The individuals in each cohort were also born between the 1930s and 1980s when standards of neonatal care were much different.

Behavioral Outcomes

A 2011 review article described a "preterm behavioral phenotype" characterized by inattention, anxiety, and social difficulties as well as a greater risk for internalizing than externalizing problems [105]. The majority of studies used to define this phenotype focused on children born very preterm, but there has been some work examining behavioral symptoms in the LPT population.

Attention Deficit Hyperactivity Disorder

Very preterm birth is a known risk factor for attention deficit hyperactivity disorder (ADHD), and attention difficulties are one of the most common issues for very preterm children [105]. Research examining ADHD in children born MLPT has yielded variable findings. A study of school-age children found a stepwise increase in prescriptions for ADHD medication with decreasing GA at birth. Children born LPT (35–36 weeks) were approximately 30% more likely to be prescribed ADHD medication than those born at term [106]. Similarly, an investigation using the Danish Psychiatric Central Register found that LPT children had a 70% increased risk of being diagnosed with hyperkinetic disorder/ADHD [107]. Conversely, a review of school and medical records for a population in Rochester, MN, revealed no significant difference in the cumulative incidence of ADHD between LPT and term children through age 19 years [108]. The rate of ADHD in both groups was within the range of ADHD in the general population. Another study of preschool-aged children with a history of LPT birth also found no difference in the incidence of ADHD compared to term controls [109].

Several groups have used parent- and/or teacher-completed behavioral rating scales to compare ADHD symptoms in MLPT versus term cohorts. These studies have also had conflicting results with some reporting no significant difference between MLPT and term children [90] and others reporting higher rates of attention and/or hyperactivity problems in the MLPT group [90, 99].

Factors that may contribute to the conflicting findings include different GA ranges (32–36 weeks vs. 34–36 weeks) and different degrees of neonatal morbidity in the study populations as well as varying ages at assessment for ADHD.

It may be that there is a subset of children born MLPT who are at increased risk for ADHD. Within the category of "moderate to late preterm," specific GA may be important given maturational changes taking place in the brain during this time period. Most studies that have found no increased risk for ADHD or inattentive symptoms have focused only on LPT infants (34–36 weeks) and have not included those born at 32–33 weeks [90, 108]. One study showed a dose effect with risk of ADHD increasing with each declining week of GA from 23 to 40 weeks suggesting that each gestational week has significance for a child's risk of ADHD [12].

Pregnancy complications and neonatal morbidities may also contribute to the risk for ADHD. One study found higher levels of ADHD symptoms in a subgroup of children whose LPT delivery was medically indicated (e.g., due to preeclampsia) rather than spontaneous although the LPT cohort as a whole did not differ significantly from controls [90]. The authors suggest that maternal hypertensive disorders may be a factor in the association between LPT birth and increased rates of ADHD. Intrauterine growth restriction has also been associated with ADHD [12]. However, in two separate studies that accounted for growth, the association between ADHD and MLPT birth appeared to be independent of growth restriction [90, 106].

Both biologic and environmental factors are thought to contribute to the development of ADHD. Psychosocial factors, such as lower maternal education and lower socioeconomic status, have been associated with ADHD symptoms in the general population [110]. Some of the available evidence suggests an association between LPT birth and ADHD symptoms that is independent of socioeconomic factors [90, 99, 106, 107]. However, the effect of preterm birth may be modified by these factors. For example, in one study, low maternal education increased the effect of MLPT birth on ADHD risk [106].

Children born MLPT may have a higher risk of ADHD than those born at term, but evidence for an association between ADHD and LPT birth is not nearly as strong as for very preterm birth. Further research is needed to clarify whether LPT birth truly is associated with higher risk of ADHD and to better define the subset within this population that is most at risk. Many studies have not accounted for specific GA, IUGR, or neonatal morbidities so these are areas to be further explored.

Autism

Increased rates of autism have been described in children born preterm [11, 111–113]. However, these children may have high rates of false-positive screens for autism due to other developmental delays or sensory impairments [113]. Available evidence suggests at least higher rates of autism symptoms in children born very preterm. This association is less clear for those born LPT though.

Children born MLPT are more likely to screen positive for autism on the Modified Checklist for Autism in Toddlers (M-CHAT) than term controls but less likely than children born very preterm [114]. They are also more likely to exhibit delayed social competence on the Brief Infant-Toddler Social and Emotional Assessment (BITSEA) in early childhood [115], but a study of school-age children found no significant difference from term children in teacher-reported social skills [116].

A population-based study which examined perinatal factors associated with a diagnosis of autism reported a 50% increased risk of autism in children born MLPT compared to those born at term [112]. In this study, the increased risk of autism both in the very and MLPT groups was mediated by specific perinatal complications such as preeclampsia, SGA, and intracranial hemorrhage. Conversely, two other studies found no difference in rates of autism between LPT and term children although rates in very preterm children were significantly higher [11, 111].

Other Behavior Problems

Findings from studies exploring a wider range of behavior problems beyond ADHD symptoms in children born MLPT have been variable. The majority of investigations in young children have been based on parent-completed behavioral rating scales rather than formal psychological or psychiatric evaluation. Multiple studies have used the Achenbach Child Behavior Checklist (CBCL) to examine behavior problems in this population in comparison to term controls. These studies have primarily defined behavior problems broadly looking at the externalizing problems and internalizing problems scales [117]. Despite using the same measure, results of investigations have been conflicting with findings ranging from significantly increased rates of both internalizing and externalizing problems [118, 119] to higher rates of internalizing problems only [90, 99] to no significant differences from controls [116]. One thing that is consistent across studies is that the majority of children have scores on the CBCL within the average range indicating that most LPT children do not have clinically significant behavior problems [90, 116, 118, 120].

Data on long-term behavioral outcomes for the LPT population is also conflicting. Some authors report an inverse correlation between higher rates of psychological disorders in adolescence and adulthood and decreasing GA at birth [11, 106]. In these studies, individuals born LPT had an approximately 50% higher rate of behavioral disorders compared to those born at term [11] as well as a higher rate of hospitalization for psychiatric disorders [106]. However, another study reported similar psychological and social functioning in adults born at 32–35 weeks compared to those born at term [121].

Late preterm birth may be a risk factor for later emotional and behavioral difficulties, but this should be considered in combination with other risk factors such as perinatal complications and psychosocial stressors. An investigation examining behavioral difficulties in very, moderate, and late preterm infants found that the degree of prematurity accounted for only a small portion (5–8%) of the variance in behavioral outcomes [122]. In this study, negative parenting style and lower socioeconomic status were also predictive of behavior problems. In multiple studies that have found an association between LPT birth and behavioral difficulties, the association persists after adjusting for social and economic variables [11, 90, 99, 119]. However, the risks are cumulative: the effects of LPT birth are greater in children who also have socioeconomic risks [106, 119].

Children born MLPT may manifest some aspects of the preterm behavioral phenotype, but current evidence suggests that it is less common, less severe, and less well-defined compared to their very preterm counterparts. Of the three aspects of the preterm behavioral phenotype (inattention, anxiety, and social difficulties), attention problems seem most prominent in MLPT children with likely a subset at greater risk for ADHD. As with developmental delays and school difficulties, LPT children may be described as being at an intermediate level of risk for behavioral disturbances. While data is not sufficient to indicate a need for screening and prevention efforts for behavioral conditions beyond usual care in MLPT children in general, those children with a combination of biological and environmental risk may warrant closer attention.

Recommendations for Follow-Up Care

- Perform routine surveillance for behavioral disorders in primary care according AAP recommendations for all children.
- Perform routine autism screening in primary care according to AAP recommendations with referral for further evaluation for children who screen positive, keeping in mind the potential higher likelihood of false-positive screens in this population.
- Consider MLPT birth along with other risk factors for behavioral disorders when targeting populations for prevention and intervention programs.

Family Outcomes

Preterm delivery is a risk factor for maternal postpartum depression. There appears to be a dose effect with higher depression scores associated with lower GA, but some research suggests that the risk is still significant for mothers of LPT infants [123]. Mothers of LPT infants have been described as experiencing more symptoms of anxiety, depression, and emotional distress than mothers of term infants and are more likely to rate their infants' behavior negatively [124, 125]. Factors contributing to emotional difficulties for mothers of MLPT infants include unexpected timing of delivery, lack of preparation for the delivery, concern regarding the infant's health, and separation from infants who require specialized neonatal care [124]. Depression can affect a mother's ability to respond to and interact with her infant which may then affect the child's development and behavior [126]. In one study, maternal psychological distress at term-corrected age was associated with increased risk of later behavior problems in preterm children (<37 weeks) [127].

Further work is needed to elucidate the role of maternal emotional distress secondary to MLPT birth in the development of behavioral symptoms in this population as well as to identify strategies to mitigate this risk. Mothers who report having more support report lower stress and are less likely to perceive their infants negatively [128]. An intervention aimed at improving the interactions between mothers and their MLPT infants was effective in reducing maternal depressive symptoms and increasing duration of breastfeeding but was not associated with improvements in maternal stress or perceptions of infant temperament [129].

Recommendations for Follow-Up Care

- Screen for maternal depression prior to NICU/nursery discharge and at well-child visits.
- Assess support systems for families of MLPT infants and link to community resources if needed.

Conclusion

While LPT birth does not carry nearly as high a risk of severe neonatal morbidities and neurodevelopmental impairment as very preterm birth, this is a population at risk for early medical complications and later developmental and behavioral difficulties.

Healthcare professionals in multiple disciplines and settings can play a role in mitigating the risks for these children. Obstetricians need to balance the risks of continuing pregnancy when the intrauterine environment may be hostile to the fetus with the risks associated with LPT birth [130] and should avoid elective deliveries <39 weeks gestation [131]. Nursery and NICU personnel should monitor for complications, avoid early discharge <48 h after birth, educate parents on risks, and arrange for appropriate follow-up [6]. Developmental screening by primary care providers is key to facilitating referrals for specialty evaluation and early intervention services. Formal NICU follow-up programs are likely to see only the small proportion of this population who have specific complications that place them at high risk for adverse neurodevelopmental outcomes. However, NICU follow-up programs may have an indirect role in ensuring providers at other levels of care understand the needs of this population.

References

1. Raju, T. N., et al. (2006). Optimizing care and outcome for late-preterm (near-term) infants: A summary of the workshop sponsored by the National Institute of Child Health and Human Development. *Pediatrics, 118*(3), 1207–1214.
2. AAP Committee on Fetus & Newborn, ACOG Committee on Obstetric Practice. (2007). *Guidelines for perinatal care*. Elk Grove Village: American Academy of Pediatrics; American College of Obstetrics and Gynecology.
3. WHO. *Preterm birth.* (2016). Retrieved from http://www.who.int/mediacentre/factsheets/fs363/en/.
4. Huppi, P. S., et al. (1998). Quantitative magnetic resonance imaging of brain development in premature and mature newborns. *Annals of Neurology, 43*(2), 12.
5. Kinney, H. C. (2006). The near-term (late preterm) human brain and risk for periventricular leukomalacia: A review. *Seminars in Perinatology, 30*(2), 81–88.
6. Engle, W. A., Tomashek, K. M., & Wallman, C. (2007). "Late-preterm" infants: A population at risk. *Pediatrics, 120*(6), 1390–1401.
7. Kugelman, A., & Colin, A. A. (2013). Late preterm infants: Near term but still in a critical developmental time period. *Pediatrics, 132*(4), 741–751.
8. Shapiro-Mendoza, C. K., et al. (2008). Effect of late-preterm birth and maternal medical conditions on newborn morbidity risk. *Pediatrics, 121*(2), e223–e232.
9. McLaurin, K. K., et al. (2009). Persistence of morbidity and cost differences between late-preterm and term infants during the first year of life. *Pediatrics, 123*(2), 653–659.
10. McGowan, J. E., et al. (2011). Early childhood development of late-preterm infants: A systematic review. *Pediatrics, 127*(6), 1111–1124.
11. Moster, D., Lie, R. T., & Markestad, T. (2008). Long-term medical and social consequences of preterm birth. *The New England Journal of Medicine, 359*, 12.
12. Sucksdorff, M., et al. (2015). Preterm birth and poor fetal growth as risk factors of attention-deficit/hyperactivity disorder. *Pediatrics, 136*(3), e599–e608.
13. Lindstrom, K., et al. (2007). Preterm infants as young adults: A Swedish national cohort study. *Pediatrics, 120*(1), 70–77.
14. Osterman, M. J., et al. (2015). Annual summary of vital statistics: 2012-2013. *Pediatrics, 135*(6), 1115–1125.
15. Committee on Fetus and Newborn. (2008). Hospital discharge of the high-risk neonate. *Pediatrics, 122*(5), 1119–1126.
16. Natarajan, G., & Shankaran, S. (2016). Short- and long-term outcomes of moderate and late preterm infants. *American Journal of Perinatology, 33*(3), 305–317.
17. Walker, M. (2008). Breastfeeding the late preterm infant. *Journal of Obstetric, Gynecologic, and Neonatal Nursing, 37*(6), 10.
18. Radtke, J. V. (2011). The paradox of breastfeeding-associated morbidity among late preterm infants. *Journal of Obstetric, Gynecologic, and Neonatal Nursing, 40*(1), 9–24.
19. Hallowell, S. G., & Spatz, D. L. (2012). The relationship of brain development and breastfeeding in the late-preterm infant. *Journal of Pediatric Nursing, 27*(2), 154–162.
20. Teune, M. J., et al. (2011). A systematic review of severe morbidity in infants born late preterm. *American Journal of Obstetrics and Gynecology, 205*(4), 374 e1–9.
21. Gianni, M. L., et al. (2015). Is nutritional support needed in late preterm infants? *BMC Pediatrics, 15*(1), 194.
22. Briere, C. E., et al. (2015). Establishing breastfeeding with the late preterm infant in the NICU. *Journal of Obstetric, Gynecologic, & Neonatal Nursing, 44*(1), 102–113. quiz E1–2.
23. Meier, P., et al. (2013). Management of breastfeeding during and after the maternity hospitalization for late preterm infants. *Clinics in Perinatology, 40*(4), 689–705.
24. Eidelman, A. I. (2016). The challenge of breastfeeding the late preterm and the early-term infant. *Breastfeeding Medicine, 11*, 99.

25. Colaizy, T. T., & Morriss, F. H. (2008). Positive effect of NICU admission on breastfeeding of preterm US infants in 2000 to 2003. *Journal of Perinatology, 28*(7), 505–510.
26. Meier, P. P., Furman, L. M., & Degenhardt, M. (2007). Increased lactation risk for late preterm infants and mothers: Evidence and management strategies to protect breastfeeding. *Journal of Midwifery & Women's Health, 52*(6), 579–587.
27. Academy of Breastfeeding Medicine. (2011). ABM clinical protocol #10: Breastfeeding the late preterm infant (34(0/7) to 36(6/7) weeks gestation) (first revision June 2011). *Breastfeeding Medicine, 6*(3), 151–156.
28. Johnson, S., et al. (2016). Eating difficulties in children born late and moderately preterm at 2 y of age: A prospective population-based cohort study. *The American Journal of Clinical Nutrition, 103*(2), 406–414.
29. Lapillonne, A., et al. (2013). Nutritional recommendations for the late-preterm infant and the preterm infant after hospital discharge. *The Journal of Pediatrics, 162*(3 Suppl), S90–100.
30. Dee, D. L., et al. (2008). Sources of supplemental iron among breastfed infants during the first year of life. *Pediatrics, 122*(Suppl 2), S98–104.
31. Berglund, S. K., Westrup, B., & Domellof, M. (2015). Iron supplementation until 6 months protects marginally low-birth-weight infants from iron deficiency during their first year of life. *Journal of Pediatric Gastroenterology and Nutrition, 60*(3), 390–395.
32. Goyal, N. K., Fiks, A. G., & Lorch, S. A. (2011). Association of late-preterm birth with asthma in young children: Practice-based study. *Pediatrics, 128*(4), e830–e838.
33. Bocca-Tjeertes, I. F., et al. (2011). Growth and predictors of growth restraint in moderately preterm children aged 0 to 4 years. *Pediatrics, 128*(5), e1187–e1194.
34. Santos, I. S., et al. (2009). Late preterm birth is a risk factor for growth faltering in early childhood: A cohort study. *BMC Pediatrics, 9*, 71.
35. Nagasaka, M., et al. (2015). Incidence of short stature at 3 years of age in late preterm infants: A population-based study. *Archives of Disease in Childhood, 100*(3), 250–254.
36. Dotinga, B. M., et al. (2016). Longitudinal growth and neuropsychological functioning at age 7 in moderate and late preterms. *Pediatrics, 138*(4), e20153638–e20153638.
37. Altman, M., et al. (2011). Neonatal morbidity in moderately preterm infants: A Swedish national population-based study. *The Journal of Pediatrics, 158*(2), 239–244. e1.
38. Bhutani, V. K., & Johnson, L. (2006). Kernicterus in late preterm infants cared for as term healthy infants. *Seminars in Perinatology, 30*(2), 89–97.
39. Wong, R. J., et al. (2006). *Neonatal jaundice and liver disease in Fanaroff and Martin's neonatal-perinatal medicine*. 46.
40. Academy of Breastfeeding Medicine Protocol Committee. (2010). ABM clinical protocol #22: guidelines for management of jaundice in the breastfeeding infant equal to or greater than 35 weeks' gestation. *Breastfeeding Medicine, 5*(2), 7.
41. Maisels, M. J., et al. (2009). Hyperbilirubinemia in the newborn infant > or =35 weeks' gestation: An update with clarifications. *Pediatrics, 124*(4), 1193–1198.
42. Maisels, M. J., et al. (2012). An approach to the management of hyperbilirubinemia in the preterm infant less than 35 weeks of gestation. *Journal of Perinatology, 32*(9), 660–664.
43. Gouyon, J. B., et al. (2012). Neonatal problems of late and moderate preterm infants. *Seminars in Fetal & Neonatal Medicine, 17*(3), 146–152.
44. Kotecha, S. J., Dunstan, F. D., & Kotecha, S. (2012). Long term respiratory outcomes of late preterm-born infants. *Seminars in Fetal & Neonatal Medicine, 17*(2), 77–81.
45. Colin, A. A., McEvoy, C., & Castile, R. G. (2010). Respiratory morbidity and lung function in preterm infants of 32 to 36 weeks' gestational age. *Pediatrics, 126*(1), 115–128.
46. Bull, M. J., & Engle, W. A. (2009). Safe transportation of preterm and low birth weight infants at hospital discharge. *Pediatrics, 123*(5), 1424–1429.
47. Boyce, T. G., et al. (2000). Rates of hospitalization for respiratory syncytial virus infection among children in medicaid. *The Journal of Pediatrics, 137*(6), 865–870.
48. Hall, C. B., et al. (2013). Respiratory syncytial virus-associated hospitalizations among children less than 24 months of age. *Pediatrics, 132*(2), 8.

49. Committee on Infectious Disease and Bronchiolitis Guidelines Committee. (2014). Updated guidance for palivizumab prophylaxis among infants and young children at increased risk of hospitalization for respiratory syncytial virus infection. *Pediatrics, 134*(2), 415–420.
50. Escobar, G. J., et al. (2010). Recurrent wheezing in the third year of life among children born at 32 Weeks' gestation or later. *Archives of Pediatrics & Adolescent Medicine, 164*(10), 8.
51. Odibo, I. N., et al. (2016). Childhood respiratory morbidity after late preterm and early term delivery: A study of Medicaid patients in South Carolina. *Paediatric and Perinatal Epidemiology, 30*(1), 67–75.
52. Haataja, P., et al. (2016). Asthma and atopic dermatitis in children born moderately and late preterm. *European Journal of Pediatrics, 175*(6), 799–808.
53. Abe, K., et al. (2010). Late preterm birth and risk of developing asthma. *The Journal of Pediatrics, 157*(1), 74–78.
54. Vrijlandt, E. J., et al. (2013). Moderately preterm children have more respiratory problems during their first 5 years of life than children born full term. *American Journal of Respiratory and Critical Care Medicine, 187*(11), 1234–1240.
55. Batshaw, M. L., Roizen, N. J., & Lotrecchiano, G. R. (2012). *Children with disabilities*. Baltimore: Paul H. Brookes Publishing Co.
56. Committee on practice and ambulatory medicine, section on ophthalmology. (2015). Visual system assessment in infants, children, and young adults by pediatricians. *Pediatrics, 137*(1), e20153596–e20153596.
57. Raffa, L., et al. (2015). Ophthalmological findings in relation to auxological data in moderate-to-late preterm preschool children. *Acta Ophthalmologica, 93*(7), 635–641.
58. Johnson, S., et al. (2015). Neurodevelopmental outcomes following late and moderate prematurity: A population-based cohort study. *Archives of Disease in Childhood-Fetal and Neonatal Edition, 100*(4), F301–F308.
59. Joint Committee on Infant Hearing. (2007). Year 2007 position statement: principles and guidelines for early hearing detection and intervention programs. *Pediatrics, 120*(4), 898–921.
60. Cristobal, R., & Oghalai, J. S. (2008). Hearing loss in children with very low birth weight: Current review of epidemiology and pathophysiology. *Archives of Disease in Childhood-Fetal and Neonatal Edition, 93*(6), F462–F468.
61. Dowley, A. C., et al. (2009). Auditory neuropathy: Unexpectedly common in a screened newborn population. *Developmental Medicine and Child Neurology, 51*(8), 642–646.
62. Xoinis, K., et al. (2007). Extremely low birth weight infants are at high risk for auditory neuropathy. *Journal of Perinatology, 27*(11), 718–723.
63. Smolkin, T., et al. (2013). Impact of gestational age on neonatal hearing screening in vaginally-born late-preterm and early-term infants. *Neonatology, 104*(2), 110–115.
64. Volpe, J. J., et al. (2011). The developing oligodendrocyte: Key cellular target in brain injury in the premature infant. *International Journal of Developmental Neuroscience, 29*(4), 423–440.
65. Laptook, A. R. (2013). Neurologic and metabolic issues in moderately preterm, late preterm, and early term infants. *Clinics in Perinatology, 40*(4), 723–738.
66. Swaiman, K. F., et al. (2011). *Ch. 16 in Swaiman's pediatric neurology: principles and practice*. Philadelphia: Elsevier Health Sciences.
67. Ment, L. R., et al. (2002). Report of the Quality Standards Subcommittee of the American Academy of Neurology and the Practice Committee of the Child Neurology Society. *Neurology, 58*, 14.
68. Ballardini, E., et al. (2014). Universal cranial ultrasound screening in preterm infants with gestational age 33–36 weeks. A retrospective analysis of 724 newborns. *Pediatric Neurology, 51*(6), 790–794.
69. Fumagalli, M., et al. (2015). Cranial ultrasound findings in late preterm infants and correlation with perinatal risk factors. *Italian Journal of Pediatrics, 41*, 65.
70. Bhat, V., et al. (2012). Utility of performing routine head ultrasounds in preterm infants with gestational age 30–34 weeks. *The Journal of Maternal-Fetal & Neonatal Medicine, 25*(2), 116–119.

71. Van Naarden Braun, K., et al. (2016). Birth prevalence of cerebral palsy: A population-based study. *Pediatrics, 137*(1), e20152872.
72. Smithers-Sheedy, H., et al. (2016). A special supplement: Findings from the Australian cerebral palsy register, birth years 1993 to 2006. *Developmental Medicine and Child Neurology, 58*(Suppl 2), 5–10.
73. Himpens, E., et al. (2008). Prevalence, type, distribution, and severity of cerebral palsy in relation to gestational age: A meta-analytic review. *Developmental Medicine and Child Neurology, 50*(5), 334–340.
74. Odd, D. E., et al. (2013). Movement outcomes of infants born moderate and late preterm. *Acta Paediatrica, 102*(9), 876–882.
75. Hirvonen, M., et al. (2014). Cerebral palsy among children born moderately and late preterm. *Pediatrics, 134*(6), e1584–e1593.
76. Petrini, J. R., et al. (2009). Increased risk of adverse neurological development for late preterm infants. *The Journal of Pediatrics, 154*, 8.
77. Spittle, A. J., et al. (2016). Neurobehaviour and neurological development in the first month after birth for infants born between 32-42 weeks' gestation. *Early Human Development, 96*, 7–14.
78. Brogna, C., et al. (2013). Prognostic value of the qualitative assessments of general movements in late-preterm infants. *Early Human Development, 89*(12), 1063–1066.
79. Crump, C., et al. (2011). Preterm birth and risk of epilepsy in Swedish adults. *Neurology, 77*, 8.
80. Sun, Y., et al. (2008). Gestational age, birth weight, intrauterine growth, and the risk of epilepsy. *American Journal of Epidemiology, 167*(3), 262–270.
81. Schonhaut, L., Armijo, I., & Perez, M. (2015). Gestational age and developmental risk in moderately and late preterm and early term infants. *Pediatrics, 135*(4), e835–e841.
82. McGowan, J. E., et al. (2012). Impact of neonatal intensive care on late preterm infants: Developmental outcomes at 3 years. *Pediatrics, 130*(5), e1105–e1112.
83. Woythaler, M., et al. (2015). Late preterm infants and neurodevelopmental outcomes at kindergarten. *Pediatrics, 136*(3), 424–431.
84. Chyi, L. J., et al. (2008). School outcomes of late preterm infants: Special needs and challenges for infants born at 32 to 36 weeks gestation. *The Journal of Pediatrics, 153*(1), 25–31.
85. Parekh, S. A., et al. (2016). Correcting for prematurity affects developmental test scores in infants born late and moderately preterm. *Early Human Development, 94*, 1–6.
86. Romeo, D. M., et al. (2010). Neurodevelopmental outcome at 12 and 18 months in late preterm infants. *European Journal of Paediatric Neurology, 14*(6), 503–507.
87. Shah, P., et al. (2016). Developmental outcomes of late preterm infants from infancy to kindergarten. *Pediatrics, 138*(2), e20153496.
88. Shapiro-Mendoza, C., et al. (2013). Enrollment in early intervention programs among infants born late preterm, early term, and term. *Pediatrics, 132*(1), e61–e69.
89. Kerr-Wilson, C. O., et al. (2012). Meta-analysis of the association between preterm delivery and intelligence. *Journal of Public Health (Oxford, England), 34*(2), 209–216.
90. Talge, N. M., et al. (2010). Late-preterm birth and its association with cognitive and socioemotional outcomes at 6 years of age. *Pediatrics, 126*(6), 1124–1131.
91. de Jong, M., Verhoeven, M., & van Baar, A. L. (2012). School outcome, cognitive functioning, and behaviour problems in moderate and late preterm children and adults: A review. *Seminars in Fetal & Neonatal Medicine, 17*(3), 163–169.
92. Baron, I. S., et al. (2011). Cognitive deficit in preschoolers born late-preterm. *Early Human Development, 87*(2), 115–119.
93. Romeo, D. M., et al. (2012). Longitudinal cognitive assessment in healthy late preterm infants. *European Journal of Paediatric Neurology, 16*(3), 243–247.
94. Baron, I. S., et al. (2009). Visuospatial and verbal fluency relative deficits in 'complicated' late-preterm preschool children. *Early Human Development, 85*(12), 751–754.
95. Mervis, C. B., Robinson, B. F., & Pani, J. R. (1999). Visuospatial construction. *American Journal of Human Genetics, 65*, 8.

96. Nepomnyaschy, L., et al. (2012). Developmental outcomes of late-preterm infants at 2 and 4 years. *Maternal and Child Health Journal, 16*(8), 1612–1624.
97. Lipkind, H. S., et al. (2012). School-age outcomes of late preterm infants in New York City. *American Journal of Obstetrics and Gynecology, 206*(3), 222 e1–222 e6.
98. Mackay, D. F., et al. (2013). Obstetric factors and different causes of special educational need: Retrospective cohort study of 407,503 schoolchildren. *BJOG, 120*(3), 297–307. discussion 307-8.
99. van Baar, A. L., et al. (2009). Functioning at school age of moderately preterm children born at 32 to 36 weeks' gestational age. *Pediatrics, 124*(1), 251–257.
100. Morse, S. B., et al. (2009). Early school-age outcomes of late preterm infants. *Pediatrics, 123*(4), e622–e629.
101. Mathiasen, R., et al. (2010). Gestational age and basic school achievements: A national follow-up study in Denmark. *Pediatrics, 126*(6), e1553–e1561.
102. Coletti, M. F., et al. (2015). One-year neurodevelopmental outcome of very and late preterm infants: Risk factors and correlation with maternal stress. *Infant Behavior & Development, 39*, 11–20.
103. Heinonen, K., et al. (2013). Late-preterm birth and lifetime socioeconomic attainments: The Helsinki birth cohort study. *Pediatrics, 132*(4), 647–655.
104. Heinonen, K., et al. (2015). Late preterm birth and neurocognitive performance in late adulthood: A birth cohort study. *Pediatrics, 135*(4), e818–e825.
105. Johnson, S., & Marlow, N. (2011). Preterm birth and childhood psychiatric disorders. *Pediatric Research, 69*(5), 8.
106. Lindstrom, K., Lindblad, F., & Hjern, A. (2009). Psychiatric morbidity in adolescents and young adults born preterm: A Swedish national cohort study. *Pediatrics, 123*(1), e47–e53.
107. Linnet, K. M., et al. (2006). Gestational age, birth weight, and the risk of hyperkinetic disorder. *Archives of Disease in Childhood, 91*(8), 655–660.
108. Harris, M. N., et al. (2013). ADHD and learning disabilities in former late preterm infants: A population-based birth cohort. *Pediatrics, 132*(3), e630–e636.
109. Rabie, N. Z., et al. (2015). ADHD and developmental speech/language disorders in late preterm, early term and term infants. *Journal of Perinatology, 35*(8), 660–664.
110. Hjern, A., Weitoft, G. R., & Lindblad, F. (2010). Social adversity predicts ADHD-medication in school children--a national cohort study. *Acta Paediatrica, 99*(6), 920–924.
111. Schendel, D., & Bhasin, T. K. (2008). Birth weight and gestational age characteristics of children with autism, including a comparison with other developmental disabilities. *Pediatrics, 121*(6), 1155–1164.
112. Buchmayer, S., et al. (2009). Can association between preterm birth and autism be explained by maternal or neonatal morbidity? *Pediatrics, 124*(5), e817–e825.
113. Stephens, B. E., et al. (2012). Screening for autism spectrum disorders in extremely preterm infants. *Journal of Developmental and Behavioral Pediatrics, 33*, 7.
114. Guy, A., et al. (2015). Infants born late/moderately preterm are at increased risk for a positive autism screen at 2 years of age. *The Journal of Pediatrics, 166*(2), 269–275. e3.
115. Johnson, S., et al. (2015). Early emergence of delayed social competence in infants born late and moderately preterm. *Journal of Developmental and Behavioral Pediatrics, 36*, 10.
116. Gurka, M., LoCasale-Croush, J., & Blackman, J. (2010). Long-term cognition, achievement, socioemotional, and behavioral development of healthy late-preterm infants. *Archives of Pediatrics & Adolescent Medicine, 164*, 6.
117. Achenbach, T. M., & Rescorla, L. A. (2001). *Manual for the ASEBA school-age forms & profiles*. Burlington, VT: University of Vermont, Research Center for Children, Youth, & Families.
118. Potijk, M. R., et al. (2012). Higher rates of behavioural and emotional problems at preschool age in children born moderately preterm. *Archives of Disease in Childhood, 97*(2), 112–117.
119. Potijk, M. R., et al. (2015). Behavioural and emotional problems in moderately preterm children with low socioeconomic status: A population-based study. *European Child & Adolescent Psychiatry, 24*(7), 787–795.

120. Boylan, J., et al. (2014). Behavioural outcomes at 3 years of age among late preterm infants admitted to neonatal intensive care: A cohort study. *Archives of Disease in Childhood-Fetal and Neonatal Edition, 99*(5), F359–F365.
121. Dalziel, S., et al. (2007). Psychological functioning and health-related quality of life in adulthood after preterm birth. *Developmental Medicine and Child Neurology, 49*, 6.
122. Shah, P. E., et al. (2013). The paradox of prematurity: The behavioral vulnerability of late preterm infants and the cognitive susceptibility of very preterm infants at 36 months post-term. *Infant Behavior & Development, 36*(1), 50–62.
123. Vigod, S. N., et al. (2010). Prevalence and risk factors for postpartum depression among women with preterm and low-birth-weight infants: A systematic review. *BJOG, 117*(5), 540–550.
124. Brandon, D. H., et al. (2011). Emotional responses of mothers of late-preterm and term infants. *Journal of Obstetric, Gynecologic, and Neonatal Nursing, 40*(6), 719–731.
125. Voegtline, K. M., & Stifter, C. A. (2010). Late-preterm birth, maternal symptomatology, and infant negativity. *Infant Behavior & Development, 33*(4), 545–554.
126. Earls, M. F. (2010). Incorporating recognition and management of perinatal and postpartum depression into pediatric practice. *Pediatrics, 126*(5), 1032–1039.
127. Gray, R. F., Indurkhya, A., & McCormick, M. C. (2004). Prevalence, stability, and predictors of clinically significant behavior problems in low birth weight children at 3, 5, and 8 years of age. *Pediatrics, 114*(3), 736–743.
128. Woodman, A. C., Mawdsley, H. P., & Hauser-Cram, P. (2014). Parenting stress and child behavior problems within families of children with developmental disabilities: Transactional relations across 15 years. *Research in Developmental Disabilities, 36C*, 264–276.
129. Ravn, I. H., et al. (2012). Effects of early mother-infant intervention on outcomes in mothers and moderately and late preterm infants at age 1 year: A randomized controlled trial. *Infant Behavior & Development, 35*(1), 36–47.
130. Edmunds, S. F., & Silver, R. M. (2013). Stillbirth reduction efforts and impact on early births. *Clinics in Perinatology, 40*(4), 611–628.
131. Ramirez, M., Ramin, S., & ACOG Committee on Practice Bulletins. (2009). ACOG practice bulletin: Clinical management guidelines for obstetrician-gynecologists. *Obstetrics and Gynecology, 114*(2), 12.

Neonatal Encephalopathy

Rudaina Banihani, Paige Terrien Church, Maureen Luther, Patricia Maddalena, and Elizabeth Asztalos

Abstract Historically, hypoxic-ischemic encephalopathy (HIE) has been used to describe any infant who is encephalopathic following birth. This terminology, however, infers etiology (hypoxia and ischemia). There is now substantial evidence demonstrating that HIE accounts for only a portion of infants presenting with neonatal encephalopathy and only accounts for 4% of the diagnosis of cerebral palsy (Badawi et al., BMJ 317:1549–1553, 1998).

Confusion over terminology has significant implications medically, developmentally, psychologically, and medicolegally. As a result, the American College of Obstetricians and Gynecologists (ACOG), in collaboration with the American Academy of Pediatrics (AAP), has issued a report clarifying the preferred term to be neonatal encephalopathy (NE) (Obstetrics and Gynecology 123:896–901, 2014). This descriptive term describes the presentation without ascribing etiology, which may take months to determine.

This chapter will focus on the terminology of NE, pathophysiology as it relates to the aspect of hypoxic ischemia, and the relationship to current management protocols. It will also discuss the evidence around outcomes and associated challenges with its interpretation.

R. Banihani (✉) · P. T. Church
Department of Paediatrics, University of Toronto, Sunnybrook Health Sciences Centre, Department of Newborn & Developmental Paediatrics, 2075 Bayview Avenue, M4-245, Toronto, ON M4N 3M5, Canada
e-mail: rudaina.banihani@sunnybrook.ca; paige.church@sunnybrook.ca

M. Luther · P. Maddalena
Sunnybrook Health Sciences Centre, Neonatal Follow-Up Clinic, 2075 Bayview Avenue, M4-245, Toronto, ON M4N 3M5, Canada
e-mail: maureen.luther@sunnybrook.ca; Patricia.maddalena@sunnybrook.ca

E. Asztalos
Department of Paediatrics, University of Toronto, Department of Newborn & Developmental Paediatrics, 2075 Bayview Avenue, M4-245, Toronto, ON M4N 3M5, Canada
e-mail: elizabeth.asztalos@sunnybrook.ca

H. Needelman, B. J. Jackson (eds.), *Follow-Up for NICU Graduates*,
https://doi.org/10.1007/978-3-319-73275-6_8

Definition and Etiologies

Neonatal encephalopathy (NE) is the recommended term used to describe a heterogeneous, clinically defined syndrome of disturbed neurologic function in the earliest days of life in an infant born at term or near term [2]. NE is manifested by a reduced level of consciousness or seizures and often accompanied by difficulty with initiating and maintaining respiration as well as by depression of tone and reflexes at birth. NE can result from a wide variety of conditions (Table 1) and often remains unexplained.

The incidence of neonatal encephalopathy depends on how the syndrome is defined, but published estimates vary between 2 and 9 per 1000 term births [6–9]. In a 2010 review, the estimated incidence of neonatal encephalopathy was 3.0 per 1000 live births, while the estimated incidence of hypoxic-ischemic encephalopathy was 1.5 per 1000 live births [10].

Historically, all neonates with encephalopathy were referred to as having hypoxia-ischemia encephalopathy (HIE) [11], but it is now known that hypoxia-ischemia is only one of many possible contributors to NE. Specifically, HIE describes a brain injury resulting from inadequate blood flow to the infant's brain as a consequence of a hypoxic-ischemic event during the peripartum period [12]. Based on the recommendations put forward by the Task Force on NE, for cases of HIE, evidence of sentinel peripartum events are required, such as cord prolapse, uterine rupture, abruptio placentae, placenta previa, maternal hypotension, breech presentation, or shoulder dystonia [13]. Clinically, however, it is often unclear whether a particular newborn's encephalopathy can be attributed to a primary hypoxic-ischemic brain injury. The underlying etiology may be genetic or metabolic with a secondary hypoxic-ischemic event. As clinical management is time sensitive with evidence supporting early therapeutic hypothermia, many advocate for the condition to be called "presumed HIE" or "apparent HIE" when the clinical features and neonatal brain injury patterns on MRI suggest that HIE is a potential factor [14]. Others favor using the nonspecific term "neonatal encephalopathy" whenever there is doubt as to the underlying mechanism of injury [15].

Despite the advances in obstetric care aimed at preventing peripartum hypoxic-ischemic events, the incidence of HIE has not declined [16]. To this extent, much of

Table 1 Most common causes of neonatal encephalopathy [3–5]

Infection: Meningitis, encephalitis, early-onset sepsis, chorioamnionitis, congenital infections
Hypoglycemia and electrolyte disturbances
Hypoxic-ischemic encephalopathy (HIE)
Stroke and thrombophilic disorders
Venous sinus thrombosis
Maternal drugs
Brain trauma
Brain malformation
Genetic metabolic disorders

the current neonatal research about HIE focuses on minimizing the extent of subsequent brain injury [13]. In the past, treatment options were limited to supportive medical therapy to maintain cardiopulmonary function and to manage seizure activity, whereas currently several treatment approaches are available to infants with HIE, and many others are being evaluated in animal models and clinical trials.

It is important to be aware of the history as inconsistent terminology impacts how the literature is interpreted and how to look at future outcomes for the infant with encephalopathy.

Pathophysiology of Neonatal Hypoxic-Ischemic Brain Injury

As much of the recent progress on NE has focused on the impact of hypoxia on the developing brain, we will now shift to focusing on HIE, a type of NE with a number of studies focused on the mechanism and pathophysiology of brain injury. The pathologic events of HIE are the result of impaired cerebral perfusion and oxygen delivery to the brain [17, 18]. These effects are complex and evolve over two phases: primary and secondary energy failures which culminate in brain injury [19].

The primary energy failure is associated with anaerobic metabolism. It's characterized by a cascade of events initiated by reductions in cerebral blood flow [19, 20], which ultimately leads to decreases in oxygen and glucose, a corresponding and significant reduction in energy (adenosine triphosphate (ATP) and increased lactate production) [21]. This primary phase is an essential prerequisite for the deleterious events that follow from acute intracellular derangements to the failure of various mechanisms responsible for cell integrity, particularly the sodium/potassium (Na/ K) pumps and those that maintain low intracellular calcium [22]. Failure in the Na/K pumps precipitates a massive depolarization of neurons due to an excessive influx of positively charged sodium ions. This leads to the release of glutamate, a prominent excitatory neurotransmitter, allowing for an additional influx of intracellular calcium and sodium [23]. Increased intracellular calcium triggers destructive pathways such as cerebral edema, ischemia, and microvascular damage with resultant necrosis and/or apoptosis [24].

The resolution of hypoxia-ischemia with reperfusion is associated with a complex cascade of mechanisms that can reverse the fall in high-energy phosphorylated metabolites and the intracellular pH as well as promote a recycling of neurotransmitters. This occurs between 6 and 48 h after the primary event. It is referred to as the secondary cerebral energy failure, also known as "delayed injury," and may continue for days to weeks. This secondary cerebral energy failure is different from the primary failure as the declines in phosphocreatine and ATP are not accompanied by brain acidosis. The severity of the secondary failure depends on the magnitude of the primary energy failure and has a prognostic value since it is associated with delayed neuronal death (apoptosis). While this time frame represents a window for therapeutic intervention, the pathophysiology of secondary energy failure is less well understood with interplay of excitatory neurotransmitter accumulation, oxidative injury, apoptosis, inflammation, derangements in growth factors, and altered protein synthesis.

The duration of time for hypoxia-ischemia to be successfully reversed and promote recovery depends on maturation, preconditioning events, substrate availability, body temperature, as well as the concurrent disease processes.

While the cerebral energy state can recover following primary energy failure, a secondary energy failure may occur at a time remote from the initiating event. This second interval of energy failure is different from the primary energy failure phase in that the declines in phosphocreatine and ATP are not accompanied by brain acidosis [19]. The presence and severity of secondary energy failure depend on the magnitude of the primary energy failure. This second phase also has an undoubtable prognostic value since it is associated with delayed neuronal death (apoptosis) during hours or days after the initiation of injury and represents a window for therapeutic intervention. However, it is important to note that the pathogenesis of secondary energy failure is not as well understood as primary energy failure. It mostly involves many pathophysiologic processes, including accumulation of excitatory neurotransmitters, oxidative injury, apoptosis, inflammation, and altered growth factors and protein synthesis [25–30].

Therapy is offered during the latent phase which is the interval between primary and secondary energy failure to reduce the risk of death or impairment among infants with hypoxic-ischemic encephalopathy. Initiating brain cooling within the first six postnatal hours following brain ischemia has been proven to reduce brain damage successfully and substantiates the presence of a therapeutic window [31, 32]. However, the precise mechanism of neuroprotection by mild hypothermia is still very unclear. The therapeutic effects of hypothermia appear to impact both primary injury and the latent phase of hypoxic-ischemic brain injury [33]. Hypothermia can lead to delayed cell death due to the suppression of many of the pathways. It also influences cell signaling cascades, which are key factors in the initiation of neuronal injury in the primary phase [33].

Clinical Presentation and Evaluation of the Neonate with Encephalopathy

A neonate presenting with encephalopathy (regardless of etiology) may have an abnormal state of consciousness (e.g., hyperalert, irritable, lethargic, or obtunded), diminished spontaneous movements, respiratory or feeding difficulties, poor tone, abnormal posturing, absent primitive reflexes, or seizure activity [34]. In clinical practice, the first signs of NE are seen around the time of birth or the first 6 h after birth. Sarnat and Sarnat [35] were the first to define this syndrome as neonatal encephalopathy following fetal distress. The clinical features and severity of encephalopathy can be distinguished into three stages of encephalopathy. Stage I, or mild encephalopathy, is associated with hyperalertness, sympathetic overdrive, normal muscle tone, and a normal EEG. Stage II, or moderate encephalopathy, is marked by obtundation or lethargy, hypotonia, multifocal seizures, and an EEG showing periodic or continuous delta activity. In Stage III, or severe encephalopathy, the infants appear stuporous and flaccid with an isoelectric or periodic EEG.

A comprehensive evaluation is recommended in all cases of NE to determine the probable cause [2]. This evaluation must include an assessment of neonatal clinical status and a thorough consideration of all factors that can potentially contribute to NE, including maternal medical history, obstetric antecedents, intrapartum factors (including fetal heart rate monitoring results and issues related to delivery), and placental pathology. A thorough maternal and family history is recommended, including a history of thromboembolic disorders, prior pregnancy loss, maternal infection, and maternal drug use. As well, blood samples need to be drawn to determine pH and base deficit.

In addition, a gross and histologic examination of the placenta and umbilical cord may provide evidence of a possible placental vascular lesion or infection, or an umbilical cord thrombosis [36]. The presence of oliguria, cardiomyopathy, or abnormal liver function tests may suggest a global hypoxic-ischemic event. Neuroimaging may provide information regarding the type and timing of brain injury [37, 38]. Metabolic derangements, unusual odors, dysmorphic features, and congenital anomalies may suggest the presence of an inborn error of metabolism or genetic disorder.

To ascribe NE to HIE necessitates a search for potential etiologies. It relies on a carefully obtained history and a thorough neurological examination [22]. There are three important features to consider in the diagnosis of an intrapartum insult leading to HIE:

1. Fetal decompensation
2. Neonatal depression at birth
3. Neonatal neurological dysfunction in the first hours of life [22]

The essential criteria suggested as prerequisites to a diagnosis of hypoxic-ischemic insult resulting in moderate or severe encephalopathy in term newborn infants include the following: documented fetal deterioration; metabolic acidosis with a cord pH of <7 or a base deficit of ≥ 12 mmol/L; early onset of encephalopathy; multi-system organ dysfunction; and exclusion of other etiology such as trauma, coagulation disorders, metabolic disorders, and genetic causes [39].

Management of NE

In the heat of the moment, it may be difficult to delineate an intrapartum event from other etiologies of NE accurately. Furthermore, the other potential etiologies of NE may result in a component of hypoxia. As a result, the management of other causes of NE often blends with that for HIE. There are two main strategies; the supportive strategy which focuses on adequate resuscitation, adequate ventilation and perfusion, careful fluid management, avoidance of hypoglycemia and hyperglycemia and treatment of seizures [40, 41], and interventional neuroprotective strategies which aim to avoid any further brain injury in these infants [42].

Supportive Management

The primary goals of supportive management include the maintenance of physiologic homeostasis and treatment of the outward manifestations of brain injury [22, 43]. Central aspects of supportive care include securing an appropriate airway and maintaining adequate circulation. Ventilatory support is often required (e.g., mechanical ventilation or continuous positive airway pressure (CPAP)). Metabolic complications, such as hypoglycemia, hypocalcemia, hyponatremia, and acidosis, frequently accompany NE and should be identified and treated. If an infection is suspected, it should be dealt with appropriately. If the history, examination or initial laboratory investigation points to an inborn error of metabolism, early treatment is crucial, and a biochemical geneticist should be consulted. The diagnosis of a severe intracranial hemorrhage should prompt consultation with a neurosurgeon to manage raised intracranial pressure from mass effect or hydrocephalus. In addition, platelet levels and coagulation function should be measured. Since the clinical syndrome evolves considerably over the first 72 h of life, management of specific complications, such as respiratory compromise or seizures, can often be anticipated.

Interventional Neuroprotective Strategies

It has been known for many decades that being cooled during hypoxia protected the brain [44–47]. Clinical reports of intact survival after prolonged cardiac arrest with accidental immersion in cold water persuaded researchers that hypothermic neuroprotection could be achieved in reality [48, 49]. In 1987, Busto et al. reported that reduction of the brain temperature by only a few degrees during ischemia would grant a protective effect in adult rats [50]. However, the debate of clinical investigators continued until 2005 when the first large randomized controlled trial established the efficacy and safety of therapeutic hypothermia in asphyxiated infants [51].

Therapeutic hypothermia within the first 6 h of age and maintained for 72 h at 33–35 °C (91.4–95.0 °F) is the only effective neuroprotective therapy currently available for treatment of NE with an associated hypoxic etiology [52]. It offers the greatest therapeutic promise, particularly in infants who present with moderate encephalopathy and without early seizures [51, 53, 54]. Infants with severe encephalopathy and/or early seizures may not gain the same benefit of therapeutic hypothermia as the extent of their injury may be greater and further advanced, with necrosis rather than apoptosis as the predominant pattern of brain injury. However, hypothermia is still considered as a reasonable choice in these infants as it decreases the multiple deleterious pathways and thus extends the “therapeutic window” for more targeted interventions such as prophylactic and/or aggressive treatment of seizures, blocking the effects of glutamate, combating inflammatory responses, inhibiting apoptosis, and the administration of growth factor [55, 56].

Therapeutic hypothermia aims to lower the temperature of the vulnerable deep brain structures, the basal ganglia, to 32–34 °C. Two methods have been evaluated in newborn infants with presumed HIE: whole body hypothermia and selective head cooling with mild systemic hypothermia. Whole body cooling relies on the core and deep brain temperatures being similar. The rationale of selective head cooling use is the creation of differential temperature gradients within the brain leading to some systemic cooling, the adverse effects of which may be minimized by selectively cooling the brain more than the body [57].

Seven large randomized controlled trials (RCTs) have now been conducted, establishing the efficacy and safety of hypothermia. All demonstrated that treatment with hypothermia improves survival and reduced the risk of composite outcome of death or major neurodevelopmental disability (cerebral palsy (CP), developmental delay and blindness) of term neonates with moderate to severe HIE at 18 months [58]. Six and seven newborns with moderate and severe HIE need to be treated to save one newborn from death or severe disability [58, 59].

This conclusion is supported by two meta-analyses of hypothermia trials involving newborns with moderate to severe neonatal encephalopathy [59–61]. In all the included trials, hypothermia was started within 6 h after birth (Table 2).

Although therapeutic hypothermia has been shown to have promising effects in clinical trials, the reduction in the composite outcome of death or disability at 18 months with therapeutic hypothermia is modest (reduced from 63.2 to 47.9%) [60]. Thus approximately half the infants who receive therapeutic hypothermia still have a major neurodevelopmental disability, and some infants with the most severe injuries may not be rescued [51]. Therefore, there is still an urgent need for other treatment options [62]. There are a variety of other potential neuroprotective treatments currently being studied to prevent the cascade of injurious effects after hypoxia-ischemia and can augment the neuroprotective effect of therapeutic hypothermia [63, 64]. Recent experimental data suggest that hypothermia extends the duration of the therapeutic window [65, 66], therefore giving more time to other neuroprotective interventions to act within an expanded therapeutic window. As an example, erythropoietin has neuroprotective properties in animal models of hypoxic-ischemic brain injury and neonatal stroke [67] and has shown promise in preliminary

Table 2 Meta-analysis of ten randomized trials of therapeutic hypothermia for term infants with HIE [61]

	Newborns with moderate to severe neonatal encephalopathy	Relative risk	Risk difference	Number needed to treat
Death or major disabilities	1344	0.75 (0.68–0.83)	−0.15 (−0.22 to −0.10)	7 (5–10)
Mortality	1478	0.75 (0.64–0.88)	−0.9 (−0.13 to −0.04)	11 (8–25)
Major disabilities	917	0.77 (0.63–0.94)	−0.13 (−0.19 to −0.07)	8 (5–14)

randomized trials for reducing brain injury and improving motor outcomes in infants with hypoxic-ischemic encephalopathy [68, 69]. Other potential therapeutic agents that have been studied include xenon (Xe), magnesium sulfate ($MgSO_4$), allopurinol (allo), topiramate, melatonin, *N*-acetylcysteine (NAC), opioids, and stem cell therapy [62, 70]. Confirmation of benefit in larger trials is needed.

Outcomes

There is a broad spectrum of neurodevelopmental outcomes following any presentation of NE. The outcomes are known to be dependent on the underlying etiology and pattern and severity of the brain injury [8, 71–74]. The neurodevelopmental effects can range from mild to significant challenges and may involve motor, visual, and cognitive functions. Both genetic and postnatal variables such as socioeconomic factors (e.g., environmental exposures and parental education) likely affect an individual's neurodevelopmental outcome following neonatal brain injury [75, 76]. The outcomes are classified and/ or described in the literature according to the different aspects and degrees of functional impairment. Often significant challenges associated with disability are the primary outcome variables and include cerebral palsy (CP), intellectual disabilities (ID), and neurosensory impairment. These outcomes are detected in early childhood, usually prior to the age of 2 years, and will significantly impact independent living in the future. Other outcomes are often described as "minor or mild" challenges and include mild neurological abnormalities, developmental coordination disorder (DCD), mild intellectual delay, learning disability, executive dysfunction, and behavioral challenges. These outcomes can be only assessed at a later age when these functions develop (e.g., executive functions) or can be reliably measured (e.g., behaviors, academic function).

Predictors of Major Neurodevelopmental Outcomes

Reliable early predictors of neurodevelopmental outcomes are very important to determine as they may both select infants for therapy and guide withdrawal of care [77]. Predictors of outcome may be clinical, neuroimaging, electrographic, and/or biochemical. It is vital to understand the effect of therapeutic hypothermia upon their predictive power.

Clinical Presentation and Grading

The discussion of outcome correlates directly to the degree of encephalopathy. Sarnat scoring of encephalopathy [35] is the most widely used clinical staging system to describe the degree of NE and is most often cited in the majority of outcome related studies [72, 73, 76]. As we mention before, the Sarnat scoring is based on the clinical presentation and severity of encephalopathy; it can be distinguished into three stages of encephalopathy. Stage I, or mild encephalopathy, is associated with hyperalertness, sympathetic overdrive, normal muscle tone, and a normal EEG. Stage II, or moderate encephalopathy, is marked by obtundation or lethargy, hypotonia, multifocal seizures, and an EEG showing periodic or continuous delta activity. In Stage III, or severe encephalopathy, the infants appear stuporous and flaccid with an isoelectric or periodic EEG [35]. The challenge, however, with the interpretation of the data on outcomes lies with the multitude of etiologies of NE, the concomitant presentation of possible HIE, and the potential confounder that genetic/metabolic cases may present to outcome studies. As the practice of clustering outcomes by the degree of encephalopathy continues, the outcomes presented reflect a combination of potential etiologies and degrees of initial encephalopathy.

There is a spectrum of outcomes associated with neonates presenting with varying degrees of encephalopathy (Fig. 1), and the pooled outcome data reflects the degree of injury rather than intervention. At the most severe degree of injury/presentation, mortality is as high as 25–50%, most of which occurs in the first week of life. Among survivors, up to 10% will appear untouched. Most (80%), however, will demonstrate findings of neurological injury, including 10–20% developing moderately serious disabilities [76, 78]. Among infants who survive moderate hypoxic-ischemic encephalopathy, 30–50% may have serious long-term complications such as cerebral palsy, intellectual disabilities, and neurosensory impairment, and 10–20% of infants have minor neurological morbidities [79, 80]. Epilepsy is identified in up to one-half of survivors from moderate to severe neonatal encephalopathy [81, 82]. Infants with mild hypoxic-ischemic encephalopathy tend to be free from serious CNS complications.

Other scoring systems have been devised to help predict an infant's subsequent risk for developing cerebral palsy or systemic morbidity [84]. One of the largest

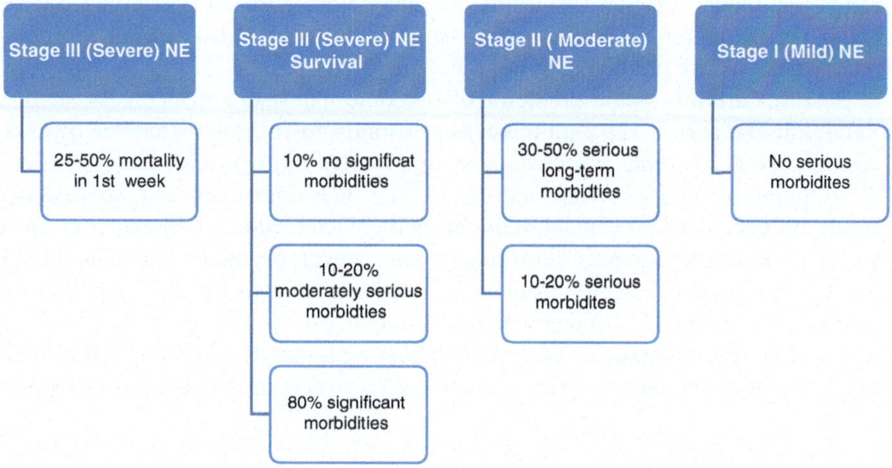

Fig. 1 Outcomes associated with varying degrees of encephalopathy [76, 78–80, 83]

studies retrospectively evaluated 365 infants with HIE and found that three clinical parameters (administration of chest compression for >1 min, onset of regular respirations >30 min after birth, and base deficit value of >16 mmol/L on any blood gas analysis within the first 4 h from birth) were predictors of severe adverse outcome (death or severe disability) [85]. Severe outcome rates with none, one, two, or all three predictors were 46%, 64%, 76%, and 93%, respectively.

A reliable predictor of long-term outcome is the neurological examination performed at the end of the first week of life [35, 73, 86]. Carefully performed serial neurological examinations provide critical information regarding the presence, extent, and evolution over time of the hypoxic-ischemic injury in the term infant. Several studies have demonstrated the reliability and prognostic ability of the neurological examination [87–96].

Recently, the assessment of general movements (GMs) has been examined in its capacity to predict long-term neurodevelopmental outcomes [97, 98]. GMs are spontaneously generated full-body movements that reflect neurological integrity. They occur as early as 10 weeks postmenstrual age and continue until approximately 4 months post-term age. Their pattern changes from preterm movements in utero or in premature infants to writhing movements which emerge at term gestation to fidgety movements at around 10–12 weeks of age. The quality of GMs is described in terms of the infant's movement variety (how variable are the movements) and complexity (is the infant using the three-dimensional space around him to move) and fluency (how smooth are the transitions between movements). These qualities as well as any potential asymmetries should be noted [98, 99]. There are several approaches to classification of the GMs, but generally the descriptive categories include normal and abnormal. Abnormal GMs; especially cramped synchronized GMs are strongly related to a later diagnosis of cerebral palsy in both the term and preterm infants [98, 100–109].

Neuroimaging abnormalities have been related to abnormal GMs in term infants with HIE. MRI lesion patterns were correlated with 1-month and 3-month GMs. Deep gray matter injury was strongly correlated with cramped synchronized GMs and abnormal motor outcomes [110]. Therefore, GMs may also be used as a complementary tool for predicting motor outcome [109, 110].

Seizures are a possible predictor of outcome, but data are inconsistent. In a longitudinal report of 129 children ages 12 months to 16.5 years (median 6 years) who survived NE, epilepsy developed in 13 (10%) [111]. Risk factors for the development of epilepsy were the occurrence of neonatal seizures, particularly status epilepticus, and neonatal brain injury on MRI. Children with epilepsy often had significant challenges and major neurodevelopmental outcomes [111]. Another report analyzed data from subjects enrolled in a therapeutic hypothermia trial and compared 127 infants who had clinical seizures during the trial with 81 infants who had no seizures [112]. After adjusting for study treatment and severity of encephalopathy, seizures were not associated with poor outcomes at 18 months of life.

Amplitude-Integrated EEG

The potential benefits of preventing further neuronal injury associated with seizures have prompted the development of continuous cerebral function monitoring like amplitude-integrated EEG (aEEG) [113]. Prolonged video/EEG monitoring has shown that 85% of all seizures were not associated with clinical manifestations [114] and that 58% of the infants with seizures persisting after treatment with phenobarbitone or phenytoin showed uncoupling of electrical and clinical seizures [115].

Several studies [116–122] have shown that outcome of infants with a presumed HIE insult can be accurately predicted from an aEEG during the first hours after birth. The predictive value of the presence of a poor background pattern for subsequent neurodevelopmental challenges at 18–24 months was assessed in these studies. Both positive and negative predictive values were slightly lower when the aEEG was assessed at 3 instead of 6 h after birth. Also, combining neurological examination with aEEG performed <12 h after birth further increases predictive accuracy from 75 to 85%. These findings were also confirmed by a meta-analysis of eight studies described by Spitzmiller and colleagues [123]; a sensitivity of 91% (95% CI 87–95) and a negative likelihood ratio of 0.09 (95% CI .06–15) for poor aEEG tracings were found to accurately predict poor outcome [123].

Brain Imaging and Outcome

Among the techniques available for imaging the newborn, brain MRI and magnetic resonance spectroscopy (MRS) are validated and well-accepted biomarkers of hypoxic-ischemic encephalopathy (HIE) severity, neurologic outcome [124–127], and treatment response following hypothermia [128–132]. Brain MRI appears to be the most sensitive in detecting periventricular white matter injury, deep gray matter lesions, arterial infarction, hemorrhage, developmental brain malformations, and other underlying causes of neonatal encephalopathy [133]. MRI provides excellent details of the brain lesions characteristic of perinatal hypoxic-ischemic injury: these lesions can be graded, and the pattern of involvement can be related to the outcome.

The distribution of injury was associated with the duration and severity of ischemia. Acute-profound asphyxia produced injury in the basal ganglia and thalamus, while partial asphyxia caused white matter injury [134, 135]. The basal ganglia-predominant pattern involves both the basal ganglia and thalamus and perirolandic cortex [37, 125, 136]. The watershed pattern predominantly affects the intravascular boundary zone of the white matter, plus cortical gray matter that is perfused by both the anterior and middle cerebral. Maximal injury in both the watershed region and basal nuclei results in the total pattern of brain injury [125, 136]. The final pattern of injury, increasingly recognized by MRI, is the "focal- or multifocal" pattern of injury: stroke or white matter injury (WMI). Recent data suggest that strokes (arterial or venous) are also associated with NE in the term newborn [137]. Many newborns with stroke have multiple risk factors for brain injury, including intrapartum complications.

The pattern of brain injury on neuroimaging conveys valuable prognostic information regarding the "pattern" of neurodevelopmental abnormalities. Patterns of injury related to the basal nuclei and abnormal signal intensity in the posterior limb of the internal capsule are both predictive of significant neurodevelopmental challenges in the motor and/or cognitive domains (CP and significant cognitive delay) [125, 138, 139]. In one study, brainstem lesions on MRI were associated with an increased risk of death [140]. In contrast, the watershed pattern is associated with cognitive deficits that are not necessarily accompanied by significant motor deficits [125] and may not be evident until after 2 years of age [125]. In survivors of NE without functional motor deficits assessed at 4 years old, the severity of the watershed-distribution injury was most strongly associated with impaired language skills [141].

Diffusion-weighted MRI can detect the presence of acute brain injury, that is, an injury that occurred within 7–10 days of the study. Thus, diffusion-weighted images can distinguish which infants with NE have suffered a significant brain injury that is associated with long-term outcome within a window of time that often includes the time of delivery [142–146]. Apparent diffusion coefficient (ADC) mapping, with ADC values calculated from the diffusion-weighted imaging sequence, shows restricted diffusion as dark areas of diminished signal intensity. In the white matter, it increases with maturation [147–149]. Hunt and colleagues [150] found that the ADC values in the posterior limb of the internal capsule were a good indicator of ischemic injury in newborns with HIE. Measuring ADC values in this brain structure may eventually be a useful and objective marker of prognosis even in normal-appearing structures [151].

MRS can detect increased lactate and decreased *N*-acetylaspartate (NAA) concentrations, which indicate derangements of the metabolic state of specific regions of the brain and portend a worse prognosis [75, 152–154]. In a meta-analysis of studies evaluating the predictive value of neonatal MRI, elevated Lac/NAA ratios in the thalamus or basal ganglia and thalamus demonstrated a pooled sensitivity of 82% and specificity of 95% for neurodevelopmental outcome [155]. However, a systematic review found that MRS was not as predictive of outcome as other MRI parameters [144].

Prediction is not infallible, however, and neurological deficits may also be found in some newborns whose brain imaging studies appear normal [156]. More subtle brain injuries associated with later neurodevelopmental deficits, such as white matter injuries or hippocampal volume loss, may only be detectable with quantitative brain imaging techniques [75, 157, 158]. Currently, no study has related neonatal MRI findings with non-disabling long-term outcomes in children such as developmental coordination disorder (DCD), mild intellectual delay, learning disability, executive dysfunction, and behavioral challenges.

Neurodevelopmental Outcomes

Cerebral Palsy (CP) and Motor Function

Perinatal asphyxia is often suspected as a cause for CP, but it likely accounts for only a small minority of cases [159, 160]. In a retrospective study using the Task Force on Neonatal Encephalopathy and Cerebral Palsy criteria of intrapartum asphyxial events sufficient to cause CP [161, 162], only 4% of children born full term who developed CP had evidence of an acute hypoxic event during labor [163]. The possibility of CP in the term survivors of hypoxic-ischemic brain injury may occur in more than one-third of affected newborns and is most common in those with a severe encephalopathy [22, 164, 165]. Overall, rates of CP in this population without therapeutic hypothermia range from 9% to 39% and depend on the included population and the severity of the insult. With current data from the seven trials examining therapeutic hypothermia, these rates appear to decrease by approximately 50% [57, 59, 166–169].

Spastic quadriparesis is the most common type of CP, although athetoid or spastic hemiparesis also occurs [170–173].

Motor impairments, such as DCD, that do not meet diagnostic criteria for CP are diagnosed in more than one-third of children with moderate encephalopathy, and in more than one-quarter of children with mild encephalopathy [174].

Vision and Hearing

One-quarter of children after moderate or severe encephalopathy have severe visual impairment [72, 76, 175]. This may be due to injury to the posterior visual pathway, including the primary visual cortex, resulting in "cortical visual impairment" (CVI) [176]. Injuries to the basal nuclei may also affect acuity, visual fields, or stereopsis (depth perception) [177, 178].

Sensorineural hearing loss, likely secondary to brainstem injury, is also seen following NE [72, 73, 76] affecting 18% of survivors of moderate encephalopathy without CP [179].

Cognition

Overall, cognitive deficits are seen in 30–50% of childhood survivors of moderate NE [180]. Intellectual performance in children with severe encephalopathy without CP is also affected [86, 173]. Recently reviewed data indicate that cognitive deficits may prominently follow HIE, even in the absence of CP [181]. This pattern is most commonly associated with the watershed pattern of injury and white matter damage, rather than the basal nuclei-predominant pattern of injury. Cognitive deficits, such as those in language and memory, may be seen, even when intellectual quotient (IQ) scores are "normal" [173].

Therefore, a comprehensive neurodevelopmental assessment of neurodevelopmental outcome in the follow-up of newborns with a history of encephalopathy must include aspects of cognition most readily assessed at school age: learning, executive function, behavior, specific language impairments, and social competence [174].

School Performance

There is a body of literature on school-age survivors that show that even in the absence of obvious neurological deficits in the newborn period, long-term functional challenges may be present [73]. Children with moderate NE were not different from controls with respect to general cognitive ability, but less proficient in language and sensorimotor domains, as well as narrative memory and sentence repetition [173]. Therefore, they are more likely to have difficulties with reading, spelling, and arithmetic or require additional school resources [73, 76, 174, 182]. They were more likely to require extra educational assessment, teaching provision, and support, even though they did not have any overt neuromotor challenges [183]. Among the more severely affected children, referred to as severe encephalopathy, memory for names, orientation, and reported everyday memory function were also significantly more impaired than for either comparison children or the moderate encephalopathy group [173].

In a cohort of school-aged children with a history of moderately severe HIE, 15–20% had significant learning difficulties, even in the absence of obvious signs of brain injury. Survivors of mild NE compared to a comparison group were not significantly more than one grade behind in reading, spelling, and arithmetic [174, 182]. Thus, all children who have moderate or severe HIE should be monitored well into school age [71, 174].

Behavioral Challenges

Behavioral difficulties, such as hyperactivity and emotional problems, should also be considered for all children affected, even in individuals without motor disability [173, 174]. Hyperactivity was more often present in children with moderate NE [173, 182, 183]. In addition, one study that used parent's observations of their child's behavior found more problems related to hyperactivity, aggression, and anxiety in a mixed group of children with NE compared to a control group [182].

There is little evidence that NE can lead to a higher risk of autism spectrum disorders (ASD) [179]. However, one study reported an unexpectedly high proportion of children with moderate and severe NE that developed ASD [184].

Therapeutic Hypothermia and Outcomes

To date, neurodevelopmental outcome following therapeutic hypothermia for HIE has only been reported to 18–24 months. School-age survivors of HIE, even those without neuromotor challenges, are at significantly greater risk of more cognitive, behavioral, and educational difficulties [173, 174, 185, 186].

In two large hypothermia trials, 23–27% of infants died before discharge from the neonatal intensive care unit (NICU), whereas the mortality rate at follow-up 18–22 months later was 37–38% [51, 187]. Neurodevelopmental assessment of infants with moderate to severe encephalopathy at 18 months revealed a mental development index (MDI) and a psychomotor development index (PDI) scores of more than 85 in 40–55% of infants, of 70–84 in 10–21% of infants, and of less than 70 in 35–39% of infants. Of those infants "disabling" CP was reported in 30%, epilepsy in 16%, blindness in 14–17%, and severe hearing impairment in 6% of survivors [187, 188].

Data regarding the long-term safety and efficacy of therapeutic hypothermia suggest that the benefit extends later into childhood. Long-term outcome data at age of 6–7 years for 190 children of those previously studied in the National Institute of Child Health and Human Development (NICHD) trial had been recently published [188]. The proportion who died or had an IQ score <70 was lower for children assigned to the hypothermia group compared with the control group (47 versus 62%), but the difference between the groups just missed statistical significance (RR 0.78, 95% CI 0.61–1.01). Therefore, Tagin and colleagues stated that "it remains appropriate for clinicians to be conservative when counseling parents about longer-term neurological function" [58]. In the Total Body Hypothermia (TOBY) trial [166], there is outcome data available of children who survived to ages 6 or 7 years; an IQ score of ≥85 was significantly higher for the hypothermia group compared with the control group (52 versus 39%, RR 1.31, 95% CI 1.01–1.71) [189]. While these results are promising, more evidence is needed to confirm the long-term benefit of therapeutic hypothermia.

Therapeutic hypothermia appears to decrease the degree of brain injury observed on MRI and magnetic resonance spectroscopy [128, 130, 132, 190]; however, the available evidence suggests that treatment with hypothermia does not affect the value of MRI for predicting outcome after neonatal encephalopathy [128, 132]. Several studies using MRI reported less severe cortical and deep gray matter injury on neonatal MRI in association with therapeutic hypothermia [128, 131, 190]. In addition, neonate treated with hypothermia showed less extension of watershed and basal ganglia/thalamus injury and more normal MRI [190].

Conclusion

Neonatal encephalopathy is the preferred term used to describe a heterogeneous, clinically defined syndrome of disturbed neurologic function in the earliest days of life in an infant born at term or near term. Neonatal encephalopathy can result from a wide variety of conditions and often remains unexplained. Hypoxic-ischemic encephalopathy is a subset of NE and may occur concomitant with other causes of NE. Therapeutic hypothermia is the treatment of choice (in the first 6 h of age) for NE with the goal of treating any component of hypoxic-ischemic injury in experienced centers.

Neurodevelopmental outcome depends on the severity of the individual presentation, the underlying etiology, and management and represents a broad spectrum. Regardless, the spectrum of outcomes is diffuse and holistic attention to all domains of development is warranted throughout childhood.

References

1. Badawi, N., et al. (1998). Antepartum risk factors for newborn encephalopathy: The Western Australian case-control study. *BMJ, 317*(7172), 1549–1553.
2. (2014). Executive summary: Neonatal encephalopathy and neurologic outcome, second edition. Report of the American College of Obstetricians and Gynecologists' task force on neonatal encephalopathy. *Obstetrics and Gynecology, 123*(4), 896–901.
3. Pierrat, V., et al. (2005). Prevalence, causes, and outcome at 2 years of age of newborn encephalopathy: Population based study. *Archives of Disease in Childhood-Fetal and Neonatal Edition, 90*(3), F257–F261.
4. Locatelli, A., et al. (2010). Antepartum and intrapartum risk factors for neonatal encephalopathy at term. *American Journal of Perinatology, 27*(8), 649–654.
5. Martinez-Biarge, M., et al. (2013). Antepartum and intrapartum factors preceding neonatal hypoxic-ischemic encephalopathy. *Pediatrics, 132*(4), e952–e959.
6. Wu, Y. W., et al. (2004). Declining diagnosis of birth asphyxia in California: 1991-2000. *Pediatrics, 114*(6), 1584–1590.
7. Lee, A. C., et al. (2013). Intrapartum-related neonatal encephalopathy incidence and impairment at regional and global levels for 2010 with trends from 1990. *Pediatric Research, 74*(Suppl 1), 50–72.
8. Thornberg, E., et al. (1995). Birth asphyxia: Incidence, clinical course and outcome in a Swedish population. *Acta Paediatrica, 84*(8), 927–932.
9. Graham, E. M., et al. (2008). A systematic review of the role of intrapartum hypoxia-ischemia in the causation of neonatal encephalopathy. *American Journal of Obstetrics and Gynecology, 199*(6), 587–595.
10. Kurinczuk, J. J., White-Koning, M., & Badawi, N. (2010). Epidemiology of neonatal encephalopathy and hypoxic-ischaemic encephalopathy. *Early Human Development, 86*(6), 329–338.
11. Sartwelle, T. P. (2009). Defending a neurologic birth injury. Asphyxia neonatorum redux. *The Journal of Legal Medicine, 30*(2), 181–247.
12. Long, M., & Brandon, D. H. (2007). Induced hypothermia for neonates with hypoxic-ischemic encephalopathy. *Journal of Obstetric, Gynecologic, and Neonatal Nursing, 36*(3), 293–298.
13. Shankaran, S. (2009). Neonatal encephalopathy: Treatment with hypothermia. *Journal of Neurotrauma, 26*(3), 437–443.
14. Volpe, J. J. (2012). Neonatal encephalopathy: An inadequate term for hypoxic-ischemic encephalopathy. *Annals of Neurology, 72*(2), 156–166.
15. Dammann, O., Ferriero, D., & Gressens, P. (2011). Neonatal encephalopathy or hypoxic-ischemic encephalopathy? Appropriate terminology matters. *Pediatric Research, 70*(1), 1–2.
16. Kumar, S., & Paterson-Brown, S. (2010). Obstetric aspects of hypoxic ischemic encephalopathy. *Early Human Development, 86*(6), 339–344.

17. Shalak, L., & Perlman, J. M. (2004). Hypoxic-ischemic brain injury in the term infant-current concepts. *Early Human Development, 80*(2), 125–141.
18. Cotten, C. M., & Shankaran, S. (2010). Hypothermia for hypoxic-ischemic encephalopathy. *Expert Review of Obstetrics & Gynecology, 5*(2), 227–239.
19. Lorek, A., et al. (1994). Delayed ("secondary") cerebral energy failure after acute hypoxia-ischemia in the newborn piglet: Continuous 48-hour studies by phosphorus magnetic resonance spectroscopy. *Pediatric Research, 36*(6), 699–706.
20. Laptook, A. R., et al. (1992). Glucose-associated alterations in ischemic brain metabolism of neonatal piglets. *Stroke, 23*(10), 1504–1511.
21. Hanrahan, J. D., et al. (1996). Cerebral metabolism within 18 hours of birth asphyxia: A proton magnetic resonance spectroscopy study. *Pediatric Research, 39*(4 Pt 1), 584–590.
22. Volpe, J. J. (2008). *Neurology of the newborn*. Philadelphia: Saunders/Elsevier.
23. Johnston, M. V., et al. (2009). Plasticity and injury in the developing brain. *Brain Dev, 31*(1), 1–10.
24. Siesjo, B. K., & Bengtsson, F. (1989). Calcium fluxes, calcium antagonists, and calcium-related pathology in brain ischemia, hypoglycemia, and spreading depression: A unifying hypothesis. *Journal of Cerebral Blood Flow and Metabolism, 9*(2), 127–140.
25. Fellman, V., & Raivio, K. O. (1997). Reperfusion injury as the mechanism of brain damage after perinatal asphyxia. *Pediatric Research, 41*(5), 599–606.
26. Mehmet, H., et al. (1994). The relationship between impaired cerebral energy metabolism and apoptosis in the cingulate gyrus of newborn piglets following transient hypoxia-ischaemia. UCL/RPMS Perinatal Brain Research Group. *Biochemical Society Transactions, 22*(4), 421s.
27. Liu, X. H., et al. (1999). Mice deficient in interleukin-1 converting enzyme are resistant to neonatal hypoxic-ischemic brain damage. *Journal of Cerebral Blood Flow and Metabolism, 19*(10), 1099–1108.
28. Mehmet, H., et al. (1994). Increased apoptosis in the cingulate sulcus of newborn piglets following transient hypoxia-ischaemia is related to the degree of high energy phosphate depletion during the insult. *Neuroscience Letters, 181*(1-2), 121–125.
29. Tan, W. K., et al. (1996). Accumulation of cytotoxins during the development of seizures and edema after hypoxic-ischemic injury in late gestation fetal sheep. *Pediatric Research, 39*(5), 791–797.
30. Gluckman, P. D., et al. (1998). Asphyxial brain injury—the role of the IGF system. *Molecular and Cellular Endocrinology, 140*(1-2), 95–99.
31. Gunn, A. J., et al. (1997). Dramatic neuronal rescue with prolonged selective head cooling after ischemia in fetal lambs. *The Journal of Clinical Investigation, 99*(2), 248–256.
32. Gunn, A. J., et al. (1999). Cerebral hypothermia is not neuroprotective when started after postischemic seizures in fetal sheep. *Pediatric Research, 46*(3), 274–280.
33. Gunn, A. J., & Thoresen, M. (2006). Hypothermic neuroprotection. *NeuroRx, 3*(2), 154–169.
34. Miller, S. P., et al. (2004). Clinical signs predict 30-month neurodevelopmental outcome after neonatal encephalopathy. *American Journal of Obstetrics and Gynecology, 190*(1), 93–99.
35. Sarnat, H. B., & Sarnat, M. S. (1976). Neonatal encephalopathy following fetal distress. A clinical and electroencephalographic study. *Archives of Neurology, 33*(10), 696–705.
36. Redline, R. W. (2005). Severe fetal placental vascular lesions in term infants with neurologic impairment. *American Journal of Obstetrics and Gynecology, 192*(2), 452–457.
37. Chau, V., Poskitt, K. J., & Miller, S. P. (2009). Advanced neuroimaging techniques for the term newborn with encephalopathy. *Pediatric Neurology, 40*(3), 181–188.
38. Barnette, A. R., et al. (2014). Neuroimaging in the evaluation of neonatal encephalopathy. *Pediatrics, 133*(6), e1508–e1517.
39. Hankins, G. D., et al. (2003). Neonatal encephalopathy and cerebral palsy: A knowledge survey of fellows of the American College of Obstetricians and Gynecologists. *Obstetrics and Gynecology, 101*(1), 11–17.
40. Shankaran, S. (2002). The postnatal management of the asphyxiated term infant. *Clinics in Perinatology, 29*(4), 675–692.

41. Stola, A., & Perlman, J. (2008). Post-resuscitation strategies to avoid ongoing injury following intrapartum hypoxia-ischemia. *Seminars in Fetal & Neonatal Medicine, 13*(6), 424–431.
42. Perlman, J. M. (2006). Intervention strategies for neonatal hypoxic-ischemic cerebral injury. *Clinical Therapeutics, 28*(9), 1353–1365.
43. Yager, J. Y., Armstrong, E. A., & Black, A. M. (2009). Treatment of the term newborn with brain injury: Simplicity as the mother of invention. *Pediatric Neurology, 40*(3), 237–243.
44. Cordey, R. (1964). Hypothermia in resuscitating newborns in white asphyxia; A report of 14 cases. *Obstetrics and Gynecology, 24*, 760–767.
45. Dunn, J. M., & Miller, J. A., Jr. (1969). Hypothermia combined with positive pressure ventilation in resuscitation of the asphyxiated neonate. Clinical observations in 28 infants. *American Journal of Obstetrics and Gynecology, 104*(1), 58–67.
46. Silverman, W. A., Fertig, J. W., & Berger, A. P. (1958). The influence of the thermal environment upon the survival of newly born premature infants. *Pediatrics, 22*(5), 876–886.
47. Ehrstrom, J., et al. (1969). Hypothermia in the resuscitation of severely asphyctic newborn infants. A follow-up study. *Annals of Clinical Research, 1*(1), 40–49.
48. Ravitch, M. M., et al. (1961). Lightning stroke. Report of a case with recovery after cardiac massage and prolonged artificial respiration. *The New England Journal of Medicine, 264*, 36–38.
49. De Villota, E. D., et al. (1973). Recovery from profound hypothermia with cardiac arrest after immersion. *British Medical Journal,* ***4***(5889), 394–395.
50. Busto, R., et al. (1987). Small differences in intraischemic brain temperature critically determine the extent of ischemic neuronal injury. *Journal of Cerebral Blood Flow and Metabolism, 7*(6), 729–738.
51. Gluckman, P. D., et al. (2005). Selective head cooling with mild systemic hypothermia after neonatal encephalopathy: Multicentre randomised trial. *Lancet, 365*(9460), 663–670.
52. Papile, L. A., et al. (2014). Hypothermia and neonatal encephalopathy. *Pediatrics, 133*(6), 1146–1150.
53. Shankaran, S., et al. (2005). Whole-body hypothermia for neonates with hypoxic–ischemic encephalopathy. *New England Journal of Medicine, 353*(15), 1574–1584.
54. Jacobs, S. E. (2005). Selective head cooling with mild systemic hypothermia after neonatal encephalopathy: Multicentre randomised trial. *The Journal of Pediatrics, 147*(1), 122–123.
55. Perlman, J. (2007). Pathogenesis of hypoxic-ischemic brain injury. *Journal of Perinatology, 27*, S39–S46.
56. Perlman, M., & Shah, P. S. (2011). Hypoxic-ischemic encephalopathy: Challenges in outcome and prediction. *The Journal of Pediatrics, 158*(2 Suppl), e51–e54.
57. Gunn, A. J., Gluckman, P. D., & Gunn, T. R. (1998). Selective head cooling in newborn infants after perinatal asphyxia: A safety study. *Pediatrics, 102*(4 Pt 1), 885–892.
58. Tagin, M. A., et al. (2012). Hypothermia for neonatal hypoxic ischemic encephalopathy: An updated systematic review and meta-analysis. *Archives of Pediatrics & Adolescent Medicine, 166*(6), 558–566.
59. Jacobs, S. E., et al. (2013). Cooling for newborns with hypoxic ischaemic encephalopathy. *Cochrane Database of Systematic Reviews, 1*, Cd003311.
60. Tagin, M. A., et al. (2012). Hypothermia for neonatal hypoxic ischemic encephalopathy: An updated systematic review and meta-analysis. *Archives of Pediatrics & Adolescent Medicine, 166*(6), 558–566.
61. Azzopardi, D. V., et al. (2009). Moderate hypothermia to treat perinatal asphyxial encephalopathy. *The New England Journal of Medicine, 361*(14), 1349–1358.
62. Robertson, N. J., et al. (2012). Which neuroprotective agents are ready for bench to bedside translation in the newborn infant? *The Journal of Pediatrics,* ***160***(4), 544–552.e4.
63. ur Rehman, S. (2011). Neuroprotection in neonatal hypoxic ischemic encephalopathy: therapeutic hypothermia and beyond. *Journal of Postgraduate Medical Institute (Peshawar-Pakistan), 25*(1).
64. Marro, P. J., & Delivoria-Papadopoulos, M. (2010). Pharmacology review. *NeoReviews, 11*(6), e311–e315.

65. O'Brien, F. E., et al. (2006). Delayed whole-body cooling to 33 or 35 C and the development of impaired energy generation consequential to transient cerebral hypoxia-ischemia in the newborn piglet. *Pediatrics, 117*(5), 1549–1559.
66. Liu, Y., et al. (2004). Topiramate extends the therapeutic window for hypothermia-mediated neuroprotection after stroke in neonatal rats. *Stroke, 35*(6), 1460–1465.
67. Rangarajan, V., & Juul, S. E. (2014). Erythropoietin: Emerging role of erythropoietin in neonatal neuroprotection. *Pediatric Neurology, 51*(4), 481–488.
68. Zhu, C., et al. (2009). Erythropoietin improved neurologic outcomes in newborns with hypoxic-ischemic encephalopathy. *Pediatrics, 124*(2), e218–e226.
69. Wu, Y. W., et al. (2016). High-dose erythropoietin and hypothermia for hypoxic-ischemic encephalopathy: A phase II trial. *Pediatrics, 137*(6).
70. Higgins, R. D., et al. (2011). Hypothermia and other treatment options for neonatal encephalopathy: an executive summary of the Eunice Kennedy Shriver NICHD workshop. *The Journal of Pediatrics, 159*(5), 851–858.e1.
71. Pin, T. W., Eldridge, B., & Galea, M. P. (2009). A review of developmental outcomes of term infants with post-asphyxia neonatal encephalopathy. *European Journal of Paediatric Neurology, 13*(3), 224–234.
72. Robertson, C., & Finer, N. (1985). Term infants with hypoxic-ischemic encephalopathy: Outcome at 3.5 years. *Developmental Medicine and Child Neurology, 27*(4), 473–484.
73. Robertson, C. M., Finer, N. N., & Grace, M. G. (1989). School performance of survivors of neonatal encephalopathy associated with birth asphyxia at term. *The Journal of Pediatrics, 114*(5), 753–760.
74. Finer, N. N., et al. (1981). Hypoxic-ischemic encephalopathy in term neonates: Perinatal factors and outcome. *The Journal of Pediatrics, 98*(1), 112–117.
75. Miller, S. P., et al. (2002). Predictors of 30-month outcome after perinatal depression: Role of proton MRS and socioeconomic factors. *Pediatric Research, 52*(1), 71–77.
76. Robertson, C. M., & Finer, N. N. (1993). Long-term follow-up of term neonates with perinatal asphyxia. *Clinics in Perinatology, 20*(2), 483–500.
77. Rutherford, M., et al. (2010). Magnetic resonance imaging in hypoxic-ischaemic encephalopathy. *Early Human Development, 86*(6), 351–360.
78. Bohr, L., & Greisen, G. (1998). Prognosis after perinatal asphyxia in full-term infants. A literature review. *Ugeskrift for Laeger, 160*(19), 2845–2850.
79. Carli, G., Reiger, I., & Evans, N. (2004). One-year neurodevelopmental outcome after moderate newborn hypoxic ischaemic encephalopathy. *Journal of Paediatrics and Child Health, 40*(4), 217–220.
80. Smith, J., Wells, L., & Dodd, K. (2000). The continuing fall in incidence of hypoxic-ischaemic encephalopathy in term infants. *BJOG, 107*(4), 461–466.
81. Brunquell, P. J., et al. (2002). Prediction of outcome based on clinical seizure type in newborn infants. *The Journal of Pediatrics, 140*(6), 707–712.
82. Legido, A., Clancy, R. R., & Berman, P. H. (1991). Neurologic outcome after electroencephalographically proven neonatal seizures. *Pediatrics, 88*(3), 583–596.
83. Shevell, M., & Miller, S. (2012). *Acquired brain injury in the fetus and newborn.* London: Mac Keith Press.
84. Thompson, C. M., et al. (1997). The value of a scoring system for hypoxic ischaemic encephalopathy in predicting neurodevelopmental outcome. *Acta Paediatrica, 86*(7), 757–761.
85. Shah, P. S., et al. (2006). Postasphyxial hypoxic-ischemic encephalopathy in neonates: Outcome prediction rule within 4 hours of birth. *Archives of Pediatrics & Adolescent Medicine, 160*(7), 729.
86. Marlow, N., & Budge, H. (2005). Prevalence, causes, and outcome at 2 years of age of newborn encephalopathy. *Archives of Disease in Childhood-Fetal and Neonatal Edition, 90*(3), F193–F194.
87. Bierman-van Eendenburg, M. E., et al. (1981). Predictive value of neonatal neurological examination: A follow-up study at 18 months. *Developmental Medicine and Child Neurology, 23*(3), 296–305.

88. Hadders-Algra, M., & Touwen, B. C. (1989). The long-term significance of neurological findings at toddler's age. *Pädiatrie und Grenzgebiete, 28*(2), 93–99.
89. Wolf, M. J., et al. (1998). Neonatal neurological examination as a predictor of neuromotor outcome at 4 months in term low-Apgar-score babies in Zimbabwe. *Early Human Development, 51*(2), 179–186.
90. Prechtl, H. F. (1997). State of the art of a new functional assessment of the young nervous system. An early predictor of cerebral palsy. *Early Human Development, 50*(1), 1–11.
91. Prechtl, H. F., & Einspieler, C. (1997). Is neurological assessment of the fetus possible? *European Journal of Obstetrics, Gynecology, and Reproductive Biology, 75*(1), 81–84.
92. Amiel-Tison, C. (2002). Update of the Amiel-Tison neurologic assessment for the term neonate or at 40 weeks corrected age. *Pediatric Neurology, 27*(3), 196–212.
93. Zafeiriou, D. I. (2004). Primitive reflexes and postural reactions in the neurodevelopmental examination. *Pediatric Neurology, 31*(1), 1–8.
94. Ricci, D., et al. (2006). Neurological examination at 6 to 9 months in infants with cystic periventricular leukomalacia. *Neuropediatrics, 37*(4), 247–252.
95. Ricci, D., et al. (2006). Sequential neurological examinations in infants with neonatal encephalopathy and low apgar scores: Relationship with brain MRI. *Neuropediatrics, 37*(3), 148–153.
96. Murray, D. M., et al. (2010). The predictive value of early neurological examination in neonatal hypoxic-ischaemic encephalopathy and neurodevelopmental outcome at 24 months. *Developmental Medicine and Child Neurology, 52*(2), e55–e59.
97. Zuk, L. (2011). Fetal and infant spontaneous general movements as predictors of developmental disabilities. *Developmental Disabilities Research Reviews, 17*(2), 93–101.
98. Hadders-Algra, M. (2004). General movements: A window for early identification of children at high risk for developmental disorders. *The Journal of Pediatrics, 145*(2 Suppl), S12–S18.
99. Groen, S. E., et al. (2005). General movements in early infancy predict neuromotor development at 9 to 12 years of age. *Developmental Medicine and Child Neurology, 47*(11), 731–738.
100. Brogna, C., et al. (2013). Prognostic value of the qualitative assessments of general movements in late-preterm infants. *Early Human Development, 89*(12), 1063–1066.
101. Chen, N., et al. (2015). Predictive value of qualitative assessment of general movements for adverse outcomes at 24 months of age in infants with asphyxia. *Zhongguo Dang Dai Er Ke Za Zhi, 17*(12), 1322–1326.
102. Cioni, G., et al. (1997). Which better predicts later outcome in full-term infants: Quality of general movements or neurological examination? *Early Human Development, 50*(1), 71–85.
103. Darsaklis, V., et al. (2011). Predictive validity of Prechtl's method on the qualitative assessment of general movements: A systematic review of the evidence. *Developmental Medicine and Child Neurology, 53*(10), 896–906.
104. Einspieler, C., & Prechtl, H. F. (2005). Prechtl's assessment of general movements: A diagnostic tool for the functional assessment of the young nervous system. *Mental Retardation and Developmental Disabilities Research Reviews, 11*(1), 61–67.
105. Ferrari, F., et al. (2002). Cramped synchronized general movements in preterm infants as an early marker for cerebral palsy. *Archives of Pediatrics & Adolescent Medicine, 156*(5), 460–467.
106. Guzzetta, A., et al. (2003). General movements detect early signs of hemiplegia in term infants with neonatal cerebral infarction. *Neuropediatrics, 34*(2), 61–66.
107. Hadders-Algra, M. (1996). The assessment of general movements is a valuable technique for the detection of brain dysfunction in young infants. A review. *Acta Paediatrica Supplement, 416*, 39–43.
108. Natarajan, G., et al. (2014). Functional status at 18 months of age as a predictor of childhood disability after neonatal hypoxic-ischemic encephalopathy. *Developmental Medicine and Child Neurology, 56*(11), 1052–1058.
109. Soleimani, F., et al. (2015). General movements as a predictive tool of the neurological outcome in term born infants with hypoxic ischemic encephalopathy. *Early Human Development, 91*(8), 479–482.

110. Ferrari, F., et al. (2011). General movements in full-term infants with perinatal asphyxia are related to basal ganglia and thalamic lesions. *The Journal of Pediatrics, 158*(6), 904–911.
111. Glass, H. C., et al. (2011). Risk factors for epilepsy in children with neonatal encephalopathy. *Pediatric Research, 70*(5), 535–540.
112. Kwon, J. M., et al. (2011). Clinical seizures in neonatal hypoxic-ischemic encephalopathy have no independent impact on neurodevelopmental outcome: Secondary analyses of data from the neonatal research network hypothermia trial. *Journal of Child Neurology, 26*(3), 322–328.
113. Hellstrom-Westas, L., Rosen, I., & Swenningsen, N. W. (1985). Silent seizures in sick infants in early life. Diagnosis by continuous cerebral function monitoring. *Acta Paediatrica Scandinavica, 74*(5), 741–748.
114. Boylan, G. B., et al. (2002). Phenobarbitone, neonatal seizures, and video-EEG. *Archives of Disease in Childhood-Fetal and Neonatal Edition, 86*(3), F165–F170.
115. Scher, M. S., et al. (2003). Uncoupling of EEG-clinical neonatal seizures after antiepileptic drug use. *Pediatric Neurology, 28*(4), 277–280.
116. Hellstrom-Westas, L., Rosen, I., & Svenningsen, N. W. (1995). Predictive value of early continuous amplitude integrated EEG recordings on outcome after severe birth asphyxia in full term infants. *Archives of Disease in Childhood-Fetal and Neonatal Edition, 72*(1), F34–F38.
117. Eken, P., et al. (1995). Predictive value of early neuroimaging, pulsed Doppler and neurophysiology in full term infants with hypoxic-ischaemic encephalopathy. *Archives of Disease in Childhood-Fetal and Neonatal Edition, 73*(2), F75–F80.
118. al Naqeeb, N., et al. (1999). Assessment of neonatal encephalopathy by amplitude-integrated electroencephalography. *Pediatrics, 103*(6 Pt 1), 1263–1271.
119. Toet, M., et al. (1999). Amplitude integrated EEG 3 and 6 hours after birth in full term neonates with hypoxic–ischaemic encephalopathy. *Archives of Disease in Childhood-Fetal and Neonatal Edition, 81*(1), F19–F23.
120. Shalak, L. F., et al. (2003). Amplitude-integrated electroencephalography coupled with an early neurologic examination enhances prediction of term infants at risk for persistent encephalopathy. *Pediatrics, 111*(2), 351–357.
121. van Rooij, L. G. M., et al. (2005). Recovery of amplitude integrated electroencephalographic background patterns within 24 hours of perinatal asphyxia. *Archives of Disease in Childhood-Fetal and Neonatal Edition, 90*(3), F245–FF251.
122. Shany, E., et al. (2006). Predictive value of amplitude-integrated electroencephalography pattern and voltage in asphyxiated term infants. *Pediatric Neurology, 35*(5), 335–342.
123. Spitzmiller, R. E., et al. (2007). Amplitude-integrated EEG is useful in predicting neurodevelopmental outcome in full-term infants with hypoxic-ischemic encephalopathy: A meta-analysis. *Journal of Child Neurology, 22*(9), 1069–1078.
124. Wu, Y. W., et al. (2006). Cerebral palsy in a term population: Risk factors and neuroimaging findings. *Pediatrics, 118*(2), 690–697.
125. Miller, S. P., et al. (2005). Patterns of brain injury in term neonatal encephalopathy. *The Journal of Pediatrics, 146*(4), 453–460.
126. Barkovich, A. J., et al. (2006). MR imaging, MR spectroscopy, and diffusion tensor imaging of sequential studies in neonates with encephalopathy. *American Journal of Neuroradiology, 27*(3), 533–547.
127. Merhar, S. L., & Chau, V. (2016). Neuroimaging and other neurodiagnostic tests in neonatal encephalopathy. *Clinics in Perinatology, 43*(3), 511–527.
128. Rutherford, M., et al. (2010). Assessment of brain tissue injury after moderate hypothermia in neonates with hypoxic-ischaemic encephalopathy: A nested substudy of a randomised controlled trial. *Lancet Neurology, 9*(1), 39–45.
129. Shankaran, S., et al. (2012). Brain injury following trial of hypothermia for neonatal hypoxic-ischaemic encephalopathy. *Archives of Disease in Childhood-Fetal and Neonatal Edition, 97*(6), F398–F404.
130. Corbo, E. T., et al. (2012). The effect of whole-body cooling on brain metabolism following perinatal hypoxic-ischemic injury. *Pediatric Research, 71*(1), 85–92.

131. Inder, T. E., et al. (2004). Randomized trial of systemic hypothermia selectively protects the cortex on MRI in term hypoxic-ischemic encephalopathy. *The Journal of Pediatrics, 145*(6), 835–837.
132. Cheong, J. L., et al. (2012). Prognostic utility of magnetic resonance imaging in neonatal hypoxic-ischemic encephalopathy: Substudy of a randomized trial. *Archives of Pediatrics & Adolescent Medicine, 166*(7), 634–640.
133. Cowan, F., et al. (2003). Origin and timing of brain lesions in term infants with neonatal encephalopathy. *The Lancet, 361*(9359), 736–742.
134. Myers, R. E. (1972). Two patterns of perinatal brain damage and their conditions of occurrence. *American Journal of Obstetrics and Gynecology, 112*(2), 246–276.
135. Myers, R. E. (1975). Four patterns of perinatal brain damage and their conditions of occurrence in primates. *Advances in Neurology, 10*, 223–234.
136. Sie, L. T., et al. (2000). MR patterns of hypoxic-ischemic brain damage after prenatal, perinatal or postnatal asphyxia. *Neuropediatrics, 31*(3), 128–136.
137. Cowan, F., et al. (2003). Origin and timing of brain lesions in term infants with neonatal encephalopathy. *Lancet, 361*(9359), 736–742.
138. Rutherford, M. A., et al. (1998). Abnormal magnetic resonance signal in the internal capsule predicts poor neurodevelopmental outcome in infants with hypoxic-ischemic encephalopathy. *Pediatrics, 102*(2 Pt 1), 323–328.
139. Okereafor, A., et al. (2008). Patterns of brain injury in neonates exposed to perinatal sentinel events. *Pediatrics, 121*(5), 906–914.
140. Martinez-Biarge, M., et al. (2011). Predicting motor outcome and death in term hypoxic-ischemic encephalopathy. *Neurology, 76*(24), 2055–2061.
141. Steinman, K. J., et al. (2009). Neonatal watershed brain injury on magnetic resonance imaging correlates with verbal IQ at 4 years. *Pediatrics, 123*(3), 1025–1030.
142. Roland, E. H., et al. (1998). Perinatal hypoxic-ischemic thalamic injury: Clinical features and neuroimaging. *Annals of Neurology, 44*(2), 161–166.
143. Alderliesten, T., et al. (2011). MR imaging and outcome of term neonates with perinatal asphyxia: Value of diffusion-weighted MR imaging and (1)H MR spectroscopy. *Radiology, 261*(1), 235–242.
144. van Laerhoven, H., et al. (2013). Prognostic tests in term neonates with hypoxic-ischemic encephalopathy: A systematic review. *Pediatrics, 131*(1), 88–98.
145. Robertson, R. L., et al. (1999). MR line-scan diffusion-weighted imaging of term neonates with perinatal brain ischemia. *American Journal of Neuroradiology, 20*(9), 1658–1670.
146. Dag, Y., et al. (2006). Clinical outcomes of neonatal hypoxic ischemic encephalopathy evaluated with diffusion-weighted magnetic resonance imaging. *Diagnostic and Interventional Radiology, 12*(3), 109–114.
147. Miller, S. P., et al. (2002). Serial quantitative diffusion tensor MRI of the premature brain: Development in newborns with and without injury. *Journal of Magnetic Resonance Imaging, 16*(6), 621–632.
148. Drobyshevsky, A., et al. (2005). Developmental changes in diffusion anisotropy coincide with immature oligodendrocyte progression and maturation of compound action potential. *The Journal of Neuroscience, 25*(25), 5988–5997.
149. Prayer, D., et al. (2001). Visualization of nonstructural changes in early white matter development on diffusion-weighted MR images: Evidence supporting premyelination anisotropy. *American Journal of Neuroradiology, 22*(8), 1572–1576.
150. Hunt, R. W., et al. (2004). Apparent diffusion coefficient in the posterior limb of the internal capsule predicts outcome after perinatal asphyxia. *Pediatrics, 114*(4), 999–1003.
151. Liauw, L., et al. (2009). Do apparent diffusion coefficient measurements predict outcome in children with neonatal hypoxic-ischemic encephalopathy? *American Journal of Neuroradiology, 30*(2), 264–270.
152. Amess, P. N., et al. (1999). Early brain proton magnetic resonance spectroscopy and neonatal neurology related to neurodevelopmental outcome at 1 year in term infants after presumed hypoxic-ischaemic brain injury. *Developmental Medicine and Child Neurology, 41*(7), 436–445.

153. Hanrahan, J. D., et al. (1999). Relation between proton magnetic resonance spectroscopy within 18 hours of birth asphyxia and neurodevelopment at 1 year of age. *Developmental Medicine and Child Neurology, 41*(2), 76–82.
154. Boichot, C., et al. (2006). Term neonate prognoses after perinatal asphyxia: Contributions of MR imaging, MR spectroscopy, relaxation times, and apparent diffusion coefficients. *Radiology, 239*(3), 839–848.
155. Thayyil, S., et al. (2010). Cerebral magnetic resonance biomarkers in neonatal encephalopathy: A meta-analysis. *Pediatrics, 125*(2), e382–e395.
156. Bax, M., Tydeman, C., & Flodmark, O. (2006). Clinical and MRI correlates of cerebral palsy: The European cerebral palsy study. *JAMA, 296*(13), 1602–1608.
157. Nagy, Z., et al. (2005). Diffusion tensor imaging on teenagers, born at term with moderate hypoxic-ischemic encephalopathy. *Pediatric Research, 58*(5), 936–940.
158. Gadian, D. G., et al. (2000). Diffusion and perfusion magnetic resonance imaging in childhood stroke. *Journal of Child Neurology, 15*(5), 279–283.
159. Perlman, J. M. (1997). Intrapartum hypoxic-ischemic cerebral injury and subsequent cerebral palsy: Medicolegal issues. *Pediatrics, 99*(6), 851–859.
160. Hankins, G. D., & Speer, M. (2003). Defining the pathogenesis and pathophysiology of neonatal encephalopathy and cerebral palsy. *Obstetrics and Gynecology, 102*(3), 628–636.
161. Morgan, M. A., et al. (2005). Neonatal encephalopathy and cerebral palsy revisited: The current state of knowledge and the impact of American college of obstetricians and gynecologists task force report. *Journal of Perinatology, 25*(8), 519–525.
162. Philopoulos, D. (2012). Application of criteria developed by the task force on neonatal encephalopathy and cerebral palsy to acutely asphyxiated neonates. *Obstetrics and Gynecology, 119*(5), 1056. author reply 1056-7.
163. Strijbis, E. M., et al. (2006). Cerebral palsy and the application of the international criteria for acute intrapartum hypoxia. *Obstetrics and Gynecology, 107*(6), 1357–1365.
164. Dixon, G., et al. (2002). Early developmental outcomes after newborn encephalopathy. *Pediatrics, 109*(1), 26–33.
165. Barnett, A., et al. (2002). Neurological and perceptual-motor outcome at 5 - 6 years of age in children with neonatal encephalopathy: relationship with neonatal brain MRI. *Neuropediatrics, 33*(5), 242–248.
166. Azzopardi, D., et al. (2008). The TOBY study. Whole body hypothermia for the treatment of perinatal asphyxial encephalopathy: A randomised controlled trial. *BMC Pediatrics, 8*, 17.
167. Zhou, W. H., et al. (2010). Selective head cooling with mild systemic hypothermia after neonatal hypoxic-ischemic encephalopathy: a multicenter randomized controlled trial in China. *The Journal of Pediatrics, 157*(3), 367–372. 372.e1-3.
168. Simbruner, G., et al. (2010). Systemic hypothermia after neonatal encephalopathy: Outcomes of neo.nEURO.network RCT. *Pediatrics, 126*(4), e771–e778.
169. Lin, Z. L., et al. (2006). Mild hypothermia via selective head cooling as neuroprotective therapy in term neonates with perinatal asphyxia: An experience from a single neonatal intensive care unit. *Journal of Perinatology, 26*(3), 180–184.
170. Kyriakopoulos, P., et al. (2013). Term neonatal encephalopathy antecedent cerebral palsy: A retrospective population-based study. *European Journal of Paediatric Neurology, 17*(3), 269–273.
171. Rosenbloom, L. (1994). Dyskinetic cerebral palsy and birth asphyxia. *Developmental Medicine and Child Neurology, 36*(4), 285–289.
172. Stanley, F. J., et al. (1993). Spastic quadriplegia in Western Australia: A genetic epidemiological study. I: Case population and perinatal risk factors. *Developmental Medicine and Child Neurology, 35*(3), 191–201.
173. Marlow, N., et al. (2005). Neuropsychological and educational problems at school age associated with neonatal encephalopathy. *Archives of Disease in Childhood-Fetal and Neonatal Edition, 90*(5), F380–F387.
174. van Handel, M., et al. (2007). Long-term cognitive and behavioral consequences of neonatal encephalopathy following perinatal asphyxia: A review. *European Journal of Pediatrics, 166*(7), 645–654.

175. Shankaran, S., et al. (1991). Acute neonatal morbidity and long-term central nervous system sequelae of perinatal asphyxia in term infants. *Early Human Development, 25*(2), 135–148.
176. Van Hof-van Duin, J., & Mohn, G. (1984). Visual defects in children after cerebral hypoxia. *Behavioural Brain Research, 14*(2), 147–155.
177. Mercuri, E., et al. (1997). Basal ganglia damage and impaired visual function in the newborn infant. *Archives of Disease in Childhood-Fetal and Neonatal Edition, 77*(2), F111–F114.
178. Mercuri, E., et al. (1997). Visual function in full-term infants with hypoxic-ischaemic encephalopathy. *Neuropediatrics, 28*(3), 155–161.
179. Lindstrom, K., et al. (2006). Teenage outcome after being born at term with moderate neonatal encephalopathy. *Pediatric Neurology, 35*(4), 268–274.
180. Dilenge, M. E., Majnemer, A., & Shevell, M. I. (2001). Long-term developmental outcome of asphyxiated term neonates. *Journal of Child Neurology, 16*(11), 781–792.
181. Gonzalez, F. F., & Miller, S. P. (2006). Does perinatal asphyxia impair cognitive function without cerebral palsy? *Archives of Disease in Childhood-Fetal and Neonatal Edition, 91*(6), F454–F459.
182. Moster, D., Lie, R. T., & Markestad, T. (2002). Joint association of Apgar scores and early neonatal symptoms with minor disabilities at school age. *Archives of Disease in Childhood-Fetal and Neonatal Edition, 86*(1), F16–F21.
183. Robertson, C. M., & Finer, N. N. (1988). Educational readiness of survivors of neonatal encephalopathy associated with birth asphyxia at term. *Journal of Developmental and Behavioral Pediatrics, 9*(5), 298–306.
184. Badawi, N., et al. (2006). Autism following a history of newborn encephalopathy: More than a coincidence? *Developmental Medicine and Child Neurology, 48*(2), 85–89.
185. Lindström, K., et al. (2006). Teenage outcome after being born at term with moderate neonatal encephalopathy. *Pediatric Neurology, 35*(4), 268–274.
186. Guillet, R., et al. (2012). Seven- to eight-year follow-up of the CoolCap trial of head cooling for neonatal encephalopathy. *Pediatric Research, 71*(2), 205–209.
187. Shankaran, S., et al. (2005). Whole-body hypothermia for neonates with hypoxic-ischemic encephalopathy. *The New England Journal of Medicine, 353*(15), 1574–1584.
188. Shankaran, S., et al. (2012). Childhood outcomes after hypothermia for neonatal encephalopathy. *The New England Journal of Medicine, 366*(22), 2085–2092.
189. Azzopardi, D., et al. (2014). Effects of hypothermia for perinatal asphyxia on childhood outcomes. *The New England Journal of Medicine, 371*(2), 140–149.
190. Bonifacio, S. L., et al. (2011). Perinatal events and early magnetic resonance imaging in therapeutic hypothermia. *The Journal of Pediatrics, 158*(3), 360–365.

Genetic Considerations in Infants with Congenital Anomalies

Elizabeth A. Sellars and G. Bradley Schaefer

Abstract Congenital anomalies or birth defects account for one of the leading causes of death in infants. When one anomaly is present, there is a 50% risk that other anomalies are present. Healthcare providers taking care of newborns should have a low threshold to look for additional anomalies in the presence of one. These anomalies may not always be identified in the hospital but may require evaluation after discharge. Close attention to the type and pattern of anomalies also provides clues to the underlying diagnosis. Genetic consultation is critical to help identify patterns and guide appropriate testing, especially in this time of ever-changing and complex diagnostic options.

Introduction

While 3–4% of babies born in the United States and Europe are found to have a major birth defect, only a subset of these infants are diagnosed with a genetic syndrome prior to discharge from the newborn nursery or NICU. Therefore, appropriate genetic follow-up is imperative for continued evaluation to work toward a diagnosis. In many cases, a congenital anomaly may not come to light in the immediate days or months after birth but may take time to be identified. It often takes just one anomaly to create a "light-bulb" moment for the proper genetic testing.

In cases where a genetic diagnosis is successfully made, follow-up is equally important in order to provide disease-specific management recommendations or to offer genetic testing in the cases in which molecular confirmation was not able to be made initially.

E. A. Sellars · G. Bradley Schaefer (✉)
Department of Pediatrics, Section of Genetics and Metabolism,
University of Arkansas for Medical Sciences, Rock, AR, USA
e-mail: SchaeferGB@uams.edu

H. Needelman, B. J. Jackson (eds.), *Follow-Up for NICU Graduates*,
https://doi.org/10.1007/978-3-319-73275-6_9

Congenital Anomalies

The vernacular term for congenital anomalies is "birth defects". Either term simply implies that something is not right at birth. These anomalies can be structural or functional. By convention, and for the purposes of this article, we will limit our discussions to structural anomalies. These can be identified during pregnancy, at birth, or later in life. They can also be a common cause of fetal deaths. The World Health Organization reports that 50% of all congenital anomalies cannot be linked to a specific cause. For many others, there are known etiologic or risk factors. Congenital anomalies can be genetic, infectious, nutritional, or environmental in origin—or some combination of these factors called multifactorial. When risk factors are identified, measures to focus on prevention can be undertaken. Examples of prevention include vaccinations, folic acid intake, and prenatal care.

The March of Dimes reports up to a 6% worldwide population risk of a major congenital anomaly in all live born infants regardless of race, location, socioeconomic status, or other demographic characterization (http://www.marchofdimes.org). There are no identified factors that can lower this baseline statistical risk. Additionally, it is thought that when one anomaly is present, there is a 50% risk that a second anomaly is present [1]. Of course, not all anomalies are obvious at birth. Therefore, there should be a low threshold to "search" for a second anomaly when one is present. This heightened awareness of the potential for additional congenital anomalies extends to the NICU graduate.

Because congenital anomalies can place a heavy burden on individuals, families, healthcare resources, and societies at large, many state health departments have birth defect registries in place to track the number and type of congenital anomalies. Since these registries have been in place for a decade or longer now, in many cases, trends can be appreciated. For instance, neural tube defects have been declining and congenital heart defects have been increasing.

With improved access to prenatal care, many birth defects are noted prenatally on a routine basis. In many cities with neonatology services, a neonatologist is "consulted" during pregnancy in order to prepare the family for any immediate needs the baby may have. For example, many congenital heart anomalies require early imaging and initiation of prostaglandins, and families can be informed of this process. Some birth centers have entire teams for fetal care that may include obstetrics, neonatology, radiology, and subspecialty services on a case-by-case basis. Additionally, the type of congenital anomaly may dictate the method of delivery. For instance, in many cases of myelomeningocele, a cesarean section is recommended to prevent rupture of any exposed membranes. In fact, many children's hospitals are creating fetal care centers where mothers can deliver babies and the babies can get immediate surgery or care required without transport. Even if a fetal care center does not exist in a given area, it is highly advisable for any child with major congenital anomalies to be born at a medical center with access to an NICU and neonatology services.

Types of Congenital Anomalies

Congenital anomalies are classified by the underlying pathologic mechanism. Four separate mechanisms have been described: malformations, deformations, disruptions, and dysplasias. The identification of congenital anomalies is not limited to the NICU setting. In fact, in many instances anomalies come to light after an infant is discharged and the immediate issues related to prematurity resolve.

A *malformation* is a morphologic defect of an organ, part of an organ, or larger region of the body resulting from an *intrinsically* abnormal developmental process. In other words, the tissue did not form correctly from the start. The majority of malformations occur in the first trimester of pregnancy. Physicians often classify malformations as "major" or "minor." Essentially, a major malformation is one in which there are significant clinical or functional implications. A "minor" malformation is one in which morbidity is not increased.

Malformations can arise from agenesis or hypogenesis of a given structure or tissue. There can also be abnormal migration of cells (as in the case of heterotopia) or whole organs (as in ectopia). Examples of these can be gray matter heterotopia seen on brain MRI or ectopic spleens noted with abdominal imaging.

Many people have minor malformations that have no functional impact. Yet minor malformations often can act as clues to search for a certain major malformation. For example, preauricular pits or dimples are minor malformations for many. While this small anomaly typically has no direct health impact on an infant, preauricular pits can signal a risk for hearing impairment or renal anomalies. Therefore, it is often recommended that any individual with three or more minor malformations have a formal evaluation for a major malformation. In the case of preauricular pits, a renal ultrasound and hearing screen should be done.

Deformations are another type of congenital malformation due to mechanical forces applied to a normally developing structure. This can include any abnormal form, shape, or position of a part of the body. The tissue is programmed normally, but external forces apply pressure. The forces can be either maternal or fetal in origin and apply direct pressure or add a constraint.

An example of a deformation due to a maternal mechanical force is uterine constraint. Uterine tumors or fibroids, a bicornate uterus, and multiple gestation pregnancy would all put a pregnancy at risk for deformations in the fetus because of fetal constraint. Examples of deformations from these etiologies would include torticollis, overlapping toes, and club foot.

A *disruption* is any morphologic defect of a tissue, organ, or larger region resulting from an extrinsic breakdown of tissue. This is a loss of normally developing cells or tissue due to any method of insult not intrinsic to the tissue.

A commonly cited example of this is amniotic band syndrome in which thin strands of tissue in the amniotic fluid impose pressure on normally developing fetal tissue. This can cause amputation of fingers, toes, and limbs. It can also cause facial clefts in severe cases. Another example of a disruption is a vascular occlusion in the developing fetus that is thought to cause gastroschisis in many instances.

The term *dysplasia* typically refers to abnormal organization of cells into tissues. This abnormality occurs later during development of the fetus and is independent of morphogenesis. Interestingly, tissues that are dysplastic in nature may predispose that tissue to cancer later in life.

A good example of a genetic syndrome that has dysplastic tissue is Beckwith-Wiedemann syndrome. This syndrome is classified as an overgrowth syndrome where parts of the body may grow asymmetrically, leading to one part being larger than the other (i.e., an arm may measure larger in circumference than the contralateral arm). Because there is a risk of nests of dysplastic fetal tissue being present in normally formed tissue, children can develop certain tumors such as Wilms tumors and hepatoblastoma. That is why there are specific recommendations for screening in childhood with frequent alpha-fetoprotein (AFP) and abdominal/kidney ultrasound imaging.

Recognizing patterns of congenital anomalies is very important to understanding the potential underlying cause. Many times, it takes several evaluations and examinations to identify additional congenital anomalies. It is important to maintain a low suspicion for a second anomaly. Recall that there is up to a 50% risk that a person has one or more additional anomalies in the presence of one congenital anomaly. Therefore, repeated evaluation by a geneticist may be necessary to put the clues together into a pattern. And these evaluations may very well be after discharge from the NICU. It is thought that 10–20% of birth defects have an identifiable pattern [2].

Patterns of Congenital Anomalies

In general, a genetic evaluation begins with early documentation of physical characteristics. Photographs may be taken to see how physical features change over time. Imaging tests are often of upmost importance in order to gather information. Skeletal surveys, echocardiograms, brain and kidney imaging are quite common to look for congenital anomalies and to build a full picture of the disorder. While much of this information is ideally gathered during the NICU stay, there are certainly many circumstances in which a study giving a critical clue is not done until after discharge. This could include things like developmental or behavioral clues not present in a neonate or physical features not easily seen in a newborn (e.g., a small coloboma of the retina seen by an ophthalmologist years later).

Anthropometric measurements are often very helpful. Specific conditions often have characteristic growth patterns. Doing standard growth measurements and then comparing these to established standards to determine normal versus abnormal growth are one of the cheapest and yet informative sets of clues in the diagnostic evaluation. Height, weight, and head circumference are relevant just as growth rates are critical. Growth rates may not be fully appreciated until a baby is discharged from the NICU. Skeletal proportionality may not be obvious until an infant is several months old.

Patterns of anomalies are grouped by common pathogenesis. A *sequence* is a pattern of multiple anomalies derived from a single known (or presumed) prior anomaly or mechanical factor. What follows is essentially a cascade of events arising from the original single anomaly. A classic example is the Robin sequence. The primary anomaly in this condition is a hypoplastic mandible. This anomaly is shared in all cases of Robin sequence. If the mandible is significantly small enough, it can posteriorly displace the tongue. Posterior displacement of the tongue prior to 9 weeks gestation (when the lateral palatine processes move across the midline to form the primary palate) can then actually impair closure of the palate by a direct mechanical obstruction in some cases. These anomalies together make up the Robin sequence.

An *association* is a nonrandom occurrence of anomalies that happen together more often than by chance. An association refers solely to statistically related anomalies (not pathogenetically or causally related). A commonly described example is VACTERL association which is a group of anomalies that are found together too often than by chance. This is an association of *v*ertebral anomalies, *a*nal atresia, *c*ardiac malformation, *t*racheal-*e*sophageal fistula, *r*enal anomalies, and *l*imb anomalies. It is important to realize that not every patient has all of these findings. It is often cited that to have a VACTERL association diagnosis, a patient has to have three characteristics or more.

Finally, *syndromes* are patterns of multiple anomalies of more than one organ system with a common etiology. Down syndrome is a well-known example. This syndrome can affect a number of organ systems such as the heart, thyroid, vision, and hearing. It is caused by an extra copy of chromosome 21—though that "extra" genetic material may be arranged in several ways. The most common arrangement is a full extra copy of chromosome 21. There are thousands of genetic syndromes and new ones discovered with greater frequency as genetic testing improves.

Newborn Screening

The purpose of newborn screening is to detect potentially fatal conditions or conditions that may affect a child's long-term health. Early detection provides a window for treatment in order to prevent death or disability. Ideally, treatment or intervention can begin before the infant displays any signs or symptoms, enabling the child to reach full potential. Millions of babies are screened each year in the United States simply with a few drops of blood from the newborn's heel. The American College of Medical Genetics recommends that states screen for 34 core disorders encompassing genetic, endocrine, and metabolic diseases. Hearing loss and congenital heart disease are also included as part of newborn screening.

Certainly, a subset of admissions to the NICU are due to the congenital disorders included on the newborn screen. In some cases, the infant becomes symptomatic prior to having results from the newborn screen. Therefore, management in the NICU centers around the diagnosis and initiation of management. As such, a NICU graduate with any disorder identified from newborn screening should either have the appropriate diagnosis at discharge or the appropriate follow-up testing in place. If there is a diagnosis, the infant should only be allowed to go home with an immediate management plan in place. These babies will need long-term follow-up by the appropriate physician. Often, this is a metabolic geneticist working closely with a metabolic nutritionist. In the first few years of life, the infant may need regular and frequent visits for diet management and growth monitoring. Frequent labwork is often necessary as well.

In the classic example of phenylketonuria (PKU), initiation of treatment should be undertaken as early as possible, ideally within the first 2 weeks of life. In order to achieve this goal, early and appropriate newborn screening should be performed, follow-up and diagnostic testing completed, and communication between family and caregivers established. As is the case of many inborn errors of metabolism, a special formula is needed. In this case, the formula has reduced or no amounts of the essential amino acid phenylalanine. The first few weeks and months require close monitoring of growth and blood phenylalanine levels. This is something that will continue beyond the NICU stay (http://www.acmg.net/docs/ACMG_PAHPKU_Guidelines_GIM_Feb2014.pdf).

Genetic Testing

The world of genetic testing is an ever-changing field with advancements in technology and interpretation of information. One of the most significant roles of a genetic consultant in the NICU, or any setting, is to help navigate this complex field. While we are working toward fast and comprehensive genetic testing, this is not yet universally available or affordable at this time. Therefore, genetic testing to obtain a diagnosis may still require careful assessment of the patient over time. A NICU graduate may need close follow-up with genetic consultants in order to continue to pursue genetic testing. It is the case for many NICU graduates that the genetic test that will provide the diagnosis was not available at the time of birth but became available months or years later.

As it stands, there is not a single test that is appropriate for all patients and situations. A karyotype is still the preferred genetic test to evaluate for Down syndrome or any disorder where aneuploidy is suspected. Gene sequencing is still the first tier for achondroplasia and Marfan syndrome or any instance where a specific syndrome is in question and the causative gene(s) is known.

Chromosomal microarrays are often the first-tier screening approach in many centers. This test utilizes microarray technology to scan the whole genome for micro-duplications and micro-deletions. Advances in this technology have allowed for identification of smaller and smaller micro-duplications and micro-deletions. Further refinements are still being developed. A microarray is a first-tier test for any infant or child with multiple congenital anomalies if the pattern of anomalies does not fit a specific syndrome with a specific test [3]. Even with advancements in this technology, this test still requires time for interpretation since so much data is being reviewed.

One of the more exciting fields of genetic testing is whole genome sequencing (WGS) and whole exome sequencing (WES). WES provides information on all the known coding regions (exomes) of the human genome. Exomic sequences account for 1–2% of the entire human genome and are considered to harbor the majority of pathogenic mutations. Whole genome sequencing is sequencing the entire 3 billion nucleotide base pairs in the human genome. Certainly it stands to reason that a subset of pathologic diseases are caused by mutations in a coding region outside of what is obtained in WES. The major hurdles in WGS is the increased cost in sequencing much more data, the ability to interpret this additional data, and the time to obtain results.

With the ever-expanding list of genetic tests that are available comes an ever more complex decision-making process. The question for the geneticist and non-geneticist alike is "What test do we order"? The complete answer to this question is beyond the scope of this review. However, several overarching principles can be discussed:

1. Cytogenetic studies (karyotype) are used much less often than previously. The major reasons for considering doing a karyotype would include:
    (a) Confirming a suspected chromosome aneuploidy (e.g., Down syndrome)
    (b) Determining the number of X and Y chromosomes which is the first bit of information needed in the workup of disorders of sexual differentiation
    (c) Checking the infant for a known familial chromosome abnormality
2. Single-locus FISH studies are also utilized much less often. They are still indicated if a clinical diagnosis is strongly suspected. For instance, if the child has all of the right features that suggest Williams syndrome, the 7p FISH can be ordered. The downside of FISH testing is that you only get a yes or no answer on the specific condition you are testing for. As the cost of chromosomal microarray has come down, it is usually better to order this instead of just FISH. That is because microarray testing will detect what any single-locus FISH test would detect, plus a whole lot more.
3. Chromosomal microarray testing currently is the most common first test to order in children with congenital anomalies. It is recognized as a true "first-tier" test in this regard.
4. Single gene sequencing is similar to FISH testing. If a specific condition is suspected that is associated with a mutation in a single gene, sequencing can be accomplished. For example, in a child with a pattern of anomalies suggestive of CHARGE syndrome, CHD7 gene sequencing could be the first test performed.
5. Next-generation sequencing (NGS) panels are becoming utilized much more often. These panels use rapid sequencing technologies to sequence multiple genes at the same time. Most NGS panels are organized in diagnostic categories. So, for instance, a child identified with a hearing loss on newborn screening may have a NGS "hearing loss panel" ordered. The standard NGS panel for hearing loss is about 120 of the most common genes known to cause hearing loss.
6. Genome-wide testing such as WES or WGS as described above currently is not used as first-level testing. However, as noted above, it may not be long before these tests supplant other tests and become a true first-tier recommendation. It is

likely not going too long before genome-wide testing becomes part of newborn screening! Currently studies are being conducted to look at the yield and ethics of such inititatives.

Genetic testing has been considered expensive and results too slow to return in order to be useful for actual management of NICU patients. Therefore, acute management decisions are often made without a diagnosis. This can lead to delays in appropriate evaluations or treatments. Since the progress of disease can be quite rapid in infants with congenital anomalies, diagnosis of any underlying genetic etiology must be timely in order to consider appropriate interventions. One of the new horizons in genetic testing is to offer rapid WGS with a turnaround time down to a matter of days [4–7]. There are places, currently, that are performing trials using this technology for acute management purposes, reporting a 26-h turnaround and over 50% diagnostic yield [4, 5, 7]. The hope is to decrease infant mortality related to a genetic disease and facilitate informed parental decision-making. While the cost of this testing may seem prohibitive, one may argue cost savings if this allows a rapid move to end-of-life care or treatment that allows an infant to move to discharge quicker.

What this means for the NICU graduate is that he or she may be part of a new horizon in genetic diagnostics. A baby may leave the NICU with a genetic diagnosis in which there is little information. We are frequently encountering test results in which the phenotype of a disease seems to lie on a continuum and boundaries between diseases are blurred. Individuals have features that overlap several diseases that were once thought distinct entities but now are known to share a common genetic cause. As a member of a medical team, it is important to help families obtain good information. The goal in genetic diagnostics is often to empower patients and families to advocate for themselves. Unfortunately, while having a genetic diagnosis is useful in many ways, there remains few treatment options specific to most genetic diagnoses. At this time, there is no way to correct a genetic abnormality.

Genetic testing has its own set of ethical issues. The best interests of the patient should always come first. It is now very standard across all genetic tests to provide informed consent. Many commercial labs have informed consent forms that must be submitted along with a specimen. Many genetic tests performed on site also have institution-specific consent forms. Several key pieces of information in the consent include discussing information related to a later-onset disease (presymptomatic testing) or information on a disease that is unrelated to the presenting symptoms. Many commercial labs allow a family or patient to "opt out" of receiving unrelated information. Informed consent should also point out diagnostic yield—or the likelihood that any given test will provide a diagnostic answer. This varies by test, and despite the great improvements in our current testing options, an answer is not always achieved. The bottom line is that genetic testing is deeply personal and can have implications for an entire family. It cannot be performed lightly or without consent from the family.

It is easy to see from these discussions that genetic testing is a complex process and part of an ever-evolving landscape. The best advice is to use genetic consultants liberally.

Outcomes

In many ways each child born with congenital anomalies or a genetic syndrome is unique. Depending on the specific anomalies or syndrome, the outcomes will vary. Structural anomalies such as gastroschisis can be repaired with surgery. Congenital hypothyroidism can be treated with medication. But infants born with the majority of congenital anomalies have an increased risk of morbidity and mortality. In the United States, at least 20% of infant deaths are caused by congenital anomalies and chromosome anomalies [6]. Many infants do not graduate from the NICU if the anomalies are incompatible with life. Others, who do get discharged from the NICU, do not make it past the first few weeks, months, or years.

One of the goals of rapid genetic testing (WES/WGS) is to get a diagnosis in order to provide prognostic information when possible. In one study of rapid WGS, palliative care was instituted more often in infants with a genetic diagnosis than in those without [5]. Diseases with a poor prognosis may allow families to feel more comfortable taking their child home for however long that may be.

For many families, taking their child home with palliative care does not mean that they do not require or wish for regular medical follow-up. Historically, the thinking has been that children in hospice will stop receiving preventative care or minor treatments. This thinking has evolved. Many children live much longer than may have been anticipated. It is important to have regular follow-up in order to modify treatment plans as indicated. For instance, medication dosages may need to be adjusted, feeding regimens may change, or small infections addressed.

While long-term outcomes for many congenital anomalies and genetic syndromes have not changed, early diagnosis and medical management of certain anomalies have. Congenital anomalies and prematurity are the leading causes of death among infants. The fiscal impact of birth defects is staggering [1]. Therefore, it is important to seek a genetic diagnosis in order to expedite appropriate treatment and to empower families to make the best possible decisions for their child.

References

1. Schaefer, G., & Thompson, J. (Eds.). (2014). *Medical genetics: An integrated approach* (1st ed.). New York: McGraw-Hill Education.
2. Carmichael, S. L. (2014). Birth defects epidemiology. *European Journal of Medical Genetics, 57*, 355–358.
3. Kearney, H. M., South, S. T., Wolff, D. J., Lamb, A., Hamosh, A., Rao, K. W., & Working Group of The American College of Medical Genetics. (2011). American College of Medical Genetics recommendations for the design and performance expectations for clinical genomic copy number microarrays intended for use in the postnatal setting for detection of constitutional abnormalities. *Genetics in Medicine, 13*, 676–679.
4. Miller, N. A., Farrow, E. G., Gibson, M., Willig, L. K., Twist, G., Yoo, B., Marrs, T., Corder, S., Krivohlavek, L., Walter, A., Petrikin, J. E., Saunders, C. J., Thiffault, I., Soden, S. E., Smith, L. D., Dinwiddie, D. L., Herd, S., Cakici, J. A., Catreux, S., Ruehle, M., & Kingsmore, S. F.

(2015). A 26-hour system of highly sensitive whole genome sequencing for emergency management of genetic diseases. *Genome Medicine, 7*, 100.
5. Petrikin, J. E., Willig, L. K., Smith, L. D., & Kingsmore, S. F. (2015). Rapid whole genome sequencing and precision neonatology. *Seminars in Perinatology, 39*, 623–631.
6. Saunders, C. J., Miller, N. A., Soden, S. E., Dinwiddie, D. L., Noll, A., Alnadi, N. A., Andraws, N., Patterson, M. L., Krivohlavek, L. A., Fellis, J., Humphray, S., Saffrey, P., Kingsbury, Z., Weir, J. C., Betley, J., Grocock, R. J., Margulies, E. H., Farrow, E. G., Artman, M., Safina, N. P., Petrikin, J. E., Hall, K. P., & Kingsmore, S. F. (2012). Rapid whole-genome sequencing for genetic disease diagnosis in neonatal intensive care units. *Science Translational Medicine, 4*, 154ra135.
7. Willig, L. K., Petrikin, J. E., Smith, L. D., Saunders, C. J., Thiffault, I., Miller, N. A., Soden, S. E., Cakici, J. A., Herd, S. M., Twist, G., Noll, A., Creed, M., Alba, P. M., Carpenter, S. L., Clements, M. A., Fischer, R. T., Hays, J. A., Kilbride, H., McDonough, R. J., Rosterman, J. L., Tsai, S. L., Zellmer, L., Farrow, E. G., & Kingsmore, S. F. (2015). Whole-genome sequencing for identification of Mendelian disorders in critically ill infants: A retrospective analysis of diagnostic and clinical findings. *Lancet Respiratory Medicine, 3*, 377–387.

Environmental Risks to NICU Outcomes

Douglas Vanderbilt, Christine Mirzaian, and Kelly Schifsky

Abstract Environmental exposures have a significant impact on high-risk infant (HRI) outcomes. While prematurity and medical severity are well understood in terms of infant survival, it is critical that clinicians understand and appreciate the impact that biological, psychological, and social risk factors have on an infant's future functioning. Taking these risk factors into a biopsychosocial framework, the chapter proposes a life course theory perspective of timeline, timing, environment, and equity to explain HRI outcomes. Specifically this chapter explores how exposures such as in utero drug exposure, parental distress and posttraumatic stress, and low socioeconomic status and low caregiver education level negatively impact infant neurodevelopment. An emphasis is placed on the strain that each of these exposures places on the caregiver-infant relationship, which is the foundation for supporting an infant's development and regulation. In discussing ways to address these environmental risk factors, methods such as breastfeeding, mental health support for NICU parents, and interdisciplinary HRI follow-up clinics are discussed and how each intervention is positioned to uniquely support this caregiver-infant dyad.

Introduction

In this chapter we will explore the environmental influences on high-risk infants (HRI), namely, those born premature or with medical risk requiring a neonatal intensive care unit (NICU) admission. The environment can be defined as factors

---

D. Vanderbilt (✉) · C. Mirzaian
Children's Hospital of Los Angeles, University of Southern California,
Los Angeles, CA, USA

Keck School of Medicine of the University of Southern California, Los Angeles, CA, USA
e-mail: dvanderbilt@chla.usc.edu

K. Schifsky
Children's Hospital of Los Angeles, University of Southern California,
Los Angeles, CA, USA

H. Needelman, B. J. Jackson (eds.), *Follow-Up for NICU Graduates*,
https://doi.org/10.1007/978-3-319-73275-6_10

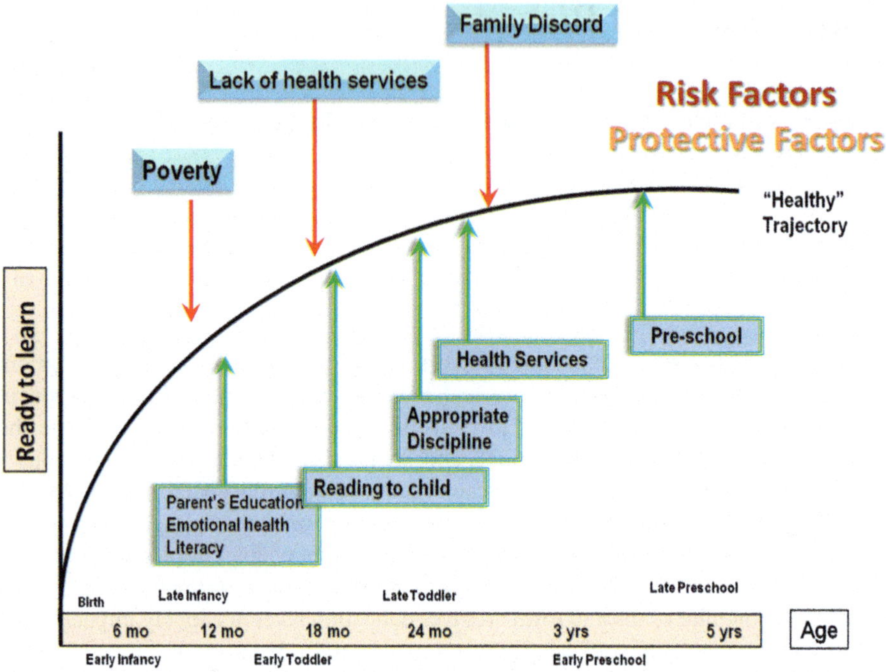

Fig. 1 Example of life course theory (CityMatCH Life Course Game [97])

that may be biological (e.g., teratogens, infections), psychological (e.g., maternal depression, traumatic stress), or social (e.g., poverty, racial health disparities). In understanding these influences, the best conceptualization of how environment affects later functioning is the life course theory (LCT) [1, 2] (see Fig. 1). This explanatory theory is summarized in these four tenets:

- "Today's experiences and exposures influence tomorrow's health. (Timeline)
- Health trajectories are particularly affected during critical or sensitive periods. (Timing)
- The broader community environment—biologic, physical, and social—strongly affects the capacity to be healthy. (Environment)
- While genetic make-up offers both protective and risk factors for disease conditions, inequality in health reflects more than genetics and personal choice. (Equity)" ([1], p. 4)

As applied to HRI, these risks can occur preconceptionally, prenatally, perinatally, or postnatally, exerting different effects depending on the susceptibility of the mother, fetus, or infant at each sensitive period. Each exposure risk then impacts the infant going forward in time to affect the physical and mental health and developmental outcomes. These biological, psychological, and social factors interact together in different ways to explain the variance in HRI outcomes for the same given medical risk such as gestational age.

One example of how LCT works in the setting of HRI is through the new focus on toxic stress [3]. Here, negative environmental exposures, known as adverse childhood experiences (ACEs), have been shown to lead to worse mental and physical health [4, 5]. The excessive stress activation seen in the setting of toxic stress is the mechanism for these experiences to create poorer outcomes. This occurs through stress shifting the biological and behavioral regulation systems to short-term survival rather than long-term development [6]. Effects are seen in epigenetic alterations, brain architecture, and chronic changes in immunity, metabolism, behavior regulation, cardiovascular function, and general growth [7, 8].

The constellation of risk may affect those with medical risk more intensely. These ACEs contribute to the toxicity of the most pervasive social risk factor of poverty on limiting the outcomes of those with medical risk [9]. The Developmental Systems Approach is a framework which posits that these risk and protective factors interact at multiple hierarchical levels to influence child development [10]. Taking a Developmental Systems Approach, these infants with medical risk have existing challenges at the level of (1) child social and cognitive competence which interacts with environmental risk at the levels of (2) family patterns of interaction and (3) family resources over time [11, 12]. Due to HRI having a relative vulnerability at the level of social and cognitive competence from their medical risk, they may be uniquely susceptible to the effects of toxic stress and social risk. These social risks affect the levels of family patterns of interaction and family resources which depress the full potential of the HRI.

Mechanism of Environmental Risk

It is well known that brain injury, through teratogenic and medical illness exposures and interruption of the supportive intrauterine environment, is a key driver of high-risk infant outcomes [13]. But if medical or biological risk predetermined outcomes, then environmental risk would not be expected to be a driver of outcomes. Thus, the role of the NICU follow-up clinic would be only to identify and accommodate the eventual or expected developmental delays. Instead, an ongoing process of mediation and modification occurs that enhances or limits the biological potential of the infant's development. The key to this modification is the parent-child relationship. Equally critical as the perspective of genetically programed and biologically constrained neuromaturational development, the environmental context is the process of co-regulation which empowers both the caregiver and child with an active process of shaping each other's responses [14]. Optimal co-regulation enhances social and cognitive competence and learning for the infant which is critical for promoting infant self-regulation for feeding, sleeping, etc. The key promoting qualities of the dyad are sensitivity and responsivity of the caregiver to the child [15, 16]. A positive dyadic relationship promotes academic and behavioral skills among preterm infants [17]. Yet, these infants often still have less optimal relationship patterns [18] despite preterm birth mothers having equal levels of sensitivity [19]. Due to the enhanced medical risk, HRI require more, not less, sensitive and responsive caregiving than infants with typical gestational age and development to achieve the same level of functioning [20]. Environmental stress from preterm birth disrupts maternal cortisol and oxytocin regulation which directly affects sensitive and responsive caregiving [21, 22]. Thus, the parent-child relationship is the common pathway for embedding social environmental risk into HRI and therefore the focus for mitigating the medical risk to long-term outcomes.

Now for the remaining sections, we can turn to examine specific environmental factors in the domains of biological, psychological, and social risks that impact the parent-child relationship.

Biological Risks

Specific biological risks, such as prenatal exposures, can impact the neurodevelopment of HRI. Many of these risks not only contribute to the cause of preterm or high-risk births but to the neurodevelopmental outcome of infants once they are born preterm or otherwise high-risk. Complicating this picture is that many of these risks co-occur with other prenatal exposures or social conditions, making isolation of the effects of single risk factors very challenging. In this section we will review the common prenatal substance exposures that could impact a high-risk infant while keeping in mind that that they rarely occur in isolation to ongoing psychological and social risk factors.

Studies have shown that cocaine use during pregnancy is strongly associated with preterm birth. In a recent meta-analysis, women who used cocaine during pregnancy had a threefold increased risk of delivering a preterm infant than those who did not use cocaine in pregnancy [23]. Women who use cocaine during pregnancy have other lifestyle factors that can adversely affect the pregnancy and later outcomes such as cigarette smoking, polysubstance (including alcohol) use, poor nutrition, lack of prenatal care, higher rates of infections, exposure to violence, low socioeconomic status, and emotional problems [24, 25]. These factors can then negatively impact the neurodevelopment of HRI via mechanisms described above. Although many studies between prenatal cocaine use and poor neurodevelopmental outcomes were attenuated when models included conditions that commonly co-occur with prenatal cocaine use [23, 26, 27], there is evidence that after controlling for these factors, there are impairments in sustained attention and behavioral self-regulation among school-aged children with prenatal cocaine exposure [28].

The adverse effects of opioid use during pregnancy are similarly confounded by additional environmental factors including polysubstance use, low socioeconomic status, and higher rates of psychiatric, nervous, and emotional disorders [24]. Even when controlling for these factors, opioid use during pregnancy is associated with adverse pregnancy outcomes including intrauterine growth restriction, oligohydramnios, preterm labor, and premature rupture of membranes [29]. Neonatal abstinence syndrome (NAS) , marked by withdrawal symptoms in the infant born to a mother who used opioids during pregnancy, is a well-known syndrome marked by a high-pitched cry, irritability, sleep-wake disturbances, alterations in tone or movement, feeding difficulties, gastrointestinal disturbances, autonomic dysregulation, poor weight gain, and occasionally seizures [30]. The long-term neurodevelopmental outcomes of infants born to mothers who used opioids during pregnancy are difficult to determine due to the above prenatal confounding factors, as well as postnatal factors such as continued maternal drug use, violence exposure, and low maternal socioeconomic status. However, there is recent evidence that, even when controlling for confounding factors, school-aged children who are exposed to opiates prenatally have lower IQ scores [31]. In addition, children with NAS have been found to be more likely to be re-hospitalized during childhood for maltreatment, assault, poisonings, and mental and behavioral disorders [32], highlighting not only the biological but significant social and psychological risks to an infant exposed to opiates in utero.

Methamphetamine use during pregnancy is associated with a higher incidence of small for gestational age neonates [33], as well as with higher rates of intrauterine fetal death, placental abruption, preterm birth, neonatal death, and infant death [34]. As with other illicit substances, direct effects of methamphetamine exposure during pregnancy have been difficult to demonstrate due to confounding factors. There is evidence that children who were prenatally exposed to methamphetamines have increased rates of attention-deficit/hyperactivity disorder, as well as increased rates of anxiety and depression [35]. However, when analysis of outcomes in children prenatally exposed to methamphetamines included an early adversity index (sum of factors including maternal postnatal substance use, extreme poverty, maternal depression, and community violence), it was found that increased rates of externalizing, rule-breaking, and aggressive behavior seen in the group exposed prenatally to methamphetamines were mediated by early adversity index scores [36]. This further highlights the confounding relationship between social risk factors and biological risk factors on developmental outcomes.

Marijuana is the most common illicit drug used in pregnancy [37]. Smoking cannabis during pregnancy is linked to lower birth weight. Further evidence supporting a relationship between smoking cannabis and other pregnancy and neonatal outcomes such as stillbirth, fetal distress, or admission to the NICU is currently unclear [38]. In terms of neurodevelopment, prenatal exposure to marijuana has not been shown to affect IQ but has been associated with deficits in problem-solving skills, sustained attention, and visual memory [39]. A recent literature review from the National Academies of Sciences concludes that there is insufficient evidence to support or refute an association between maternal cannabis smoking and later cognitive/academic outcomes in the offspring [38]. The report noted that attributing later childhood outcomes to prenatal exposures is particularly difficult as attempts to control for a child's environment may be insufficient to detect potentially subtle differences in the family and neighborhood environments of women who smoke cannabis during pregnancy and those who do not [38].

Cigarette smoking during pregnancy is one of the most prevalent and preventable causes of adverse pregnancy outcomes [40] including miscarriage, fetal growth restriction, ectopic pregnancy, and preterm labor [41]. Maternal smoking during pregnancy has not been found to lower the IQ of offspring when controlling for maternal IQ and education [42]. However, increased rates of impulsivity, attention problems, and negative and externalizing behaviors have been identified in children prenatally exposed to tobacco [39].

Alcohol is a teratogen and use during pregnancy is the leading preventable cause of birth defects and intellectual and neurodevelopmental disabilities [43]. Fetal alcohol spectrum disorders (FASD) is an overarching term that includes fetal alcohol syndrome (FAS), partial fetal alcohol syndrome (PFAS), alcohol-related birth defects (ARBD), and alcohol-related neurodevelopmental disorder (ARND) [44]. Diagnostic criteria for FAS include at least two of the characteristic facial anomalies (short palpebral fissures, thin vermilion border of the upper lip, and a smooth philtrum), as well as findings in all three of the following categories: prenatal and/or postnatal growth deficiency, brain abnormalities (deficient brain growth, abnormal brain morphogenesis, or abnormal neurophysiology), and neurobehavioral impairment (marked by cognitive or behavioral impairment in children greater than 3 years of age and developmental delay in children less than 3 years of age). Documentation of prenatal alcohol exposure is not necessary to make the diagnosis of FAS. The diagnosis of PFAS can be made with or without documentation of prenatal alcohol exposure, and criteria are the same as for FAS, though not all need to be met in order to make the diagnosis. ARBD is diagnosed based on documentation of prenatal alcohol exposure as well as one or more specific major malformations understood to be a result of prenatal alcohol exposure, such as specific cardiac defects. ARND requires documented prenatal alcohol exposure and neurobehavioral impairment [44]. A separate diagnosis but with some overlapping criteria is neurobehavioral disorder associated with prenatal alcohol exposure (ND-PAE), which is included in the *Diagnostic and Statistical Manual of Mental Disorders*, Fifth Edition [45]. Diagnostic criteria for ND-PAE include confirmed prenatal alcohol exposure and impaired neurocognitive function (such as intellectual disability or memory), impaired self-regulation (such as issues with attention or impulse control), and impaired adaptive function (such as problems with communication or daily living skills). These impairments must affect function and not be attributable to other teratogens or genetic or medical conditions. While FAS leads to specific CNS dysfunctions that can be considered primary disabilities, secondary disabilities associated with FAS such as mental health problems, disrupted school experience, trouble with the law, and alcohol and drug problems can be reduced when various protective factors are present, such as living in a stable, nurturing, and good quality home [46].

Psychological Risks: Parental Distress (Anxiety/Depression and Traumatic Stress)

Within a biopsychosocial framework, parental mental health has a significant impact on a premature infant's neurodevelopment. Preterm birth itself is associated with many traumatic experiences [47], not only for the infant, such as the trauma of preterm birth, numerous invasive procedures, and separation from parents, but for the parents as well, including the appearance of their sick infant, the threat of potentially losing their child, and the uncertainty of their role as a parent in the NICU setting. These traumatic experiences have led to parental reports of increased distress during their child's NICU stay as compared to parents of full-term, healthy infants [48–52]. Parents report feelings of anxiety, sleep disturbance, and avoidance of medical visits due their experience of caring for a child in a highly technological environment such as the NICU [47, 53]. A substantial number of NICU parents, in fact, surpass the clinical cutoff scores for depression, anxiety, and PTSD, both in the immediate aftermath of the NICU course and for well over a year or more later [53–59].

Major depressive disorder, at baseline, is most prevalent during childbearing years and is more common in women than men, affecting approximately 8.5% of women [60]. Postpartum depression affects approximately 10–15% of women of healthy babies [53]. Thus, it is understandable that, when this already vulnerable population faces the distress of having a child in the NICU, these women have a greater risk for postpartum depression, with up to 39–40% of NICU mothers meeting the cutoff for PPD at 1 month postpartum [51, 53]. This, too, is compounded by the potential for a pre-existing risk present prior to delivery, either from the stress of a complicated pregnancy or a mother's own history of traumatic experiences and struggle with depression, which may have contributed to the preterm birth itself [61]. Two studies of NICU mothers showed that these mothers have increased rates of postpartum depression (PPD) as compared to mothers of healthy infants [51, 53]. Identifying those with PPD is critical because PPD has been shown to have a direct influence on infant neurodevelopment. Infants of mothers with PPD are known to be at an increased risk for lower cognitive scores, disruptive behavior, difficulties with feeding and growing, and inappropriate utilization of the healthcare system [62–64].

Many parents report symptoms of traumatic stress when their child is first admitted to the NICU. However, a sizable number may continue to experience these symptoms beyond 30 days, warranting a full diagnosis of PTSD [53]. Several studies reported higher rates of PTSD in NICU mothers [51, 54, 65, 66]. As seen with PPD, the effect of maternal PTSD on the developing infant can be significant, placing them at increased risk for poorer health outcomes and lower scores on developmental testing [47, 53]. Further, mothers with PTSD display more controlling and less sensitive behaviors which may exacerbate behavioral problems seen in the premature population [66, 67].

The development of a strong and nurturing mother-infant relationship is often stressed and even disrupted when an infant spends a significant period in the NICU [3, 47, 62]. Maternal depression and traumatic stress are known to further disrupt this dyad, leading to insecure and distorted attachment [68]. Additional studies indicate that the impact of maternal depression on this dyad may be unique, and even greater, in the premature population, with one study showing that mothers experiencing significant psychological distress display less positive perceptions of their child at 1 year of life adjusted [69]. A healthy dyad is critical for the developing infant and is discussed in further detail above.

Social Risks

There are several studies that highlight that some social risk factors are just as important, if not more important, than medical factors in predicting neurodevelopmental outcome in high-risk, preterm infants.

Vohr et al. [70] demonstrated in a group of extremely low birthweight (ELBW) infants that having a mother with less than a high school education was independently associated with significant cognitive impairment. Even when controlling for medical factors, white race was associated with a decreased risk of significant cognitive impairment. In contrast, many medical factors, including surfactant administration, antenatal steroids, and necrotizing enterocolitis, were not significantly associated with cognitive impairment in the multivariate analysis.

Similar results were found in a separate cohort of very low birth weight (VLBW) infants who were serially evaluated until 8 years of age using measures of vocabulary and IQ [71]. Multivariate analysis including medical factors revealed that residence in a two-parent household, higher levels of maternal education, absence of special services, and absence of significant central nervous system (CNS) injury were associated with increases in vocabulary scores over time. Similarly, increasing age, higher levels of maternal education, residence in a two-parent household, and absence of significant CNS injury were associated with an improvement in visual IQ scores over time.

Finally, in a third cohort of VLBW infants [72], maternal demographic factors such as white race, maternal age, and private insurance contributed 3.6% of the variance to the model predicting unimpaired outcome; in comparison, absence of neonatal morbidities (such as chronic lung disease, sepsis, NEC, IVH grade 3 or 4) contributed 5.5% of the variance in the model.

Overall, the literature in ELBW and VLBW infants supports the concept that environmental factors, including maternal education, race, and socioeconomic status, can have an equally important impact on neurodevelopmental outcome in high-risk, preterm infants as many medical factors. This highlights the importance of interventions to support the proximal environment (mother, family, and home) where the infant will reside in order to optimize neurodevelopmental outcomes.

Addressing Environmental Risks

Preventive interventions must target family patterns of interaction. Support that encourages and fosters a strong parent-child relationship early in a child's development may help the child maximize their developmental potential by minimizing external stressors and allowing the child to adapt and adjust when faced with biologic constraints. Premature infants show enhanced cognitive development in the short to medium term but no improvements in motor outcomes with generic early intervention services [73]. Additionally heterogeneous and inconsistent outcome effects are seen when looking at a variety of interventions in hospital, bridging hospital to home and then in the community [11]. What many of these interventions neglect is how the psychological and social environment shape these outcomes and the dramatic role interventions can have on these outcomes. Utilizing a team of people, including medical professionals, to help families establish this support network is critical. Focusing on the infant-caregiver dyad within a Developmental Systems Approach (DSA) may be helpful going forward to target-specific interventions. The DSA recognizes that, as infants develop and demonstrate their social and cognitive competence, there are sets of fundamental domains that assert a reciprocal influence on this competence [11]. By directly supporting family patterns of interaction (e.g., socioemotional connectedness) and the family resources (e.g., financial resources) available to these families, clinicians can optimize the relationship between the child and family and, in turn, the child's learning environment and development. We will now outline a few interventions that take these dynamics into consideration.

Numerous studies have found an association between breast milk and improved neurodevelopmental outcomes in term [74–76] and at-risk preterm and low birth weight [76–78] infants. Evidence suggests that the beneficial effects of breast milk on cognitive development may be even larger in low birth weight, preterm infants than in normal birth weight, term infants [74, 76] emphasizing a greater need to promote the provision of breast milk in this vulnerable population. Current recommendations to support breastfeeding in preterm and sick infants include facilitation of early, continuous, and prolonged skin-to-skin contact (kangaroo mother care [KMC]), early initiation of breastfeeding, and mothers' access to breastfeeding support throughout the whole hospital stay [79]. Kangaroo mother care (KMC) has not only been shown to increase rates of breastfeeding among very low birth weight infants but to significantly reduce morbidity and mortality [80]. In a follow-up study of a randomized trial of KMC versus traditional care, it was found that at 20 years, the KMC group continued to have reduced mortality, and individuals had less aggressive drive, were less impulsive and hyperactive, and had less antisocial behavior [81]. One theory as to why KMC is so effective is that it promotes bonding between the infant and mother [82].

The Family Nurture Intervention extends the skin-to-skin concept in the NICU hospital setting. Here the focus is on creating a cycle of attentive caregiving by the mother through positive reinforcement of the emotional connection and co-regulation between the mother and premature infant. The components include calming sessions, a scent cloth exchange, verbal soothing, eye contact, and skin-to-skin holding in addition to family-based support sessions. Positive outcomes have been shown in maternal caregiving, EEG coherence and power, lower rates of maternal depression and anxiety, and improved toddler attention, language, and cognition [83–89].

Another intervention that directly targets the trauma mothers experience in the NICU is targeted cognitive behavioral therapy, as laid forth by Shaw et al. [90]. In this six-session intervention, the mother receives emotional support of her distress with psychoeducation about PTSD, cognitive restructuring to address traumatic thoughts, anxiety-focused muscle relaxation, and creation of a trauma narrative. These have been shown to reduce maternal trauma symptoms and depression at

1 month and prevent symptoms at 6 months which removes a barrier to optimal dyadic functioning [91, 92].

Finally, a preterm infant intervention by Borghini [93] utilized three components. A therapist helped parents observe the infant's interactions and enhance their sensitivity and responsiveness to their child. The intervention used videotaping of sessions, and parents received feedback on their actions and reflected on their feelings. This intervention was found to reduce maternal PTSD symptoms at 4 and 12 months of age and improve maternal-infant interactions at 4 months [93]. In fact many interventions that include components such as psychosocial support for mothers with preterm infants were found to promote outcomes by lowering maternal anxiety, depressive symptoms, and self-efficacy [94]. What these interventions share is the understanding that the immediate caregiving environment of the maternal-infant dyad is critical to promoting outcomes. These early patterns of interaction carry forward to optimal parenting skills that lead to improved academic and school-age functioning [95, 96].

Summary

This chapter provides research and sound evidence as to why thinking toward "NICU outcomes" must be comprehensive in nature. If we only focus on medical interventions and outcomes, we fail to address the critical impact that the environment, including exposures, infant-parent relationships, and caregiver mental health, has on a high-risk infant's development. The NICU and follow-up clinics must work to create a mindset and atmosphere that supports medical staff and families in identifying and addressing these environmental influences, with both evidenced-based modalities and a mind toward compassion and comprehensive care.

References

1. Fine, A., Kotelchuck, M. (2010). *Rethinking MCH: The MCH life course model as an organizing framework*. US DHHS, HRSA, Maternal and Child Health Bureau. Retrieved November 28, 2016, from http://www.hrsa.gov/ourstories/mchb75th/images/rethinkingmch.pdf.
2. Halfon, N., & Hochstein, M. (2002). Life course health development: An integrated framework for developing health, policy and research. *Milbank Quarterly, 80*(3), 433–479.
3. Garner, A. S., Shonkoff, J. P., & Committee on Psychosocial Aspects of Child and Family Health; Committee on Early Childhood, Adoption, and Dependent Care; Section on Developmental and Behavioral Pediatrics. (2012). Early childhood adversity, toxic stress, and the role of the pediatrician: translating developmental science into lifelong health. *Pediatrics, 129*(1), e224–e231. https://doi.org/10.1542/peds.2011-2662. Retrieved from http://pediatrics.aappublications.org/content/pediatrics/129/1/e232.full.pdf on 11/28/16.
4. Centers for Disease Control and Prevention. (2009). *Adverse childhood experiences study*. Centers for Disease Control and Prevention Web Site. Retrieved February 23, 2009, from http://www.cdc.gov/nccdphp/ace.

5. Felitti, V. J., Anda, R. F., Nordenberg, D., et al. (1998). The relationship of adult health status to childhood abuse and household dysfunction. *American Journal of Preventive Medicine, 14*(4), 245–258.
6. Shonkoff, J. P. (2016). Capitalizing on advances in science to reduce the health consequences of early childhood adversity. *JAMA Pediatrics, 170*(10), 1003–1007.
7. Lupien, S. J., McEwen, B. S., Gunnar, M. R., & Heim, C. (2009). Effects of stress throughout the lifespan on the brain, behaviour and cognition. *Nature Reviews Neuroscience, 10*(6), 434–445.
8. Miller, G. E., Chen, E., & Parker, K. J. (2011). Psychological stress in childhood and susceptibility to the chronic diseases of aging: Moving toward a model of behavioral and biological mechanisms. *Psychological Bulletin, 137*(6), 959–997.
9. Parker, S., Greer, S., & Zuckerman, B. (1988). Double jeopardy: The impact of poverty on early child development. *Pediatric Clinics of North America, 35*(6), 1227–1240.
10. Guralnick, M. J. (2001). A developmental systems model for early intervention. *Infants Young Child, 14*, 1–18.
11. Guralnick, M. J. (2012). Preventive interventions for preterm children: Effectiveness and developmental mechanisms. *Journal of Developmental and Behavioral Pediatrics, 33*, 352–364.
12. Guralnick, M. J. (2011). Why early intervention works: A systems perspective. *Infants Young Child, 24*(1), 6–28.
13. Aylward, G. P. (2005). Neurodevelopmental outcomes of infants born prematurely. *Journal of Developmental and Behavioral Pediatrics, 26*(6), 427–440. Review. PubMed PMID: 16344661.
14. Sameroff, A. (2010). A unified theory of development: A dialectic integration of nature and nurture. *Child Development, 81*(1), 6–22. https://doi.org/10.1111/j.1467-8624.2009.01378.x.
15. Bowlby, J. (1969). *Attachment and loss: Vol. 1. Attachment*. New York: Basic Books.
16. Feldman, R. (2007). Parent-infant synchrony and the construction of shared timing; physiological precursors, developmental outcomes, and risk conditions. *Journal of Child Psychology and Psychiatry, 48*(3–4), 329–354.
17. Boyce, L. K., Cook, G. A., Simonsmeier, V., & Hendershot, S. M. (2015). Academic outcomes of very low birth weight infants: The influence of mother–child relationships. *Infant Mental Health Journal, 36*, 156–166.
18. Feldman, R., & Eidelman, A. I. (2006). Neonatal state organization, neuromaturation, mother-infant interaction, and cognitive development in small-for-gestational-age premature infants. *Pediatrics, 118*(3), e869–e878.
19. Bilgin, A., & Wolke, D. (2015). Maternal sensitivity in parenting preterm children: A meta-analysis. *Pediatrics, 136*, 2014–3570.
20. Jaekel, J., Pluess, M., Belsky, J., & Wolke, D. (2015). Effects of maternal sensitivity on low birth weight children's academic achievement: A test of differential susceptibility versus diathesis stress. *Journal of Child Psychology and Psychiatry, 56*, 693–701.
21. Habersaat, S., Borghini, A., Nessi, J., Pierrehumbert, B., Forcada-Guex, M., Ansermet, F., & Müller-Nix, C. (2014). Posttraumatic stress symptoms and cortisol regulation in mothers of very preterm infants. *Stress Health, 30*(2), 134–141.
22. Samuel, S., Hayton, B., Gold, I., Feeley, N., Carter, C. S., & Zelkowitz, P. (2015). Maternal mental health moderates the relationship between oxytocin and interactive behavior. *Infant Mental Health Journal, 36*(4), 415–426.
23. Gouin, K., Murphy, K., Shah, P. S., & Knowledge Synthesis group on Determinants of Low Birth Weight and Preterm Births. (2011). Effects of cocaine use during pregnancy on low birthweight and preterm birth: a systematic review and meta analyses. *American Journal of Obstetrics and Gynecology, 204*, 340.e1-2.
24. Bauer, C. R., Shankaran, S., Bada, H. S., et al. (2002). The maternal lifestyle study: Drug exposure during pregnancy and short-term maternal outcomes. *American Journal of Obstetrics and Gynecology, 186*, 487–495.
25. Henretic, F. M. (2009). Toxins. In W. B. Carey, A. C. Crocker, W. L. Coleman, E. R. Elias, & H. M. Feldman (Eds.), *Developmental-behavioral pediatrics* (4th ed., pp. 314–324.0). Saunders Elsevier: Philadelphia, PA.

26. Hunt, H., Brodsky, N. L., Roth, H., et al. (2005). School performance of children with gestational cocaine exposure. *Neurotoxicology and Teratology, 27*, 203–211.
27. Messinger, D. S., Bauer, C. R., Das, A., et al. (2004). The maternal lifestyle study: Cognitive, motor, and behavioral outcomes of cocaine-exposed and opiate-exposed infants through three years of age. *Pediatrics, 113*, 1677–1685.
28. Ackerman, J. P., Riggins, T., & Black, M. M. A. (2010). review of the effects of prenatal cocaine exposure among school-aged children. *Pediatrics, 125*, 554–565.
29. Maeda, A., Bateman, B. T., Clancy, C. R., et al. (2014). Opioid abuse and dependence during pregnancy: Temporal trends and obstetrical outcomes. *Anesthesiology, 121*, 1158–1165.
30. Kochrlakota, P. (2014). Neonatal abstinence syndrome. *Pediatrics, 134*, e547–e561.
31. Nygaard, E., Moe, V., Slinning, K., & Walhovd, K. B. (2015). Longitudinal cognitive development of children born to mothers with opioid and polysubstance use. *Pediatric Research, 78*(3), 330–335.
32. Uebel, H., Wright, I. M., Burns, L., et al. (2015). Reasons for rehospitalization in children who had neonatal abstinence syndrome. *Pediatrics, 136*, e811–e820.
33. Nguyen, D., Smith, L. M., LaGasse, L. L., et al. (2010). Intrauterine growth of infants exposed to prenatal methamphetamine: Results from the infant developmental, environment, and lifestyle study. *The Journal of Pediatrics, 157*, 337–339.
34. Gorman, M. G., Orme, K. S., Nguten, N. T., et al. (2014). Outcomes in pregnancies complicated by methamphetamine use. *American Journal of Obstetrics and Gynecology, 211*, 429.e1–7.
35. LaGasse, L. L., Derauf, C., Smith, L. M., et al. (2012). Prenatal methamphetamine exposure and childhood behavior problems at 3 and 5 years of age. *Pediatrics, 129*, 681–688.
36. Eze, N., Smith, L. M., LaGasse, L. L., et al. (2016). School-aged outcomes following prenatal methamphetamine exposure: 7.5-year follow-up from the infant development, environment, and lifestyle study. *The Journal of Pediatrics, 170*, 34–38.
37. Metz, T. D., & Stickrath, E. H. (2015). Marijuana use in pregnancy and lactation: A review of the evidence. *American Journal of Obstetrics and Gynecology, 213*(6), 761–778.
38. National Academies of Sciences, Engineering, and Medicine. (2017). *The health effects of cannabis and cannabinoids: The current state of evidence and recommendations for research*. Washington, DC: The National Academies Press.
39. Behnke, M., & Smith, V. C. (2013). Prenatal substance abuse: Short- and long-term effects on the exposed fetus. *Pediatrics, 131*, e1009–e1024.
40. Institute of Medicine (US) Committee on Understanding Premature Birth and Assuring Healthy Outcomes. (2007). In R. E. Behrman & A. S. Butler (Eds.), *Preterm birth: Causes, consequences, and prevention*. Washington, DC: National Academies Press.
41. Practice Committee of the American Society for Reproductive Medicine. (2012). Smoking and infertility: A committee opinion. *Fertility and Sterility, 98*(6), 1400–1406.
42. Batty, G. D., & Der Gm Deary, I. J. (2006). Effect of maternal smoking during pregnancy on offspring's cognitive ability: Empirical evidence for complete confounding in the US national longitudinal survey of youth. *Pediatrics, 118*(3), 943–950.
43. Williams, J. F., Smith, V. C., & The Committee on Substance Abuse. (2015). Fetal alcohol spectrum disorders. *Pediatrics, 136*(5), e1395–e1406.
44. Hoyme, H. E., Kalberg, W. O., Elliott, A. J., et al. (2016). Updated clinical guidelines for diagnosing fetal alcohol spectrum disorders. *Pediatrics, 138*(2), e20154256.
45. American Psychiatric Association. (2013). *Diagnostic and statistical manual of mental disorders* (5th ed.). Arlington, VA: American Psychiatric Association.
46. Streissguth, A. P., Barr, H. M., Kogan, J., & Bookstein, F. L. (1996). *Understanding the occurrence of secondary disabilities in clients with fetal alcohol syndrome (FAS) and fetal alcohol effects (FAE): Final report*. Seattle, WA: University of Washington Publication Services.
47. Vanderbilt, D., & Gleason, M. (2011). Mental health concerns of the premature infant through the lifespan. *Pediatric Clinics of North America, 58*, 815–832.
48. Brandon, D. H., Tully, K. P., Silva, S. G., Malcolm, W. F., Murtha, A. P., Turner, B. S., et al. (2011). Emotional responses of mothers of late-preterm and term infants. *Journal of Obstetric, Gynecologic, and Neonatal Nursing, 40*(6), 719–731.

49. DeMier, R. L., Hynan, M. T., Harris, H. B., & Manniello, R. L. (1996). Perinatal stressors as predictors of symptoms of posttraumatic stress in mothers of infants at high-risk. *Journal of Perinatology, 16*(4), 276–280.
50. Singer, L. T., Salvator, A., Guo, S., Collin, M., Lilien, L., & Baley, J. (1999). Maternal psychological distress and parenting stress after the birth of a very low-birth-weight infant. *JAMA, 281*(9), 799–805.
51. Vanderbilt, D., Bushley, T., Young, R., & Frank, D. A. (2009). Acute posttraumatic stress symptoms among urban mothers with newborns in the neonatal intensive care unit: A preliminary study. *Journal of Developmental and Behavioral Pediatrics, 30*(1), 50–56.
52. Voegtline, K. M., Stifter, C. A., & The Family Life Project Investigators. (2010). Late-preterm birth, maternal symptomatology, and infant negativity. *Infant Behavior & Development, 33*(4), 545–554.
53. Lefkowitz, D. S., Chiara Baxt, C., & Evans, J. R. (2010). Prevalence and correlates of posttraumatic stress and postpartum depression in parents of infants in the neonatal intensive care unit (NICU). *Journal of Clinical Psychology in Medical Settings, 17*(3), 230–237.
54. Feeley, N., Zelkowitz, P., Cormier, C., Charbonneau, L., Larcoix, A., & Papageorgiou, A. (2011). Posttraumatic stress among mothers of very low birthweight infants at 6 months after discharge from the neonatal intensive care unit. *Applied Nursing Research, 24*(2), 114–117.
55. Garfield, L., Holditch-Davis, D., Carter, C. S., McFarlin, B. L., et al. (2015). Risk factors for postpartum depressive symptoms in low-income women with very low-birth-weight infants. *Advances in Neonatal Care, 15*(1), E3–E8.
56. Holditch-Davis, D., Bartlett, T. R., Blickman, A. L., & Miles, M. S. (2003). Posttraumatic stress symptoms in mothers of premature infants. *Journal of Obstetric, Gynecologic, and Neonatal Nursing, 32*(2), 161–171.
57. Miles, S. M., Holditch-Davis, D., Schwartz, T. A., & Scher, M. (2007). Depressive symptoms in mothers of prematurely born infants. *Journal of Developmental and Behavioral Pediatrics, 28*(1), 36–44.
58. Poehlmann, J., Schwichtenberg, A. J., Bolt, D., & Dilworth-Bart, J. (2009). Predictors of depressive symptom trajectories in mothers of preterm or low birth weight infants. *Journal of Family Psychology, 23*(5), 690–704.
59. Shaw, R. J., Bernard, R. S., DeBlois, R., Ikuta, L. M., Ginzburg, K., & Koopman, C. (2009). The relationship between acute stress disorder and posttraumatic stress disorder in the neonatal intensive care unit. *Psychosomatics, 50*(2), 131–137.
60. National Institute of Mental Health: 12-month prevalence of major depressive episode among US adults (2015). *Courtesy of SAMHSA.* Data based on results from the 2015 National Survey on Drug Use and Health. Retrieved January, 4 2017, from https://www.nimh.nih.gov/health/statistics/prevalence/nsduh-mde-2015-graph-logo_148151.pdf.
61. Greene, M., Rossman, B., Patra, K., et al. (2015). Depressive, anxious, perinatal post-traumatic distress in mother of very low birth weight infants in the NICU. *Journal of Developmental and Behavioral Pediatrics, 36*(5), 362–370.
62. Hynan, M. T., Mounts, K. O., & Vanderbilt, D. L. (2013). Screening parents of high-risk infants for emotional distress: Rationale and recommendations. *Journal of Perinatology, 33*, 748–753.
63. Murray, L., Fiori-Cowley, A., Hooper, R., et al. (1996). The impact of postnatal depression and associated adversity on early mother-infant interactions and later infant outcome. *Child Development, 67*, 2512.
64. Sharp, D., Hay, D. F., Pawlby, S., et al. (1995). The impact of postnatal depression on boys' intellectual development. *Journal of Child Psychology and Psychiatry, 36*, 1315.
65. Kim, W. J., Lee, E., Kim, K. R., Namkoong, K., Park, K. S., & Rha, D. W. (2015). Progress of PTSD symptoms following birth: a prospective study of mothers of high-risk infants. *Journal of Perinatology, 35*(8), 575–579.
66. Pierrehumbert, B., Nicole, A., Muller-Nix, C., Forcada-Gue, M., & Amsermet, F. (2003). Parental post-traumatic reactions after premature birth: Implications for sleeping and eating problems in the infant. *Archives of Disease in Childhood. Fetal and Neonatal Edition, 88*(5), F400–F404.

67. Muller-Nix, C., Forcada-Guex, M., Pierrehumbert, B., et al. (2004). Prematurity, maternal stress and mother-child interactions. *Early Human Development, 79*, 145.
68. Peohlmann, J., & Fiese, B. H. (2001). The interaction of maternal and infant vulnerabilities on developing attachment relationships. *Development and Psychopathology, 13*(1), 1–11.
69. Holditch-Davis, D., Santos, H., Levy, J., White-Traut, R., O'Shea, T. M., Geraldo, V., & David, R. (2015). Patterns of psychological distress in mothers of preterm infants. *Infant Behavior & Development, 41*, 154–163. https://doi.org/10.1016/j.infbeh.2015.10.004. Epub 2015 Oct 22. PubMed PMID: 26495909; PubMed Central PMCID: PMC4654120.
70. Vohr, B. R., Wright, L. L., Dusick, A. M., et al. (2000). Neurodevelopmental and functional outcomes of extremely low birth weight infants in the National Institute of Child Health and Human Development Neonatal Research Network, 1993-1994. *Pediatrics, 105*, 1216–1226.
71. Ment, L. R., Vohr, B., Allan, W., et al. (2003). Change in cognitive function over time in very low-birth-weight infants. *Journal of the American Medical Association, 289*, 705–711.
72. Gargus, R. A., Vohr, B. R., Tyson, J. E., et al. (2009). Unimpaired outcomes for extremely low birth weight infants at 18 to 22 months. *Pediatrics, 124*, 112–121.
73. Spittle, A. J., Orton, J., Doyle, L. W., & Boyd, R. (2007). Early developmental intervention programs post hospital discharge to prevent motor and cognitive impairments in preterm infants. *Cochrane Database of Systematic Reviews, 18*, CD005495.
74. Anderson, J. W., Johnstone, B. M., & Remley, D. T. (1999). Breastfeeding and cognitive development: A meta-analysis. *The American Journal of Clinical Nutrition, 70*, 525–535.
75. Kramer, M. S., Aboud, F., Mironova, E., Promotion of Breastfeeding Intervention Trial (PROBIT) Study Group, et al. (2008). Breastfeeding and child cognitive development: New evidence from a large randomized trial. *Archives of General Psychiatry, 65*(5), 578–584.
76. Quigley, M. A., Hickley, C., Carson, C., et al. (2012). Breastfeeding is associated with improved child cognitive development: A population-based cohort study. *The Journal of Pediatrics, 160*, 25–32.
77. Vohr, B. R., Poindexter, B. B., Dusick, A. M., et al. (2006). Beneficial effects of breast milk in the neonatal intensive care unit on the developmental outcome of extremely low birth weight infants at 18 months of age. *Pediatrics, 118*, e115–e123.
78. Vohr, B. R., Poindexter, B. B., Dusick, A. M., et al. (2007). Persistent beneficial effects of breast milk ingested in the neonatal intensive care unit on outcomes of extremely low birth weight infants at 30 months of age. *Pediatrics, 120*, e953–e959.
79. Nyqvist, K. H., Häggkvist, A. P., Hansen, M. N., et al. (2013). Expansion of the baby-friendly hospital initiative ten steps to successful breastfeeding into neonatal intensive care: Expert group recommendations. *Journal of Human Lactation, 29*(3), 300–309.
80. Conde-Agudelo, A., & Díaz-Rossello, J. L. (2016). Kangaroo mother care to reduce morbidity and mortality and low birthweight infants. *Cochrane Database of Systematic Reviews, 8*, CD002771. https://doi.org/10.1002/14651858.CD002771.pub4.
81. Charpak, N., Tessier, R., Ruiz, J. G., et al. (2017). Twenty-year follow-up of kangaroo mother care versus traditional care. *Pediatrics, 139*(1), e20162063.
82. Tessier, R., Cristo, M., Velez, S., et al. (1998). Kangaroo mother care and the bonding hypothesis. *Pediatrics, 102*(2), e17.
83. Welch, M. G., Hofer, M. A., Brunelli, S. A., Stark, R. I., Andrews, H. F., Austin, J., Myers, M. M., & Family Nurture Intervention (FNI) Trial Group. (2012). Family nurture intervention (FNI): Methods and treatment protocol of a randomized controlled trial in the NICU. *BMC Pediatrics, 12*, 14. https://doi.org/10.1186/1471-2431-12-14. Erratum in: BMC Pediatr. 2012;12:107.
84. Welch, M. G., Hofer, M. A., Stark, R. I., Andrews, H. F., Austin, J., Glickstein, S. B., Ludwig, R. J., Myers, M. M., & FNI Trial Group. (2013). Randomized controlled trial of Family Nurture Intervention in the NICU: assessments of length of stay, feasibility and safety. *BMC Pediatrics, 13*, 148.
85. Welch, M. G., Myers, M. M., Grieve, P. G., Isler, J. R., Fifer, W. P., Sahni, R., Hofer, M. A., Austin, J., Ludwig, R. J., Stark, R. I., & FNI Trial Group. (2014). Electroencephalographic activity of preterm infants is increased by family nurture intervention: A randomized controlled trial in the NICU. *Clinical Neurophysiology, 125*(4), 675–684.

86. Welch, M. G., Firestein, M. R., Austin, J., Hane, A. A., Stark, R. I., Hofer, M. A., Garland, M., Glickstein, S. B., Brunelli, S. A., Ludwig, R. J., & Myers, M. M. (2015). Family nurture intervention in the neonatal intensive care unit improves social-relatedness, attention, and neurodevelopment of preterm infants at 18 months in a randomized controlled trial. *Journal of Child Psychology and Psychiatry, 56*(11), 1202–1211.
87. Welch, M. G., Halperin, M. S., Austin, J., Stark, R. I., Hofer, M. A., Hane, A. A., & Myers, M. M. (2016). Depression and anxiety symptoms of mothers of preterm infants are decreased at 4 months corrected age with family nurture intervention in the NICU. *Archives of Women's Mental Health, 19*(1), 51–61.
88. Hane, A. A., Myers, M. M., Hofer, M. A., Ludwig, R. J., Halperin, M. S., Austin, J., Glickstein, S. B., & Welch, M. G. (2015). Family nurture intervention improves the quality of maternal caregiving in the neonatal intensive care unit: Evidence from a randomized controlled trial. *Journal of Developmental and Behavioral Pediatrics, 36*(3), 188–196.
89. Myers, M. M., Grieve, P. G., Stark, R. I., Isler, J. R., Hofer, M. A., Yang, J., Ludwig, R. J., & Welch, M. G. (2015). Family nurture intervention in preterm infants alters frontal cortical functional connectivity assessed by EEG coherence. *Acta Paediatrica, 104*(7), 670–677.
90. Shaw, R. J., Brecht, C. J., St John, N., et al. (2013). Prevention of postpartum traumatic stress in mothers with preterm infants: Manual development and evaluation. *Issues in Mental Health Nursing, 34*(8), 578–586.
91. Shaw, R. J., St John, N., Lilo, E. A., Jo, B., Benitz, W., Stevenson, D. K., & Horwitz, S. M. (2013). Prevention of traumatic stress in mothers with preterm infants: A randomized controlled trial. *Pediatrics, 132*(4), e886–e894.
92. Shaw, R. J., St John, N., Lilo, E., Jo, B., Benitz, W., Stevenson, D. K., & Horwitz, S. M. (2014). Prevention of traumatic stress in mothers of preterms: 6-month outcomes. *Pediatrics, 134*(2), e481–e488. https://doi.org/10.1542/peds.2014-0529.
93. Borghini, A., Habersaat, S., Forcada-Guex, M., Nessi, J., Pierrehumbert, B., Ansermet, F., & Müller-Nix, C. (2014). Effects of an early intervention on maternal post-traumatic stress symptoms and the quality of mother-infant interaction: the case of preterm birth. *Infant Behavior & Development, 37*(4), 624–631. Epub 2014 Sep 15. PubMed PMID: 25222614. https://doi.org/10.1016/j.infbeh.2014.08.003
94. Benzies, K. M., Magill-Evans, J. E., Hayden, K. A., & Ballantyne, M. (2013). Key components of early intervention programs for preterm infants and their parents: A systematic review and meta-analysis. *BMC Pregnancy and Childbirth, 13*(Suppl. 1), S10.
95. Treyvaud, K., Doyle, L. W., Lee, K. J., Ure, A., Inder, T. E., Hunt, R. W., & Anderson, P. J. (2016). Parenting behavior at 2 years predicts school-age performance at 7 years in very preterm children. *Journal of Child Psychology and Psychiatry, 57*, 814–821.
96. Wolke, D., Jaekel, J., Hall, J., & Baumann, N. (2013). Effects of sensitive parenting on the academic resilience of very preterm and very low birth weight. *Journal of Adolescent Health, 53*(5), 642–647.
97. City MatCH The Life Course Game. (2017). For questions about copyright, please contact Mark Law at mlaw@unmc.edu or 402-552-9500. Retrieved April 3, 2017, from www.citymatch.org:8080/lifecoursetoolbox/documents/LCGamePPTTemplate.ppt.

Part IV
Issues in Clinic Organization

HRIF Clinic Organization: A Statewide Approach

Anne DeBattista

Abstract There are a great deal of variability in the structure of neonatal follow-up clinics and no clear standards. In California, there is an established Title V program referred to as California Children's Services (CCS) which has a High-Risk Infant Follow-Up program as a diagnostic service to ensure NICU graduates are connected to medical specialty practices and early intervention services at the earliest signs of difficulty or developmental delay. Key to the program is a dedicated coordinator who makes certain that the child accesses follow-up services. Medical eligibility is determined by risk factors for neurodevelopmental disabilities. HRIF evaluations include assessments of complex care needs, growth, development and the neurological examination. Many ill children need EI services, but states are variable in their criteria required to access these services and many do not accept an "at-risk" eligibility category. A recommended follow-up schedule is included.

High-Risk Infant Follow-Up (HRIF) programs are designed to periodically evaluate children who have identified risk factors for developmental disabilities in order to track outcomes and facilitate early intervention services at the first sign of delay. Although the importance of providing HRIF to children who received neonatal intensive care services has been recognized for decades, currently, there is not a national standard of care for the way in which these services are provided. There are variations in the risk factors used to identify children needing follow-up. The timing of evaluations varies, along with recommendations for the ages at which children should be followed and for how long they should be followed. Different assessment protocols and procedures are used to evaluate children. Lastly, the data collection and the way in which feedback about population outcomes is provided back to neonatal care units are not standardized. The purpose of this chapter is to describe one state's approach to creating a standard for High-Risk Infant Follow-Up.

A. DeBattista (✉)
Developmental-Behavioral Pediatrics, Stanford Children's Health,
750 Welch Road, Suite 212, Palo Alto, CA 94304 USA
e-mail: adebattista@stanfordchildrens.org

H. Needelman, B. J. Jackson (eds.), *Follow-Up for NICU Graduates*,
https://doi.org/10.1007/978-3-319-73275-6_11

Nationally, there are states that support HRIF programs through Title V Maternal and Child Health Block (MCHB) Grant Programs, a longstanding federal-state partnership designed to improve access to high-quality preventive and rehabilitative care, reduce infant mortality, and provide family-centered, community-based systems of coordinated care for children with special healthcare needs. California is one state that has an established Title V program, referred to as California Children's Services (CCS), which has been in existence since the polio epidemic. In order to be a CCS-approved center, hospitals must comply with program standards and quality metrics. The overall CCS program provides access to high-quality specialty medical care for specific chronic and disabling conditions and requires California residents to meet medical criteria and financial eligibility. To be financially eligible, a family's adjusted gross income must be less than $40,000, or their out-of-pocket medical expenses for the child are expected to be more than 20% of the family income. There is no financial eligibility requirement for diagnostic or specific rehabilitative services.

The CCS HRIF program was established in 1979 as a diagnostic service to identify infants who might develop CCS-eligible conditions after discharge from a CCS-approved neonatal intensive care unit (NICU). CCS program standards require that each approved NICU identify infants at risk and ensure that they receive follow-up after discharge, either through their own organized HRIF program or a written agreement for provision of these services by another CCS-approved HRIF program. The following HRIF program description is based on this CCS model of care.

Medical Eligibility for CCS High-Risk Infant Follow-Up (Prematurity and Beyond)

Medical eligibility is multifaceted in the CCS HRIF programs (Attachment A). Firstly, children must have been admitted and determined eligible for CCS coverage in a CCS-approved NICU or PICU or had a CCS-eligible medical condition. Secondly, they must have at least one of the following risk factors.

1. The birth weight was less than or equal to 1500 g *or* the gestational age at birth was less than 32 weeks.
2. The birth weight was more than 1500 g and the gestational age at birth was 32 weeks or more, *and* one of the following documented criteria was met during the NICU stay:

   (a) pH less than 7.0 (on an umbilical cord blood sample or a blood gas obtained within 1 h of life) or an Apgar score of less than or equal to 3 at 5 min, or an Apgar score of less than 5 at 10 min

   (b) An unstable infant manifested by hypoxia, acidemia, hypoglycemia, and/or hypotension requiring pressor support

   (c) Persistent apnea which required caffeine or other stimulant medication for the treatment of apnea at discharge

(d) Required oxygen for more than 28 days of hospital stay and had radiographic finding consistent with chronic lung disease
(e) Infants placed on extracorporeal membrane oxygenation (ECMO)
(f) Infants who received inhaled nitric oxide greater than 4 h and/or treatment during hospitalization with sildenafil or other pulmonary vasodilatory medications for pulmonary hypertension
(g) History of observed clinical or electroencephalographic (EEG) seizure activity or receiving antiepileptic medication(s) at time of discharge
(h) Evidence of intracranial pathology, including but not limited to intracranial hemorrhage, white matter injury including periventricular leukomalacia, cerebral thrombosis, cerebral infarction or stroke, congenital structural central nervous system (CNS) abnormality, or other CNS problems associated with adverse neurologic outcomes
(i) Other documented problems that could result in a neurologic abnormality, such as:

- History of central nervous system infection
- Documented sepsis
- Bilirubin at excessive levels concerning for brain injury as determined by NICU medical staff
- History of cardiovascular instability as determined by NICU medical staff due to sepsis, congenital heart disease, patent ductus arteriosus (PDA), necrotizing enterocolitis, and other documented conditions
- Clinical history and/or physical exam findings consistent with neonatal encephalopathy

These risk factors for HRIF medical eligibility are associated with future developmental delays and disabilities. Children with these risk factors often have multiple medically complex conditions requiring specialty services and have high needs for care coordination.

Enrollment in the High-Risk Infant Follow-Up Program

It takes a village to identify and enroll an infant in HRIF, and the process is most successful if started in the hospital. After a child is determined to meet medical eligibility for enrollment in HRIF, physicians, nurses, social workers, and developmental specialists should take every opportunity to explain and support the family's engagement in this "usual and customary" continuation of care after discharge from the hospital. The first appointment in an HRIF clinic is scheduled between 4 and 8 months adjusted age and should be listed in the hospital discharge plan that is copied to the primary care provider. If families receive information about the HRIF appointment prior to discharge, they have the opportunity to understand their child's risk factors from someone with whom they have built a relationship over the course of the hospital admission. Additionally, parents will understand the HRIF program as a support service designed to help them closely monitor their child's developmental progression and needs. After discharge, it can be disturbing for parents to hear for the first time from a stranger on the phone, or by letter, that their child is at risk for developmental delay. While many parents rely on their primary care provider to advise them on their child's development, the amount of time allotted for a visit in the primary care office is insufficient to fully evaluate (rather than screen) a child's developmental skill progression.

A dedicated and conscientious HRIF Coordinator is key to the successful identification of the HRIF-eligible infant and enrollment in HRIF services. The HRIF Coordinator can be a pediatrician or neonatologist, pediatric nurse practitioner, nurse specialist, psychologist, social worker, physical therapist, or occupational therapist. CCS requires specific levels of education and experience to become paneled as an approved provider for the hospital-administered CCS HRIF program. The specific responsibilities of the coordinator are to serve as the primary person coordinating and connecting the child to HRIF services. This entails participating in NICU discharge planning and multidisciplinary rounds, ensuring identification of HRIF-eligible clients according to HRIF eligibility criteria, ensuring the NICU discharge planning process includes referrals to the County CCS Program or CCS Regional Office for authorization of HRIF services. Having the parents complete the application process for CCS HRIF services before their child is discharged from the hospital better ensures that CCS HRIF services will be authorized. This CCS authorization allows for care coordination, team conference, and social work services to be billed and reimbursed. These important services for complex care management are not otherwise generally reimbursed by third-party payers.

High-Risk Infant Follow-Up Team

The compilation of the HRIF team can vary among programs. CCS recognizes the members of the HRIF team to include a social worker, physicians, pediatric nurse practitioners, nurse specialists, developmental specialists, psychologists, physical and occupational therapists, an audiologist, and an ophthalmologist. It is the Coordinator's responsibility to facilitate communication about the patient among the team members and oversee the clinic flow of professionals and patients.

HRIF Clinic Visit Schedule: Assessment Over Time

The first 3 years of life is an important time for brain growth and development. In the CCS HRIF program, children are evaluated at least three times before age 3 years. The first visit is targeted to occur between 4 and 8 months adjusted age. The second visit is between 12 and 16 months adjusted age, and the third visit is between

18 month and 3 years. Although one main purpose of the HRIF program is to identify delays early and connect children to necessary intervention services, long-term outcome studies suggest that many of the morbidities of NICU graduates may not appear until later in the school years [1, 2]. Children with HRIF risk factors are at higher risk for learning disabilities, executive function problems, and behavior disorders [1, 3, 4]. It is time to consider a new standard of care for HRIF clinic visit schedules that includes an evaluation during the early school years.

HRIF Program Services

Comprehensive Care Coordination

Prior to the HRIF clinic appointment, the Coordinator does a comprehensive chart review of the medical record and prepares a summary of the child's history for the HRIF team. The HRIF Coordinator, in collaboration with the medical provider (MD or PNP), ensures the child is connected to appropriate specialty medical services for conditions identified in the NICU or during the HRIF visit. The Coordinator oversees arrangements for interpreter services if necessary. The Coordinator participates in the case conference discussion of assessment findings and in the development of the treatment recommendations. The coordinator organizes educational resources and necessary forms to ensure connection to early intervention services or public or private therapies. At the end of the clinic visit, the Coordinator is prepared to meet with families to provide detailed information about any referrals. Families are encouraged to call the Coordinator for assistance with any barriers they encounter when accessing services.

Psychosocial Assessment and Support

Children who have the HRIF medical risk factors also have parents who are at higher risk for mental health conditions secondary to the birth experience and/or the intensive medical experience of their child [5]. Additionally, these high-risk infants are at risk for vulnerable child syndrome [6]. For these reasons, one purpose of the High-Risk Infant Follow-Up program targets the well-being of the child and the parents. The CCS HRIF mandate includes the provision of a psychosocial assessment that evaluates the parent's recovery from the NICU experience and need for support services or counseling. This can be provided by a CCS paneled social worker or nurse. The psychosocial assessment includes addressing the need for community resources for financial support, including enrollment in SSI, food stamps, Women Infants and Children.

Medical Assessment

The medical provider leads the HRIF visit. In order to accomplish the goals of care during the clinic visit, it is important to review the medical history. Eliciting any parental concerns about the child's development will also guide the assessment and recommendations. It is more effective to wait to address parental concerns until the end of the visit when they can be informed by the evaluation. Gathering an interval history is important for ensuring follow-up with medical subspecialty care and informing the evaluation with regard to health status, surgeries, or additional hospital admissions. The comprehensive medical assessment includes follow-up or identification of neurosensory impairments and a careful assessment of neurological status, growth, and potential feeding issues.

High-risk infants are at increased risk for neurosensory impairments including hearing loss, visual impairments, and cerebral palsy. The first HRIF visit must report the status of hearing and vision follow-up. Children treated in the NICU must have another audiology evaluation between 24 and 36 months secondary to the increased risk for potential progressive sensorineural hearing loss. More frequent hearing tests are needed for infants with specific risk factors (persistent pulmonary hypertension, congenital exposure to toxoplasmosis or cytomegalovirus). An audiology evaluation is also needed if a speech and language delay is detected. Premature infants who have had retinopathy of prematurity require regular ophthalmologic follow-up once the retinal specialist has completed care. Infants born prematurely and those who have incurred brain injuries in the neonatal period are at higher risk for myopia, strabismus, visual field defects, and cortical visual impairment.

The CCS HRIF mandate states that a careful neurological assessment be performed by a physician or pediatric nurse practitioner at each visit to detect any early signs of cerebral palsy. The neurologic evaluation should include an age-appropriate evaluation of muscle tone, range of motion, deep tendon reflexes, presence of persistent primitive reflexes, and asymmetry of movement, abnormal posturing, and atypical movements. Evaluation of cranial nerves, sensory perception to touch, and oral motor sensitivity based on feeding history should also be included. Dysphagia and oral motor aversion are common in high-risk infants.

Growth patterns are monitored using a standardized longitudinal growth curve. For children born prematurely, both adjusted and chronologic age measurements are plotted. It is important to evaluate individual growth curves informed by feeding practices, presence of illness, and genetic influences to determine whether additional intervention is needed to optimize growth. Weight for length, or the measure of proportion, is a better indicator of adequate caloric intake than weight or length alone. Head circumference that fails to progress can be associated with problems with the developing brain or prematurely fused skull bones. In turn, a rapidly increasing head circumference (not associated with similar catch-up growth in weight and height) can be associated with the development of a post-hemorrhagic hydrocephalus. Both deviations in head circumference would likely require further assessment, often with imaging studies.

Developmental Assessment

The CCS HRIF program mandates the use of a standardized assessment measure to identify delays and disorders as early as possible in the first 3 years of life. The medical clinician, nurse specialist, psychologist, physical/occupational therapist, or developmental specialist may administer the developmental assessment. An assessment measure provides a norm-referenced score allowing comparison to same-age individuals. For children born prematurely, scores for both adjusted age and chronological age are produced, thus allowing examiners to track a child's progress moving toward developmental skills consistent with chronologic age peers. Clinicians counsel families to adjust age level developmental expectations for their prematurely born children by subtracting the number of weeks that the child was born prematurely from the child's chronological age [7]. However, the evidence that supports this clinical recommendation is difficult to find. One recent systematic review concluded that there is evidence to support adjusting age expectations until age 12 months [8]. Another systematic review reported that there is evidence to support adjusting age expectations until age 2 years for infants born at 35 weeks gestational age, but not for those born at earlier gestational ages [9]. Long-term outcome studies report continued lower mean neurodevelopmental and behavioral scores into adulthood, suggesting that a large proportion of individuals born prematurely may never fully catch up over time [10–12]. Findings to date indicate that further research is needed to support the policy and practice of adjusting age expectations in prematurely born children [8, 9, 13].

Children are required to have a norm-referenced assessment consistent with the Bayley Scales of Infant and Toddler Development-III (BSID) at least by the third HRIF visit. The BSID is an individually administered measure of developmental function of infants and toddlers. It is designed to identify delays, strengths, and weaknesses and provide a method for monitoring developmental progress [14, 15]. The BSID is currently considered the gold standard assessment measure for evaluating developmental outcomes in research and is often cited in the medical literature [16, 17]. Some programs use a combination of briefer assessment measures over the first two visits because of the shorter attention spans of infants and toddlers. Areas of social emotional development and adaptive behavior are often evaluated through the history and additional screening measures (see Table 1 for an example of evaluation measures). A systematic and careful assessment of development informs the issue specific and anticipatory guidance provided to the family, as well as the need for early intervention services.

Connection to Appropriate Community-Based Early Intervention Services

Early intervention (EI) services provide experiences designed to facilitate neuroplasticity and aid in the progression of developmental skills from birth to 3 years of age [18]. Neuroplasticity is a process of recovery from brain injury and progression

Table 1 HRIF schedule of evaluations

CCS HRIF core visit intervals	4–8 months adjusted age	12–16 months adjusted age	18–36 months adjusted age
Evaluators	Physician or pediatric nurse practitioner Social worker Dietician, as needed OT or PT, as needed	Physician or pediatric nurse practitioner Social worker, as needed Dietician, as needed OT or PT, as needed	Physician or pediatric nurse practitioner Psychologist for Bayley Scales of Infant Development Social worker, as needed OT or PT, as needed Dietician, as needed
Assessment measures	Capute Scales Ages and Stages Social Emotional	Capute Scales Bayley Scales Gross Motor Domain Ages and Stages Social Emotional	Bayley Scales of Infant and Toddler Development-III MCHAT
Special requirements	Report results of hearing screening Report follow-up of vision testing		Plan for subsequent follow-up: Clients of CCS Medical Therapy Program should be referred to neurology when graduating from HRIF Audiology between 24 and 36 m

CCS supports three clinic visits up to age 3 years that need to be scheduled at the appropriate intervals. Visits between or beyond these intervals should be recorded as "additional visits"

toward age-appropriate skills [19]. Early intervention services are often comprised of a combination of private and public community resources. Private resources include services that are purchased by the family or obtained using medical insurance, such as physical, occupational, or speech therapy. Public services are regulated and supported through local and federal government.

Public Early Intervention Service Delivery Through CCS Medical Therapy Program

Public services are paid for by tax revenue. California has another CCS program that does not require families to be financially eligible, the Medical Therapy Program (MTP). The CCS Medical Therapy Program provides free physical and occupational therapy to children at specific public school sites. Medical eligibility is specific to chronic musculoskeletal and neurological conditions, with the exception of children under the age of 3 years at risk for cerebral palsy who have two neurological findings on exam [20]. One purpose of documenting a careful neurological examination in HRIF clinic is to facilitate eligibility determination for the CCS MTP and early connection to physical and occupational therapies. Access to early treatment with physical and occupational therapy can help at-risk children develop motor skills [21].

Public Early Intervention Service Delivery Through IDEA: Part C

Federal legislation for early intervention services is outlined under the Individuals with Disabilities Education Act Part C (IDEA Part C) for children ages birth to 3. Although these services are mandated at the federal level, the implementation of regulations and services provided varies widely [18, 22–27]. Less than a handful of states currently provide EI services based on the eligibility category of "at risk." Access to services by this category provides an avenue for premature infants with biologic risk factors to obtain EI services that begin in the first year of life. Without it, children need to have an established developmental disability or have developmental scores "delayed enough" to qualify. The percentage of delay needed to qualify for Part C eligibility varies by state. Additionally, for children born prematurely, scores based on expectations for adjusted age (higher scores) are used to calculate the level of delay. This results in children born prematurely needing to be more delayed than term born peers in order to qualify for EI services designed to help them catch up to same-age peers. By the time many qualify, they are in their second year of life, leaving little time to receive services before they age out of their eligibility for services at age 3.

While the scientific literature suggests that receipt of developmental interventions shortly after birth results in better brain development and outcomes [21, 28], in 2012, less than 1% of the pediatric population in California received Part C services in the first year of life [29]. Although exact numbers of children born prematurely are not specified, about 10% of the pediatric population was born prematurely, suggesting a significant gap in service receipt in their first year of life [30]. Other evidence for this gap was provided in a recent report which noted that a large percentage of children in California was not referred to EI services in the first year of life, although they were presumably eligible by developmental scores reflecting percentage of delay [27]. Additionally, on a national scale, only 53% of the most vulnerable of premature infants, born at less than 28 weeks gestation, received any early intervention by 18 months adjusted age [31]. With large proportions of children born prematurely not closing the gap, the functionality of the referral process is critical to ensure these children receive necessary interventions.

The Big Picture: Site- and Population-Based Outcome Information

Challenges continue to plague the NICU clinician's ability to connect medical risks with neurodevelopmental outcomes. There are HRIF program models that employ personnel actively working in the NICU to contribute to the HRIF clinic service provision. This approach provides an opportunity for families to reconnect with the clinicians who saved their infant's life. It also helps clinicians appreciate individual outcomes of neonates they cared for. However, many NICU clinicians feel ill prepared to expand their scope of practice to the assessment and treatment of the host of developmental and behavioral disorders common in this high-risk population. While there are HRIF programs involved with network randomized controlled trials or observational studies that produce outcomes shared with the NICU staff, the ability to evaluate local and regional population outcomes has been limited. In California, the Children's Medical Services (CMS) Branch in collaboration with the California Children's Services (CCS)/California Perinatal Quality Care Collaborative (CPQCC) High-Risk Infant Follow-Up (HRIF) Quality of Care Initiative (QCI) has developed a web-based reporting system for CCS HRIF-eligible children. The CCS HRIF program submits data from each visit to CPQCC using an online, web-based reporting system. This newly assembled statewide data repository is designed to identify improvement opportunities for NICUs in the reduction of long-term morbidity. Each CCS NICU will have access to aggregated HRIF outcome data specific to their site. Papers are just beginning to be published using this statewide data [27, 32].

Appendix

Attachment A

UPDATED 01/2016

CCS HRIF PROGRAM MEDICAL ELIGIBILITY CRITERIA

Data should be collected on infants/children under three years of age who meet California Children's Services (CCS) HRIF medical eligibility criteria and who met CCS medical eligibility criteria for Neonatal Intensive Care Unit (NICU) care OR had a CCS eligible medical condition at some time during their stay in a CCS-approved NICU, even if they were never a CCS client. Infants are medically eligible for the HRIF Program when the infant:

Met CCS medical eligible criteria for NICU care, in a CCS-approved NICU, regardless of length of stay, *(as per Number Letter 05-0502, Medical Eligibility in a CCS-approved NICU).* **NOTE:** Medical eligibility includes neonates who require direct admit to a CCS-approved PICU, who are never admitted to a CCS-approved NICU, but who otherwise meet all medical eligibility criteria for HRIF services.

OR

Had a CCS eligible medical condition in a CCS-approved NICU, regardless of length of stay, *(as per California Code of Regulations, Title 22, Section 41800 through 41872, CCS Medical Eligibility Regulations).*

AND MET ONE OF THE FOLLOWING:

Birth weight ≤ 1500 grams or the gestational age at birth < 32 weeks.

OR

Birth weight > 1500 grams and the gestational age at birth ≥ 32 weeks and one of the following criteria was met during the NICU stay:

A. pH less than 7.0 on an umbilical blood sample or a blood gas obtained within one hour of life) or an Apgar score of less than or equal to three at five minutes or an Apgar score less than 5 at 10 minutes.

B. An unstable infant manifested by hypoxia, acidemia, hypoglycemia and/or hypotension requiring pressor support.

C. Persistent apnea which required caffeine or other simulant medication for the treatment of apnea at discharge.

D. Required oxygen for more than 28 days of hospital stay and had radiographic finding consistent with chronic lung disease (CLD).

E. Infants placed on extracorporeal membrane oxygenation (ECMO).

F. Infants who received inhaled nitric oxide greater than four hours, and/or treatment during hospitalization with sildenafil or other pulmonary vasodilatory medications for pulmonary hypertension.

G. History of observed clinical or electroencephalograhic (EEG) seizure activity or receiving antiepileptic medication(s) at time of discharge.

H. Evidence of intracranial pathology, including but not limited to, intracranial hemorrhage (grade II or worse), white matter injury including periventricular leukomalacia (PVL), cerebral thrombosis, cerebral infarction or stroke, congenital structural central nervous system (CNS) abnormality or other CNS problems associated with adverse neurologic outcome.

I. Clinical history and/or physical exam findings consistent with neonatal encephalopathy.

J. Other documented problems that could result in neurologic abnormality, such as: history of CNS infection, documented sepsis, bilirubin at excessive levels concerning for brain injury as determined by NICU medical staff, history of cardiovascular in stability as determined by NICU medical staff due to: sepsis, congential heart disease, patent ductus arteriosus (PDA), necrotizing enterocolitis, other documented conditions.

HRIF Program Referral Process:

Communication is between the CCS-approved NICU and HRIF Program.

1. The discharging/referring NICU/Hospital will refer eligible infants to the HRIF Program at the time of discharge to home, and complete the "Referral/Registration (RR) Form" via the web-based HRIF-QCI Reporting System.
2. The discharging/referring NICU/Hospital or HRIF Program will submit a Service Authorization Request (SAR) to local CCS Office. (Service Code Group [SCG] 06, should be requested). http://www.dhcs.ca.gov/services/ccs/cmsnet/Pages/SARTools.aspx
3. The discharging/referring NICU/Hospital will send a copy of the Discharge Summary to the HRIF Program.

Medical eligibility for the HRIF Program is determined by the County CCS Program or Regional Office staff. The CCS Program is also required to determine residential eligibility. As the HRIF Program is a diagnostic service, there is no financial eligibility determination performed at the time of referral to CCS. However, insurance information shall be obtained by CCS. An infant or child is eligible for the HRIF Program from birth up to 3 years of age.

References

1. Chyi, L. J., et al. (2008). School outcomes of late preterm infants: Special needs and challenges for infants born at 32 to 36 weeks gestation. *The Journal of Pediatrics, 153*(1), 25–31.
2. Peterson, B. S., et al. (2000). Regional brain volume abnormalities and long-term cognitive outcome in preterm infants. *JAMA, 284*(15), 1939–1947.
3. Loe, I. M., et al. (2011). Behavior problems of 9 to 16 year old preterm children: Biological, sociodemographic, and intellectual contributions. *Early Human Development, 87*(4), 247–252.
4. Loe, I. M., et al. (2012). Executive function skills are associated with reading and parent-rated child function in children born prematurely. *Early Human Development, 88*(2), 111–118.
5. Shaw, R. J., et al. (2013). Parental coping in the neonatal intensive care unit. *Journal of Clinical Psychology in Medical Settings, 20*(2), 135–142.
6. De Ocampo, A. C., et al. (2003). Caretaker perception of child vulnerability predicts behavior problems in NICU graduates. *Child Psychiatry and Human Development, 34*(2), 83–96.
7. Pediatrics, A.A.o. *Preemie Milestones*. (2013). [Cited last *Updated12/3/20 Source Supporting you and your preemie: Milestone guidelines for premature babies* (Copyright © 2008 American Academy of Pediatrics © Copyright 2013 American Academy of Pediatrics. All rights reserved. Retrieved from http://www.healthychildren.org/English/ages-stages/baby/preemie/Pages/Preemie-Milestones.aspx.
8. D'Agostino, J. A. (2010). An evidentiary review regarding the use of chronological and adjusted age in the assessment of preterm infants. *Journal for Specialists in Pediatric Nursing, 15*(1), 26–32.
9. Wilson, S. L., & Cradock, M. M. (2004). Review: Accounting for prematurity in developmental assessment and the use of age-adjusted scores. *Journal of Pediatric Psychology, 29*(8), 641–649.
10. Luu, T. M., et al. (2011). Evidence for catch-up in cognition and receptive vocabulary among adolescents born very preterm. *Pediatrics, 128*(2), 313–322.
11. Hack, M. (2009). Adult outcomes of preterm children. *Journal of Developmental and Behavioral Pediatrics, 30*(5), 460–470.
12. Saigal, S., et al. (2007). Comparison of current health, functional limitations, and health care use of young adults who were born with extremely low birth weight and normal birth weight. *Pediatrics, 119*(3), e562–e573.
13. D'Agostino, J. A., et al. (2013). Provider use of corrected age during health supervision visits for premature infants. *Journal of Pediatric Health Care, 27*(3), 172–179.
14. Albers, C. A., & Grieve, A. J. (2007). Test review: Bayley, N. (2006). Bayley Scales of infant and toddler development—Third edition. *Journal of Psychoeducational Assessment, 25*(2), 180–190.
15. Mackin, R., Fadel, N. B., Feberova, J., et al. (2017). ASQ3 and/or the Bayley-III to support clinicians' decision making. *PLoS One, 12*(2), e0170171.
16. Hintz, S. R., et al. (2005). Changes in neurodevelopmental outcomes at 18 to 22 months' corrected age among infants of less than 25 weeks' gestational age born in 1993–1999. *Pediatrics, 115*(6), 1645–1651.
17. Vohr, B. R., et al. (2000). Neurodevelopmental and functional outcomes of extremely low birth weight infants in the National Institute of Child Health and Human Development Neonatal Research Network, 1993–1994. *Pediatrics, 105*(6), 1216–1226.
18. Rosenberg, S. A., et al. (2013). Part C early intervention for infants and toddlers: Percentage eligible versus served. *Pediatrics, 131*(1), 38–46.
19. Cicchetti, D., & Cohen, D. J. (2006). Developmental psychopathology. In D. Cicchetti & D. J. Cohen (Eds.), *Developmental neuroscience* (Vol. 2, p. 875). Hoboken, NJ: Wiley.
20. Services, C.D.o.H.C. (2017). *Medical therapy program*. Retrieved from http://www.dhcs.ca.gov/services/ccs/Pages/MTP.aspx.
21. Spittle, A., et al. (2009). Early developmental intervention programs post hospital discharge to prevent motor and cognitive impairments in preterm infants [Systematic Review]. *Cochrane Database of Systematic Reviews, 3*, 3.

22. Center, D. A. (2007). *Individuals with Disabilities Education Act (IDEA)—Part C National Trend Data.* Table C2 number and percent of population served, by age group (0–1, 1–2, 2–3): 1998 through 2007. IDEA 618 Data Tables. Retrieved from http://www.ideadata.org/PartCData.asp.
23. Dunst, C. J., & Bruder, M. B. (2002). Valued outcomes of service coordination, early intervention, and natural environments. *Exceptional Children, 68*(3), 368–375.
24. Mott, D., & Dunst, C. J. (2006). Use of presumptive eligibility for enrolling children in Part C Early Intervention. *Journal of Early Intervention, 29*(1), 22–31.
25. Rosenberg, S. A., Zhang, D., & Robinson, C. C. (2008). Prevalence of developmental delays and participation in early intervention services for young children. *Pediatrics, 121*(6), e1503–e1509.
26. Wang, C. J., et al. (2009). Factors influencing the enrollment of eligible extremely-low-birth-weight children in the Part C Early Intervention program. *Academic Pediatrics, 9*(4), 283–287.
27. Tang, B. G., et al. (2012). Missed opportunities in the referral of high-risk infants to early intervention. *Pediatrics, 129*(6), 1027–1034.
28. Feldman, R., Rosenthal, Z., & Eidelman, A. I. (2014). Maternal-preterm skin-to-skin contact enhances child physiologic organization and cognitive control across the first 10 years of life. *Biological Psychiatry, 75*(1), 56–64.
29. OSEP. (2013). *2012 Part C child count and services data.* Retrieved from https://explore.data.gov/Education/2012-IDEA-Part-C-Child-Count-and-Settings/dg4k-psxe.
30. KIDS COUNT Data Set. (2014). A.E.C. Foundation (Eds.). Retrieved from http://datacenter.kidscount.org.
31. Hintz, S. R., et al. (2008). Community supports after surviving extremely low birth weight, extremely preterm birth: Special outpatient services in early childhood. *Archives of Pediatrics, 162*(8), 748–755.
32. Fuller, M. R. (2015). *Factors associated with high risk infant follow-up attendance, in digital USD citation*. San Diego: UCSD.

Personnel in the NICU Follow-Up Clinic

Cheryl A. Milford

Abstract The staff in the high-risk neonatal infant follow-up clinic is critical to its success in caring for the children and their families. This chapter defines the personnel that are required, the training and expertise they need, and their roles in the clinic. Types of clinics and their settings are also discussed within the context of the types of evaluation and therapy that can be offered. Recommendations for optimal and minimum personnel will be specified.

Abbreviations

HRIFU	High-risk infant follow-up
NICU	Neonatal intensive care unit

Introduction

The personnel who staff the high-risk infant follow-up clinic provide the foundation for the program to deliver high-quality services. In 2004, Vohr et al. noted that developmental screening and assessment are essential for this population to monitor neonatal intensive care unit (NICU) interventions and their outcomes. The data collected from the assessments supports policy discussions and public health initiatives to improve outcomes. The American Academy of Pediatrics (2008) in its policy statement on the discharge of high-risk infants stated, "High-risk infants should be enrolled in a follow-up clinic that specializes in the neurodevelopmental assessment through early childhood." (p. 1119) [1], indicating that infants of lower gestational

C. A. Milford (✉)
Cheryl Milford Consulting, Orange County, CA, USA

National Perinatal Association, Lonedell, MO, USA
e-mail: cheryl.milford@gmail.com

H. Needelman, B. J. Jackson (eds.), *Follow-Up for NICU Graduates*,
https://doi.org/10.1007/978-3-319-73275-6_12

age and birth weights are surviving in greater numbers and require follow-up with providers who can provide both physical and developmental assessments. In 2014, Petra et al. observed that high-risk follow-up clinics not only provide assessments to support optimal developmental outcomes for infants and their families but are necessary for the education of medical and psychology trainees in the work with these infants as they mature.

Meeting all of these goals stated in the literature requires the recruitment of professionals from different disciplines with specialized training in assessing and providing intervention services to high-risk infants and their families. Such professionals have the skills to assist families in understanding their child's medical and developmental needs, provide referrals to needed resources, and provide anticipatory guidance for ongoing developmental and medical issues.

The Rationale for Specially Trained Personnel

All of the professionals in the high-risk infant follow-up (HRIFU) clinic should have training, expertise, and experience in working with NICU graduates and their families. HRIFU professionals need a comprehensive understanding of the impact of preterm birth, neurological disorders, and perinatal asphyxia on both the short-term and long-term developmental trajectories of the child [2]. While initially the family is significantly impacted by the NICU environment, the family resources and mental health status of the parent on developmental outcomes become more significant as the child matures [3]. Many times working with a mentor with significant HRIFU experience facilitates the integration of new staff. Professionals providing services in HRIFU clinics must be aware of the current literature regarding the impact of these factors and issues to provide competent assessments and interventions to the infants and their families. Professionals require competence in knowledge and skills concerning the emotional and psychological effects of the NICU hospitalization on the parents and other family members [3]. The mental health of the parent significantly influences the developmental and mental health outcomes of the child [3, 4]. HRIFU professionals must also be current and competent in evidence-based literature on standard of care policies and protocols for HRIFU clinics [3].

The NICU is a stressful environment and family members continue to process the experience after discharge. If the professionals in the HRIFU have not worked in the NICU, they will need significant time and opportunity to observe and receive training to be able to support the transition from the NICU to home and outpatient services [3]. Professionals who have the necessary experience working with families and infants in the NICU can also receive training to work in the HRIFU clinic. If these professionals are working in both the NICU and HRIFU, they provide continuity of care for the infants and families. Such an approach can also help with decreasing attrition rates of families from the clinic [5].

Professionals across disciplines who are trained in working with NICU infants and NICU graduates and their families should also be skilled in working with other disciplines in collaborative practice to support the assessment and treatment goals that occur in high-risk infant follow-up clinics [6]. While all team members working in the clinic are defined professionally by their discipline, they must have the skills and ability to respect the collaborative practice approach land be open to discussion and feedback [7]. The experience and training these professionals offer is the structural integrity of the HRIFU clinic and empowers families to provide optimal care to their high-risk children. Family-centered care tenets of respect, empowerment, advocacy, and parental decision-making that began during the NICU hospitalization should continue into the HRIFU program. Working with families to make appointments that fit within their schedule, providing assessment results and recommendations to families, and then developing a plan of care with them facilitate a positive and caring relationship with the professionals and encourage return visits [5, 6].

The Commitment to the Team by the Institution

The HRIFU clinic is an integral part of the perinatal-neonatal services offered by the hospital or healthcare system [8]. For academic institutions, it is required for the training of pediatric residents and neonatal fellows. The HRIFU clinic provides the institution with the ability to evaluate the outcomes of graduates from its NICU, engage in quality improvement initiatives around care practices, and analyze data regarding sequela of medical conditions treated in the NICU [9].

The HRIF clinic is not a revenue stream for the institution. Without grants or endowments, this program is generally not able to meet its expenses. It is important for hospital and healthcare system administrators to understand that the role of the clinic is supporting the work of the perinatal and neonatal services and continuity of care is important. The professionals who staff the HRIFU clinic are highly trained and experienced. Their compensation packages reflect that training and experience. Once the NICU has determined their HRIFU criteria and visit schedule, then developing a schedule for dates, times, and location of the clinic. Continuity of care is supported by the approach of HRIFU staff providing services in both the NICU and HRIFU; however some institutions have professionals that only work in the HRIFU clinic [5, 10].

Whatever approach the institution or healthcare system takes, the administration needs to be knowledgeable and respectful of the role of the HRIFU clinic in its perinatal-neonatal services. Appropriate funding, space, and resources must be provided and updated as necessary to maintain standard of care services.

Personnel by Discipline, Competencies, and Training

Vohr et al. [9] outline the types of assessment and therapy services needed for developmental assessment of high-risk infants. Physical growth and medical assessment and referral for treatment are essential. The other areas for assessment and treatment include cognitive, fine, and gross motor skills, sensory processing, and adaptive, behavioral, and language abilities. Memory and learning skills become important as the child matures. For developmental concerns, together with the family, HRIFU can collaborate with the local early intervention services to facilitate timely referral for educational and developmental therapy services. Medical issues should be referred to appropriate subspecialists with the family's knowledge and consent. The authors recommended follow-up of high-risk children until the age of 8 years [9]. The expertise in assessment and treatment in these areas requires several disciplines.

Medical

The medical team can include neonatologists, neonatal nurse practitioners, pediatric nurse practitioners, and developmental pediatricians. The medical team assesses the child's medical condition, growth parameters, and ongoing medical needs. HRIFU medical professionals need to be respectful and cognizant of the medical professionals in the child's medical home. A report to the primary care provider and consultation before referral for subspecialist services helps to maintain an effective working relationship with other providers. Prescriptions for medications and services are also completed by this team member, often in consultation with the primary care provider. Neonatal medical team members are important for supporting the family in their interactions with medical specialists. Providing anticipatory medical guidance regarding the child's medical needs as a result of their NICU hospitalization and ongoing conditions is important for the family so they can be competent healthcare consumers for their child. As the child matures, pediatric nurse practitioners and developmental pediatricians are the appropriate medical professionals to work with the child and his or her family [5]. For academic centers, pediatric residents and neonatal and perinatal fellows also participate as part of the training.

Nursing

In some programs, NICU nurses complete pre-visit and post-visit phone calls with families. They complete an extensive chart review, and they report this information to the HRIFU professional team at the onset of the visit. During the clinic visit, they complete growth parameter assessment and take a history of immunizations, pediatric visits, and hospitalizations since the last HRIFU visit. It is appropriate for pediatric nurses to complete these activities after the first year of life. Nurses who work in the NICU can be the continuity of care from the NICU to the first year of developmental assessment. This continuity of care can also assist with decreasing attrition from the HRIFU clinic. Nurses develop caring and supportive relationships with families during their neonatal hospitalization. Parents feel more comfortable and familiar with the nurses. They may be more willing to attend clinic to continue that supportive and important relationship. Nurses should have training in the impact of prematurity and medical complications during the neonatal period on the long-term physical and medical needs of the high-risk infant and child [4, 5].

Mental Health

Families who have experienced a NICU hospitalization often have many psychological and resource needs. Social workers assist the families with obtaining the needed resources to care for their children. They work together with the family to contact community agencies and government services based on an assessment of the family's resource needs. Perinatal social workers have the appropriate training and expertise to provide these services in the HRIFU clinic. Psychologists as part of the team support the assessment and treatment of psychosocial, developmental, and perinatal mental health complications such as postpartum depression, anxiety, and posttraumatic stress disorder, usually related to the NICU experience. Psychologists with training in neurodevelopmental and long-term developmental assessment can also complete standardized assessment tools. Psychologists in the HRIFU clinic must have the requisite knowledge, expertise, and skills in perinatal mental health complications and mental health issues that arise from NICU hospitalizations for all members of the family. Social workers and psychologists should have training in infant mental health concepts, trauma-informed care, and the impact of the NICU experience on family dynamics, functioning, and mental health [3, 11].

Developmental Therapy

Occupational therapists with training in NICU care provide fine motor, sensory processing assessment and treatment to support the sensory motor, sensory integration, and adaptive skills for the high-risk child and the family. Physical therapists with NICU care training provide gross motor, muscle tone, and balance assessment and treatment for the child and family. Developmental therapists can be early childhood educators or neonatal nurses who have training, expertise, and skills in developmental assessment and intervention. It is essential for developmental therapists working in the HRIFU clinic to have knowledge of the long-term impact of prematurity and critical illness in the neonatal period on the development of the sensory and motor systems. An understanding of brain development for preterm infants is critical to their work in the HRIFU program. The national occupational and physical therapy associations have competencies developed for therapists working in the NICU and high-risk follow-up clinic. The institution's administration should hire developmental therapists who meet these competencies or are committed to meeting these competencies through ongoing training in order to provide standard of care in the clinic setting. In addition, there is a national organization for neonatal therapists which supports training and education for all therapists working in these settings. For further information, visit their website: neonataltherapists.com.

Dietician

Physical growth and nutrition are ongoing issues for the high-risk infant after discharge from the NICU. Families have numerous concerns about nutrition and growth, especially in the first year of life. The neonatal dietician has many responsibilities in the NICU but should be available for consultation to the HRIFU clinic. Dietician consultations are necessary for infants with growth concerns, both too slow and too advanced weight, head circumference, or length for either chronological or corrected age. In addition, HRIFU professionals, specifically medical and nursing, should collaborate with the dietician to develop instructional information for HRIFU families. These materials generated from this collaboration should provide information on optimal caloric intake, transitions to solid foods, and healthy food choices by the age of the child [9, 12].

Administrative Support

The HRIFU administration should include a medical director and a clinic coordinator. The medical director is responsible for the medical policies, procedures, and schedules for the medical team. The clinic coordinator is a HRIFU professional responsible for the policies, procedures, and scheduling related to allied health professional services and nursing. The clinic coordinator works closely with the medical director to change clinic practices by being vigilant to the evidence-based literature, family needs, and community resources. The HRIFU clinic requires a clinic assistant who manages appointment scheduling, insurance, and billing for each visit. This individual answers the clinic phone, distributes clinic reports to subspecialist and primary care providers, and greets families when they arrive for their visit [5]. The clinic assistant with background as a NICU unit secretary is highly desirable for their knowledge of working with families and NICU culture.

Evaluation and Intervention Models

When providing care in the HRIFU clinic, the methods by which the various disciplines evaluate and provide therapy should be evidence-based and tailored to the particular institution and community [7]. There are three models for evaluation and intervention [13]. A discipline is an area of knowledge and expertise gained through education, training, and experience. The three models are multidisciplinary, interdisciplinary, and transdisciplinary. Each can work in clinical and research settings.

Multidisciplinary

This is the first model that emerged, and it is most commonly used in clinical practice in healthcare settings. Each discipline works within their scope of practice and collaborates with the rest of the team to develop an evaluation and care plan (e.g., referral to early intervention or clinical practices in healthcare settings) after they have completed their work with the child and family. Multidisciplinary teams respect and value the expertise of each discipline in caring for the high-risk child and their family [7, 13]. This approach can be time-consuming and confusing for families if the multidisciplinary team does not collaborate and communicate effectively. Multidisciplinary teams in intervention programs meet to discuss their findings and their treatment plans, but they do not collaborate or develop plans together. In clinic settings, each discipline discusses their findings and plan. Families are left to work with each discipline independently and often without regular communication among the team.

Interdisciplinary

Often multidisciplinary teams evolve into interdisciplinary teams for the purpose of efficiency and cost containment. Interdisciplinary teams work to educate each other about their areas of expertise and support other team members in learning about their theoretical foundations, skills, and approaches to evaluation and treatment. In this way, the HRIFU team can engage in evaluation and therapy with the ability to understand the other discipline and learn some of its skills as well as respecting and valuing other disciplines. The discipline of infant mental health emerged from interdisciplinary research and practice [13]. Infant mental health professionals come from all of the disciplines in HRIFU and communicate with common theoretical constructs, language, and assessment approaches that facilitate communication and collaboration in working together with the family.

In HRIFU clinics, interdisciplinary teamwork can be observed with developmental therapists and mental health professionals. Nurses and dieticians also often work together with interdisciplinary approaches and services. All HRIFU professionals need to gain skills in administrative and patient retention activities to support the viability of the clinic [7, 13]. Many HRIFU clinics have evolved into interdisciplinary teams to provide optimal services within limited budgets and resources. Intervention programs often have the developmental therapists learn skills from each other to provide services with one professional. This professional can develop an ongoing relationship with the family, and the family is not overwhelmed by the numbers of professionals providing services to them.

Transdisciplinary

This evaluation and therapy model is seen much less in healthcare settings. Transdisciplinary teams are defined by [13] as "… work using a shared conceptual framework, drawing together discipline-specific theories, concepts and approaches to address a common problem (p. 355)." Discipline scope of practice is enhanced by learning other team members' theoretical foundations, skills sets, and expertise in specific areas of assessment and treatment. A single team member can therefore be the conduit with the family. An example of a transdisciplinary approach is the Neurorelational Framework developed by Lillas and Turnbull in 2009. The framework utilized the theoretical models from neuroscience, psychology, education, attachment, and occupational therapy to develop a part-to-whole training, assessment, and treatment plan. In the Neurorelational Framework, all disciplines work together across all diagnostic categories, and each team member can see the whole picture with regard to the assessment and intervention needs of the child and family [6].

The Neurorelational Framework transdisciplinary model is currently being implemented in juvenile courts, mental health, and early intervention programs [6]. This approach utilizes established theoretical constructs and integrates them into a new assessment and intervention approach that not only utilizes a common language, but a sharing of skills and expertise to support each discipline in understanding and integrating the lens of other disciplines into their practice.

The above models of care are determined by the institution based on its mission, goals, and financial and professional resources. The process of deciding which model to utilize should occur in a collaboration between administrators and NICU providers with thoughtful respect for each member's expertise and concerns. The structure of the HRIFU program emerges out of this collaboration.

Clinic Models

The vast majority of HRIFU clinics are part of hospitals and healthcare systems [1, 8, 14, 15]. Within academic healthcare systems, the HRIFU clinic is a training site for pediatric residents and neonatal and perinatal fellows as well as a clinical service. Understanding families' preferences for days and times of appointments leads to efficacy of resource utilization. Frequent reminders of appointments, pre- and post-visit calls from nursing, and/or developmental therapy and short waiting times at the visit support families' willingness to continue their attendance.

HRIFU services are important for the optimal outcomes for high-risk infants and their families.

HRIFU programs have also initiated Internet-based parent questionnaires to support developmental screening of high-risk infants and children [10, 16]. These questionnaires were helpful in gathering data, but were not valid and reliable with standardized neurodevelopmental assessment tools. This research demonstrated that high-risk (less than 1500 grams birthweight, less than 32 weeks of gestational age, and/or significant social risk factors) infants and children require regular visits to the HRIFU clinic for accurate neurodevelopmental assessments [10]. The data from the questionnaires can be used as part of the interview process and assessment of the child. Some HRIFU programs utilized the AQS or PEDS tools which are valid and reliable, but are not specific to the NICU graduate population.

Developmental screening and assessment can also be conducted by early intervention professionals in community agencies. These are often completed as government-authorized high-risk infant monitoring programs such as early intervention high-risk infant tracking programs and public health nursing mother-baby tracking programs [12, 15]. The children are usually seen by a team of developmental therapists, educators, and/or nurses. If the child has been evaluated and found to require therapy services, these are provided by the early intervention agency or are contracted to other therapists or agencies [6].

Comprehensive HRIFU services require medical, nursing, and mental health services as well as developmental therapy. These services are available in the HRIFU clinics that are operated in hospitals and healthcare systems [9, 15]. Early intervention providers and HRIFU clinics should collaborate in regarding their assessments and treatment plans.

All children should receive developmental screening at specific well-child visits as part of developmental surveillance during their first 3 years of life in the primary care setting [17]. This screening is especially important for the high-risk child. If the primary care healthcare providers have concerns about the child's developmental screening results, they should contact medical development treatment centers or the appropriate early intervention agency, the developmental specialists, or the HRIFU clinic for an appointment. If there are pressing concerns, the HRIFU clinic should be able to address the concerns in a timely fashion. All of the providers can support the family and child best by maintaining communication and collaboration in their assessment and treatment goals and plans. Sharing assessments and intervention plans among providers with the family's permission decreases the chances of overlap and duplication of resources and services.

All of these settings and models can be part of a continuous and comprehensive high-risk infant follow-up, developmental assessment, and intervention. It takes effort and dedication to develop relationships and collaboration with other providers; often joint educational opportunities and conferences can support the process and encourage professionals to network and develop new professional relationships within their community.

Recommendations

HRIFU programs should have the available expertise of multiple disciplines. These include:

- Physician and/or physician extender (e.g., neonatologist or developmental pediatrician or neonatal or pediatric nurse practitioner)
- Nurse
- Social worker
- Psychologist
- Dietician
- Lactation consultant
- Occupational therapist
- Physical therapist
- Case managers
- Clinic coordinator
- Administrative assistant [12]

With such a team, the family and infant remain in the same place and the providers come to them. Sufficient physical space for families and space for provider consultation are necessary. Booking appointments can be based on the number of providers to assess the child at that visit and the age distribution of that day's schedule. Each child and family may not need to see each provider at every visit. The determination of provider services can be made based on level of assessment and intervention needs, age, and family concerns.

It is clear, however, that such a large team may be impossible to organize based on limited professional and financial resources. The HRIFU team can be more limited, with an interdisciplinary model of services. As discussed previously, the team can gain skills from each other and consultants to be able to provide optimal evaluation and therapy services. It is important within such a structure to have other professionals available to consult in situations where the team does not have the expertise to address the situation independently. Examples of consultation professionals include social work, psychology, dietician, and medical subspecialist. HRIF clinics can operate successfully with the following disciplines:

- Neonatologist/developmental pediatrician or neonatal or pediatric nurse practitioner
- Nurse
- Social worker or psychologist
- Occupational or physical therapist, developmental specialist, or developmental psychologist
- Clinic coordinator
- Administrative assistant [18]

It is important to maintain as comprehensive a team as possible [18]. HRIFU clinics need the resources that give them the opportunity to coordinate care from the

NICU to the community to support continuity of care so that high-risk infants and their families receive all the services they require [12]. It is fundamental to have HRIFU team members who interact regularly with families. This encourages an ongoing relationship with the HRIFU clinic and professionals. In addition, it decreases attrition from the program. Data for decision-making in the institution is stronger and more reliable when it is from ongoing follow-up and not just one or two visits [9].

Many clinics operate with minimal staff, many of whom are often part time [12]. While this is less than optimal, it is certainly better than no HRIFU program. This means being efficient and organized is critical. Patra et al. [11] found in their research that Wednesdays were best days for parents to attend clinic in this community. They also found that early morning and late afternoon appointments were often missed. This research helps HRIFU clinics to think about analysis of their own appointment patterns and alter them accordingly to their specific community transportation patterns and community activities. This facilitates utilized valuable HRIFU resources in the most clinically effective and fiscally responsible manner [11].

Summary

High-risk infant follow-up necessitates the commitment of the institution to its purpose, goals, and operation. These programs provide vital data on patient outcomes, quality improvement initiates, and continuity of care for infants and families. In addition, academic institutions utilize these programs to train pediatric residents and neonatal and perinatal fellows and allied health professionals. The HRIFU programs often require grants or endowments in addition to insurance reimbursement for them to meet their expenses. If insurance reimbursement is the only source of revenue, it is usually unrealistic to expect the program to be fiscally neutral. Understanding the financial difficulties of HRIFU by hospital and healthcare system administration is essential to ongoing resource allocation of the institution.

To provide HRIFU services entails the recruitment of specifically trained professionals or professionals willing to obtain the necessary training. These professionals have the knowledge, expertise, and skills in evaluation and treatment of high-risk children and their families. Professionals in the HRIFU clinic must demonstrate competencies within their discipline's scope of practice as well as the ability to respect, value, and learn from the other disciplines that are part of the team. Team members need to be flexible with the ability to be highly organized and knowledgeable about current literature and practices.

Many HRIFU clinics employ a multidisciplinary approach to services provision. Each discipline completes their assessment and makes recommendations for the evaluation and therapy care plan. The team meets and collaborates on the care plan for the child and the family.

More recently, HRIFU teams have evolved into interdisciplinary models of services. The disciplines not only respect and value the other disciplines but also gain knowledge, expertise, and skills from them. In this model, services are provided within a comprehensive and seamless evaluation and therapy process, with each team member competent in working within their scope of practice and supporting the role and expertise of the other team members. Research is indicating that this model may be a more effective approach to HRIFU [1, 11].

The most common setting for HRIFU clinics is in hospitals and healthcare systems. The disciplines needed for evaluation and treatment are also usually part of the NICU team and the pediatric program. For training purposes, academic institutions must maintain HRIFU clinics. Hospitals and healthcare systems need to make the commitment to HRIFU programs when they offer perinatal/neonatal services.

Early intervention and primary healthcare providers are partners in the physical and developmental surveillance of high-risk infants. All of these professionals support families in caring for their high-risk children and assist them in integrating them into their culture and community. It is vital that the HRIFU program work collaboratively with these providers as part of the goals of the clinic.

It is crucial that the most comprehensive team of HRIFU specifically trained personnel are recruited and retained in their program. The personnel of the high-risk follow-up clinic are the foundation of the program. They need to be chosen carefully and respectfully and receive compensation that is appropriate to their training, expertise, and experience. The institution's financial commitment to the program must reflect these requirements.

High-risk follow-up supports optimal outcomes of children and their families. It also optimizes the quality of life for these individuals and their communities. Trained, passionate, and dedicated personnel are the key to this process.

References

1. Pula-Phillips, L., & McGrath, J. M. (2012). Follow-up care for the neonatal intensive care unit graduate. *Newborn & Infant Nursing Reviews, 12*(4), 182–183.
2. Casiano, R. G. M., Gaspardo, C. M., & Linhares, M. B. M. (2016). Prematurity, neonatal health status, and later child behavioral/emotional problems: A systematic review. *Infant Mental Health Journal, 37*(3), 274–288.
3. Purdy, I. B., Craig, J. W., & Zeanah, P. (2015). NICU discharge planning and beyond: Recommendations for parental psychosocial support. *Journal of Perinatology, 25*, s24–s28.
4. Ballantyne, M., Stevens, B., Guttmann, A., Willan, A. R., & Rosenbaum, P. (2013). Maternal and infant predicators of attendance at neonatal follow-up programmes. *Child: Care, Health and Development, 40*(2), 250–258.
5. Purdy, I. B., & Melwak, M. A. (2012). Who is at risk? High-risk infant developmental follow-up. *Newborn & Infant Nursing Reviews, 12*(4), 221–226.
6. Lillas, C., & Turnbull, J. (2009). *Infant/child mental health, early intervention, and relationship-based therapies: A neurorelational framework for interdisciplinary practice*. New York, NY: W.W. Norton.
7. Liberati, E. G., Gorli, M., & Scaratti, G. (2016). Invisible walls within multidisciplinary teams: Disciplinary boundaries and their effects on integrated care. *Social Science & Medicine, 150*, 31–39.
8. American Academy of Pediatrics Committee on Fetus and Newborn. (2008). Policy statement on the discharge of the high-risk infant. *Pediatrics, 122*, 1119–1126.

9. Vohr, B., Wright, L. L., Hack, M., Aylward, G., & Hirtz, D. (2004). Follow-up care of high-risk infants. *Pediatrics, 114*(Supplement Part 2 of 3), 1377–1397.
10. Belfort, M. B., Santo, E., & McCormick, M. C. (2012). Using parent questionnaires to assess neurodevelopment in former preterm infants: A validation study. *Pediatric and Perinatal Epidemiology, 27*, 199–207.
11. Patra, K., Greene, M. M., Perez, B., & Silvestri, J. M. (2014). Neonatal high-risk follow-up clinics: How to improve attendance in very low birth weight infants. *E-Journal of Neonatology Research, 4*(1), 3–13.
12. Gong, A., Johnson, V. R., Livingston, J., Matula, K., & Duncan, A. F. (2015). Newborn intensive care survivors: A review and a plan for collaboration in Texas. *Maternal Health, Neonatology and Perinatology, 1*, 24. https://doi.org/10.1186/s40748-015-0025-2.
13. Choi, B. C. K., & Pak, A. W. P. (2006). Multidisciplinary, interdisciplinary and transdisciplinary in health research, services, education and policy: 1. Definitions, objectives, and evidence of effectiveness. *Clinical and Investigative Medicine, 29*(6), 351–364.
14. Orton, J. L., McGinley, J. L., Fox, L. M., & Spittle, A. J. (2015). Challenges of neurodevelopmental follow-up for extremely preterm infants at two years. *Early Human Development, 91*, 689–694.
15. Walker, K., Holland, A. J. A., Halliday, R., & Badawi, N. (2012). Which high-risk infants should we follow-up and how should we do it? *Journal of Paediatrics and Child Health, 48*, 789–793.
16. Lakshmanan, A., Santo, E., McCormick, M. C., & Balfort, M. (2014). Parental preference and ability to participate in web-based developmental screening and surveillance: Preliminary evidence for preterm infants after NICU discharge. *Clinical Pediatrics, 53*(13), 1278–1284.
17. American Academy of Pediatrics Committee on Children with Disabilities. (2001). Developmental surveillance and screening of infants and young children. *Pediatrics, 108*(1), 192–196.
18. Waruingi, W., Iyer, S., & Collin, M. (2015). Improving health care usage in a very low birth weight population. *World Journal of Pediatrics, 11*(3), 239–244.

Financing NICU Follow-Up Programs

Lauren A. Boyd

Abstract There is limited information available for public review outside of anecdotal reports and published surveys about NICU follow-up clinic revenues versus expenditures in part because this is considered confidential business information. In survey responses, both private and academic follow-up programs report that they obtain funding from a variety of sources including hospital and/or department, clinical revenue from insurance billing, research grants, and other state and government programs. Obtaining adequate funding is often cited as a concern for these clinics. Insurance issues, travel distance, and travel costs have been cited by families as the most frequent reasons for noncompliance with clinic attendance in studies of follow-up appointment compliance. Family-centered follow-up programs will need to demonstrate their worth to their shareholders in the community and local hospitals. Clinic programs will need to consider the cost to families in their structure and development of their follow-up program.

Financing a NICU Follow-Up Program

While follow-up clinics of NICU graduates are established as a standard of care, surveys of follow-up programs have consistently shown that obtaining adequate funding for follow-up clinics is challenging (B. Tang, personal communication, August 15, 2015; [1–3]). There are no national criteria to help risk stratify which NICU graduates should be evaluated in a NICU clinic versus followed in their medical home. However, because itemized hospital and clinic budgets are typically considered proprietary business information, there is limited information available for public review outside of anecdotal reports and published surveys about clinic revenues vs. expenditures. Reimbursement for clinical care provided both in the intensive care unit and the outpatient follow-up clinic will vary based on contracts negotiated

L. A. Boyd, MD (✉)
Department of Pediatrics, Developmental and Behavioral Pediatrics, Loyola University Chicago Stritch School of Medicine, Loyola University Health System, Maywood, IL, USA
e-mail: laboyd@lumc.edu

H. Needelman, B. J. Jackson (eds.), *Follow-Up for NICU Graduates*,
https://doi.org/10.1007/978-3-319-73275-6_13

by the hospital or health network with both private and public insurers. Cost of clinical follow-up includes tangible costs such as electricity, clinical equipment to obtain vital signs, salaries for front desk clinical staff, physician and therapist salaries, malpractice insurance, and cost of renting or maintaining clinic rooms.

Payments from insurers may also be bundled into general hospital or healthcare system funding in population-based health plans that do not reimburse for care under a fee for service model. Funding difficulties impact both families and clinics. Most centers that offer level II (specialty care) and level III (subspecialty neonatal intensive care units) care report that funding for their follow-up programs comes from a variety of sources (B. Tang, personal communication, August 15, 2015; [1, 4]), and most programs (63%) report that costs exceed revenue (B. Tang, personal communication, August 15, 2015).

Initial neonatal ICU care is provided at both private hospitals and academically affiliated institutions. Private clinics from level III NICUs reported that their funding primarily comes from the hospital (59%) with supplementation from the state funding (23%) and grant funding (20%). Clinics with an academic affiliation reported the majority of their funding comes from the NICU department (42–51%) and hospital (31–62%), with supplementation from state funding (22%) [1, 4]. More than a third of level II and level III NICU (37%) reported when surveyed that they do not receive reimbursement for developmental assessments (B. Tang, personal communication, August 15, 2015). Most clinics in published surveys report that they do bill a family's insurance for the follow-up evaluation [2].

Neonatology fellowship programs are required to provide training to their fellows in neonatal follow-up care in the current American College of Graduate Medical Education guidelines. The requirement for fellows to participate in follow-up might influence department budgets. A variety of other learners including nurse practitioner students, speech/language therapy students, occupational therapy students, physical therapy students, pediatric residents, neurology residents, and medical students may participate in follow-up clinics, but surveys have not captured whether their presence leads to any additional funding sources for the actual clinic.

A variety of factors including visit length, visit frequency, and the number of providers seeing and billing for their evaluation of the infant or child at each visit all impact the cost of follow-up clinics to hospitals/clinics/institutions and families. There is not yet a set of national guidelines or clear evidence favoring a particular team composition or interval for follow-up. If a child is participating in a clinical trial or research project, the timing (how often and at what age) and content of follow-up evaluations may be standardized and often paid for by the research trial or research group. Each state may also set requirements for neonatal follow-up including up to what age a child should be followed. Costs to families may include travel costs to a regional center, missed wages from employment absence, insurance co-pays, and deductibles including decreased coverage if providers are out of network. A study of follow-up clinic attendance from Rush University Medical Center, an urban academic center in Chicago, found that families reported insurance (typically out of network vs. no insurance) and travel time to the clinic as the most frequent reasons given for not attending the 2-year follow-up clinic visit [6]. A Virginia study of noncompliance with NICU follow-up clinic attendance found that families cited travel costs and travel distance most frequently [4]. Families who obtained an exception to keep a critically ill extremely premature neonate at an out-of-network hospital may not be successful in obtaining a gap exception to return to the same institution for follow-up care after the child is discharged.

Choice of developmental assessment tools (direct evaluation of a child's skills vs. parent report) and choice of assessment team (psychologist, speech therapist, occupational therapist, physical therapist, nutritionist) both influence length of the appointment as well. Physician evaluation and management (E/M) Current Procedural Terminology (CPT) codes for outpatient visits are selected either by the complexity of the history, physical examination, or physician face-to-face time with the patient and family if more than 50% of the visit is spent in counseling and care coordination. In a fee for service model of reimbursement, higher levels of outpatient coding are associated with higher reimbursement and may also be used to measure physician productivity. In a managed care model or a model where a hospital receives bundled payments for caring for a large group of patients, reimbursement does not change regardless of visit length or the complexity of the issues addressed during the visit. Some insurance plans will require prior authorization for any therapy evaluations such as speech/language or occupational therapy. Care coordination may be helpful to assist families with prior authorizations and referrals needed for these appointments especially in families with low health literacy.

Most common developmental assessments, both screeners such as the Bayley Scales of Infant and Toddler Screener, and comprehensive assessments such as the Battelle Developmental Inventory, Mullen Scales of Early Learning, and the Bayley Scales of Infant and Toddler Development, Third Edition, increase the administration time required as children increase in age. Many centers will change team composition depending on the corrected age of the child. For example, a team might initially see a recently discharged infant with speech pathologist to address feeding, then subsequent visits will focus more on motor development during the first year of life, and the therapist evaluator will be a physical therapist or occupational therapist. If the child continues to be followed after 12 months, speech and language concerns may become more evident between 12 and 30 months. In children without obvious delays who are without parent concerns about development and do not have chronic medical conditions, there can be discussion of what to code for the evaluation that will meet insurance coverage requirements for an evaluation. Direct assessment during the clinic visit requires establishment of rapport with an infant or young child for results to be valid. Though trained therapists and clinicians are skilled at working with young children, warm-up time does need to be part of a visit for evaluation to be valid. Strategies for professionals to provide this complex care have not been standardized.

Neonatal follow-up clinics have costs to the institution and family associated with time required for developmental assessment of young children, complexity of medical needs in NICU graduates, and need for multidisciplinary assessment. Nutrition and weight gain are important issues especially in the first few months after NICU discharge, but often insurance will not pay for nutrition evaluations though they might be offered by other stakeholders such as early intervention. Some clinics will bill their developmental evaluations through the Early Intervention Program if their providers are credentialed by the state's Early Intervention Program and have authorization to perform evaluations in the clinic setting.

Across the United States, some NICU teams have collaborated with other stakeholders to provide for family-centered, comprehensive collaborative care for high-risk NICU graduates. The University of Chicago Comer Children's Hospital and University of Kansas Medical Center at Kansas City have both created a neonatal medical home that provides comprehensive care including well visits, sick visits, telephone triage after hours, and developmental follow-up until age 5 for high-risk NICU graduates. This collaborative model might reduce readmission to the hospital by providing close follow-up for NICU graduates by a team very familiar with child's comorbidities with the opportunities to address questions and concerns in a timely matter before a child becomes critically ill.

Nebraska developed a statewide program that began in 2000 coordinated through the Munroe-Meyer Institute at the University of Nebraska Medical Center in collaboration with the Nebraska Department of Health and Human Services and Nebraska Department of Education and most hospitals with at least level II NICUs called Tracking Infant Progress Statewide (TIPS). The TIPS program risk stratifies all babies who were admitted to neonatal intensive care units in Nebraska for several days or more into three risk categories at discharge: lowest risk, medium risk, and highest risk. Lowest-risk infants are initially followed up by questionnaire. Medium-risk children are evaluated by an interdisciplinary team in a clinic setting by one of the five statewide clinic teams, usually consisting of a developmental pediatrician, nurse, developmental specialist, and family members. Highest-risk infants are referred to early intervention to determine eligibility for services. Children can shift tiers at any point in the process. Care coordination for medical problems remains in the medical home. Evaluations are state funded and families are not billed for services [5].

California has created the California Perinatal Quality of Care Collaborative and California Children's Services High Risk Infant Follow-Up Quality of Care Initiative. California has mandated a set of eligibility criteria for NICU babies and mandated a timeline for follow-up. The state requires clinical information be abstracted into a web-based data reporting system up until age 36 months, and the state reimburses for Medicaid evaluations. California has not mandated which developmental assessment tool may be used but does offer suggestions (http://www.dhcs.ca.gov/services/ccs/Pages/HRIF.aspx#current2).

Reduced rates of reimbursement and the drive for evidence-based practice have created the need for better understanding and research as to what quality patient-centered, evidence-based neonatal follow-up programs should look like. Because state resources, hospital resources, and distance that families must travel to neonatal centers vary, neonatal follow-up programs will need to create systems of follow-up care that address the needs of their vulnerable graduates. Clinicians that care for these high-risk children should take into account their institutional costs and the costs to the families to create a follow-up program that is able to serve the needs of the community in which they practice.

References

1. Bockli, K., Andrews, B., Pellerite, M., & Meadow, W. (2014). Trends and challenges in United States neonatal intensive care units follow-up clinics. *Journal of Perinatology, 34*, 71–74. https://doi.org/10.1038/jp.2013.136. Published online 31 October 2013.
2. Kuppala, V. S., Tabangin, M., Haberman, B., Steichen, J., & Yolton, L. (2012). Current state of high-risk infant follow-up care in the United States: Results of a national survey of academic follow-up programs. *Journal of Perinatology, 32*, 293–298. https://doi.org/10.1038/jp.2011.97. Published online 14 July 2011.
3. Lantos, J. (2010). Cruel calculus: Why saving premature babies is better business than helping them thrive. *Health Affairs (Millwood), 29*, 2114–2117.
4. Harmon, S., Conaway, M., Sinkin, R., & Blackman, J. (2013). Factors associated with neonatal intensive care follow-up appointment compliance. *Clinical Pediatrics, 52*(5), 389–396.
5. Jackson, B. J., & Needelman, H. (2007). Building a system of child find through a 3-tiered model of follow-up. *Infants and Young Children, 20*(3), 255–265.
6. Patra, K., Greene, M., Perez, B., & Silvestri, J. (2014, Winter). Neonatal high-risk follow-up clinics: How to improve attendance in very low birth weight infants. *e-Journal of Neonatology Research, 4*(1), 3–13.

Issues in Neurodevelopmental Testing of Infants Born Prematurely: The Bayley Scales of Infant Development Third Edition and Other Tools

Glen P. Aylward

> *"Developmental assessment is the practice of measuring children's strange behavior in strange settings with strange people for the briefest period of time possible."*
>
> *Urie Brofenbrenner, 1977*

Abstract Neurodevelopmental assessment of infants and toddlers consists of administration of specific tasks, observation of behavior, and use of caregiver report. It is best done in a serial fashion. Issues that affect neurodevelopmental assessment include secular changes where mean scores increase over time, the need to balance thoroughness of evaluation with pragmatic issues such as length of the test, and what functions can be assessed at specific ages. Prediction is often a long-term goal, but it is difficult due to the tests used, the infant's current level of function, and the effects of protective or risk environmental factors. There are various levels of follow-up available for infants born prematurely, ranging from mailings/phone interviews to full assessment by a multidisciplinary team. The Bayley Scales have been the reference standard and have undergone several revisions, the latest being the Bayley-III. There is concern that Bayley-III scores are inflated. Numerous attempts at making these scores more comparable to earlier Bayley-II have been minimally successful. The best solutions available for the Bayley-III are to use a control group who receive the same test and use different categorical cutoffs to indicate severity of developmental delays. Other issues that are discussed include correction for prematurity, the relationship between ASD and prematurity, and how best to define neurodevelopmental impairment.

G. P. Aylward (✉)
Pediatrics and Psychiatry, Southern Illinois University School of Medicine,
Springfield, IL, USA
e-mail: gaylward@siumed.edu

H. Needelman, B. J. Jackson (eds.), *Follow-Up for NICU Graduates*,
https://doi.org/10.1007/978-3-319-73275-6_14

General Issues in Developmental Testing

Developmental assessment consists of three components: (1) administration of specific tasks, (2) observation, and (3) use of caregiver report. Despite opinions to the contrary, there is no "gold standard" in developmental testing. This is due to factors such as variability in the infant and toddler's behavior, concerns about item sampling, difficulty adhering to standardization procedures, and the fact that development is dynamic and uneven. As a result we have *reference standards* where a test serves as a reference standard for comparison purposes, but is not absolute or as precise as a laboratory value. Instead of sensitivity and specificity, co-positivity and co-negativity are recommended: co-positivity indicating that both measures under study are positive for developmental delays and co-negativity meaning both measures indicate absence of delay.

A second issue in neurodevelopmental testing involves the secular changes in test norms that occur over time; this referred to as the Flynn effect [1]. Basically, the average test score inches upward at the rate of 0.3–0.5 points per year, thereby increasing by as much as 5 points per decade; this is equivalent to 1/3 of a standard deviation. Stated differently, a test normed in 1995 would currently have a mean score of approximately 110 versus 100, twenty years ago. This phenomenon has an impact on longitudinal studies, as the mean scores over a two-decade time span since norms were published could feasibly increase by two-thirds of a standard deviation.

The question arises in developmental testing regarding how much is too much. A balance must be struck between thoroughness of the assessment and what is practical. Essentially the conceptual does not necessarily translate into the pragmatic. It is true that a larger number of items increase reliability, but the long duration of the test might reduce validity of the findings. For example, the Bayley-III [2] is estimated to take >90 min to complete at 13 months and older. Obviously, there are very few situations where a 13-month-old would have to sustain performance for 1.5 h. Similarly, items on the Bayley-III such as imaginative play or multi-scheme play yield valuable information regarding the child's early abstracting abilities and accommodative cognitive skills, but they are time-consuming and difficult to elicit in a clinical situation. Therefore, failure of these items could be due to a variety of causes.

What we can assess varies by age, evolving from neurologic → motor → sensorimotor → cognitive [3]. This will have an impact on the breadth of assessment at any given age. For example, it is very difficult to determine the cognitive ability level in an infant with significant upper extremity motor impairment early on. This becomes easier as the child ages and motor and cognitive skills diverge. Along these lines, there is a difference among a *skill*, a *function* and an *integrated functional unit* [3]. A skill could be conceptualized as reaching for or grasping an object. A function would involve reaching, grasping, and purposefully bringing the item to the infant's mouth. An integrated functional unit incorporates reaching and attempting to grasp an object but having to overcome a barrier such as a clear plastic box in order to do so. This behavior is superimposed on reasoning and learning and is a precursor to intelligence. Related is the concept of canalized behaviors. These behaviors (e.g., smiling, babbling, reaching) are "pre-wired" [4] and have been conceptualized to be based on experience-expectant synapses. These behaviors are simple, do not require complex neural integration, and therefore are not easily disrupted by medical/biological issues such as being born prematurely. Many components of early developmental tests are composed of canalized functions, and this would have a strong impact on prediction. Complex behaviors are most likely based on integrated neural interconnections that are more susceptible to insult. However, it is often not possible to assess these behaviors at early ages, and the effects of potential damage are essentially silent until such behaviors emerge.

Disruption/Insult

Being born early causes disruption in brain development and disordered maturation. The lower the gestational age at birth, the greater the immaturity and vulnerability of the infant's central nervous system. Add to this insult due to issues such as hypoxemic-ischemic encephalopathy or intraventricular hemorrhage, which involve the necrotic-apoptotic continuum, release of glutamate, free radicals, and other inflammatory processes. The combination is sometimes referred to as encephalopathy of prematurity, and it will affect white and gray matter in the areas with greatest rapidity and complexity of developmental events and in certain cell types and regions of the brain (e.g., preoligodendrocytes, subplate neurons). This results in microscopic glial scars, neuronal/axonal damage, and disrupted cell-to-cell interactions [5]. Thus there are multiple risk factors that include being born at extremely or very preterm (EPT/VPT) gestational age, neonatal complications, and other conditions such as being small for gestational age (SGA) [6]. In reaction to these factors, which often overlap, are the processes of reorganization and recovery, which will vary in the individual child. Therefore, disruption, injury, reorganization, recovery, as well as genetics will work in a complex matrix to make prediction of later developmental function of an individual infant or toddler difficult.

Prediction

A favorite quote attributed to the late Maureen Hack is "prediction is hard, especially about the future." She was spot on. While many developmental test authors eschew prediction as a purpose of early testing, in actuality, it is one of the main reasons that this testing is employed in research and other clinical endeavors. There are several, specific issues that may affect prediction.

Tests Used

The choice of the test employed will have an impact on prediction. For example, if a measure of receptive vocabulary such as the Peabody Picture Vocabulary Test-IV [7] is employed, this may predict later receptive vocabulary, but would not necessarily measure higher-order, more complex verbal skills. In fact, naming vocabulary (which is more advanced than receptive one-word skills) tends to be spared in those born preterm. Similarly, if the test primarily assesses sensorimotor functioning without considering learning or reasoning, predictive utility is compromised. Learning and reasoning require higher-level cognitive abilities, and these are functions necessary for later intelligence. As the toddler acquires a greater breadth of representative skills, tests become better at measuring these more complex integrated functions. Age at time of testing is also important, because there may be quirks in tests at specific ages such as test floors, ceilings, and item gradients (where a small difference in raw scores could translate into a substantial difference in scaled scores).

Child's Current Level of Functioning

If an infant or toddler shows significant developmental impairment, unfortunately there is better prediction due to a ceiling being placed on the child's abilities. Conversely, if a child is found to be at an advanced level of developmental functioning, the likelihood of average later functioning or better is increased. Unfortunately, prediction is poorest in children with moderate delays, and these are the children who perhaps comprise the largest proportion of infants born at medical/biological risk and who may benefit the most from intervention services.

Environment

Environment can be a risk or protective factor and will strongly influence prediction. Environmental influences become particularly evident from 18 to 24 months onward, although there is evidence that they exert effects even earlier. Verbal and cognitive functions are the domains most affected by environmental influences. Moreover, many children born prematurely are at both environmental and medical/biological risk, essentially placing them at "double jeopardy." Social risk, family capital, parent educational level, and the child's access to educational enrichment are particularly powerful factors [8].

Follow-Up Protocols

Depending on the purpose of follow-up, namely, research, clinical, or both, different protocols are available [9, 10]. These vary in terms of level of assessment as well as frequency of follow-up contact. The degree of medical/biological risk of the infant (e.g., born EPT, having severe HIE) will also have an impact on the selection of a specific follow-up model.

Level 1

This may consist of a telephone interview by a member of the NICU staff. Level 1 follow-up may use a questionnaire such as the ASQ-3 (in part or all subscales) [11], PEDS:DM, or other screens that can be completed by interview without direct administration of tasks. This technique is useful in maintaining contact with the family while monitoring the child's development. However, this method would not be sufficient for data collection or clinical purposes without the use of additional measures.

Level 2

Level 2 involves hands-on screening using screening tools such as the Bayley Infant Neurodevelopmental Screener [12], the Capute Scales [13], or the Bayley-III Screener [2]. The benefit of this level of follow-up is that these screenings could be administered rather quickly, thereby decreasing cost. However, the data derived from screeners are less precise or detailed, and although they have clinical value, they are less useful for research purposes.

Level 3

This level of follow-up consists of a comprehensive assessment on a single visit, often at 18–24 months using a test such as the Bayley-III. This protocol is often used in research studies such as those run by the NICHD Neonatal Research Network. While the assessment is comprehensive, it is not serial and therefore may be affected by the generally uneven course of development.

Level 4

This protocol incorporates serial assessments by a multidisciplinary team using an instrument such as the Bayley-III, as well as other discipline-specific evaluation tools. While thorough and serial, the cost and time required for repeated, detailed assessment are often prohibitive.

Perhaps the best approach is a combination of these levels. An initial screening phone contact can be implemented early on, followed by regularly scheduled hands-on screenings; these periodically interspersed with more detailed evaluations such as the Bayley-III. This would provide serial, longitudinal data, continuous monitoring, and periodic scores derived from a reference standard for research purposes. For example, phone screening could be employed at 3 months, hands-on screening at 6 and 12 months, and more detailed developmental assessment at 24 or 36 months. This approach can be tailored to meet the needs of individual centers or collaborative, multicenter studies, and it could also be responsive to the degree of prematurity, with more detailed and frequent assessment being given to those born very preterm (VPT) or extremely preterm (EPT).

Age at Testing and Prediction of School-Age Functioning

There is debate regarding how long to follow children born prematurely. School age is often a desired end point particularly for research studies, but cost and loss to follow-up are significant concerns. Moreover, there is a "signal-to-noise" phenomenon whereby the direct effects of an early, medical/biological risk factor such as being born EPT (signal) become clouded by subsequent experiences and environmental factors (noise). Conversely, more subtle, high-frequency/low-severity dysfunctions such as ADHD, learning disabilities, behavioral concerns, or executive dysfunction [14] may not be apparent earlier. Because of these issues, the following two age ranges often are selected as acceptable follow-up end points to provide gross estimates of current and later school-age function.

18–24 Months

At this age the ability to predict school-age function improves in comparison to earlier ages. Cognitive and motor functions diverge, while language and reasoning skills develop. Intelligence (IQ) per se cannot be measured because the tests are developmental in nature with IQs being generated by tests administered to children older than 2 years (preferably 3 or older). The direct effect of an earlier event such as hypoxic-ischemic encephalopathy on outcome is still more apparent than at later ages, although environmental effects also become more influential.

3–4 Years

IQ, executive function, pre-academic readiness, visual motor integrative skills, and verbal/nonverbal discrepancies can be assessed. However, environmental effects become stronger, and some IQ tests have weak test floors at early ages, thereby inflating scores at the lower end. This was the case with the Stanford-Binet-IV where a raw score of zero at age 3 yielded a standard score in the 70s. Because testing at this age is closer in time to later school-age performance, the correlations between scores obtained in this age range and later functioning are greater.

At both ages, refusals are problematic. Refusals could be evident throughout the testing, be present only in assessment of certain areas such as language or motor function, or occur when the test items become too difficult (ceiling effect). It is estimated that occasional test refusals occur to some degree in 40% of children. There is a greater likelihood of refusals in children from lower socioeconomic households or those at biological risk (e.g., born EPT). This raises an interesting question in regard to premature infants: are lower scores due to noncompliant test behaviors true deficits or both [3]?

Tests

A selected listing of both developmental and intelligence tests is found in Table 1. The listing is not exhaustive, but includes the main infancy and early childhood tests that are found in published follow-up studies. The focus is up through early school age and does not include intelligence tests whose beginning age is 6 years or above.

Issues with the Bayley Scales

The Bayley Scales changed significantly from the original Bayley Scales (BSID; [15]) and the Bayley Scales of Infant Development-II (Bayley-II; [16]) to the Bayley Scales of Infant and Toddler Development (Bayley-III; [2]). In the former two versions, a Mental Developmental Index (MDI) and a Psychomotor Developmental Index (PDI) were produced. With the introduction of the Bayley-III, Cognitive, Language (Receptive Communication and Expressive Communication), Motor (Fine Motor and Gross Motor) subscales replaced the MDI/PDI concept; caregiver-completed Adaptive and Social-Emotional scales were also included. The MDI was split into the Cognitive and Language subscales (*r*'s = 0.60–0.67 for MDI and Cognitive composite; *r*'s = 0.71–0.87 for MDI and Language composite). The PDI was transformed into the Motor subscales and the correlations ranged from 0.60 to 0.65.

Comparison of the BSID to the Bayley-II, both administered to 200 children (1–42 months; mean age 15.5 months), produced a BSID-II MDI and a PDI that

Table 1 Developmental and intelligence tests used in infancy and early childhood

Test	Date of publication	Age range	Comments
Griffiths (developmental)	1996; 2016	0–2; 2–8 years	Used primarily in UK; five scales
Mullen Scales (developmental)	1995	0–68 months	Visual receptive/expressive; language receptive/expressive
Bayley Scales of Infant and Toddler Development-III (developmental)	2006	16 days to 42 months	Cognitive, language (expressive, receptive), motor (fine, gross), social-emotional, adaptive
Differential Ability Scales-II (DAS-II) (Intelligence)	2007	2.6–3.5 up to 4 subtests Goes up to 17.11	Originally developed from the British Ability Scales
Stanford-Binet V for Early Childhood (intelligence)	2005	2–5 years; can also be applied to 6 and 7 year olds	Five scales; can also derive an abbreviated Battery IQ (ABIQ), based on two subtests
Wechsler Preschool and Primary Scale of Intelligence-IV (WPPSI-IV) (intelligence)	2012	2.6–7.7	FSIQ, VIQ, PIQ
Kaufman Assessment Battery for Children-2 (KABC-2) (intelligence)	2004	3–18	Sequential, Simultaneous Processing, Mental Processing Index
NEPSY-II (neuropsychological)	2007	3–4	Not IQ but various functions

were 12 and 7 points lower than the BSID MDI and PDI, respectively [16]. This trend was reversed in a later comparison of 102 children (mean age 14.8 months) who were given the Bayley-II and Bayley-III. Here the Bayley-III Cognitive Composite was 6 points higher than the MDI, while the Motor Composite was 8 points higher [2]. Other studies indicated that the Cognitive score was up to 10 points higher and the Motor by as much as 18 points in comparison to the MDI and PDI [17]. Of note is the fact that the Bayley-III norms included 10% of "at-risk" infants and toddlers (the premise being that mixed sampling reflects the diversity found in the so-called normal population) [18, 19]. This method could inflate the norms and decrease sensitivity. This situation would be even more problematic if the at-risk children were not evenly distributed across all age ranges of the normative sample.

The Bayley-II/Bayley-III discrepancies result in two problems: (1) in longitudinal research studies, the Bayley-III yields higher scores than the Bayley-II raising concerns in regard to comparisons of different cohorts, that is, differences in scores attributable to improvements in medical care, to test issues, or to a combination of both possibilities; and (2) clinically, the Bayley-III scores may be inflated and lead to under-identification of children needing intervention services [18]. Of concern is the fact that differences between Bayley-II and Bayley-III scores are particularly problematic at the lower end of the normal distribution [20].

This raises the question as to whether the Bayley-II might be too conservative [21]. However, even if it was the case, this does not explain why, when using the Bayley-III, some control populations have received mean scores greater than 1 SD above the mean [22] while children born EPT received scores that were in the average range. This has been arguably called the "reverse Flynn effect," a term that does not clarify the conundrum.

Attempts to Reconcile Bayley Scores

There have been numerous attempts to make the scores obtained on the Bayley-II and Bayley-III more comparable. These include:

- Combine Bayley-III Cognitive and Language scores into a Composite score (CB-III). Moore et al. [23, 24] found that the Composite was still 7 points higher than the MDI and the inflated combined score was more strongly influenced by the Language versus the Cognitive Composite score.
- Moore et al. [23, 24] used nonlinear regression to produce a predicted MDI from Bayley-III scores. Although there was some improvement in comparability, the algorithm resulted in a SD of 23 points. A least squares regression [20] was used to convert Bayley-II to Bayley-III scores. Greater discrepancies were found at the lower end of the continuum (e.g., a Bayley-II MDI of 60 would convert to a Bayley-III cognitive score of 87; an MDI of 110 would convert to a Bayley-III cognitive score of 117). The resultant differences in scores could range from 7 to 27 points, again with greater differences in the lower range of functioning.
- Use of a developmental quotient score [25] was attempted, but developmental age is an imprecise measure, and standard deviations are not the same at each age. For example, at 36 months a cognitive raw score of 69 yields a composite score of 90 but an age equivalent of 28 months (8-month delay). This equals a DQ of 78. At 18 months a cognitive raw score of 48 again yields a composite score of 90, an age equivalent of 16 months (2-month delay) and a resultant DQ of 89. This comparison reveals raw scores that produce the same standard scores could have markedly different DQs due to marked variations in age equivalents. Moreover, DQs are totally dependent on the numerator of the ratio, which is derived from the test that is administered. The test, in turn, may or may not have acceptable item gradients, and this will affect the validity or accuracy of the numerator.
- The use of different cutoffs for categorization [26] has been proposed. A cognitive score of <70 is typically used to indicate significant neurocognitive impairment; a motor score of <70 is considered indicative of significant motor impairment. However, given the apparent inflation of the Bayley-III scores, cutoff scores of 80 or even 85 versus 70 have been recommended. This appears to be the most parsimonious approach to handling the apparent inflation in Bayley-III test scores.

Going Forward

One way to address the Bayley-III concerns is to prospectively employ a normative, full-term comparison group [18, 27]. If the full-term controls show no change in scores but the premature infants do, then the difference could be attributable to factors found in the preterm group. However, if scores for both groups improve or decline, there is an increased likelihood of problematic psychometric issues. The normative group could also be compared to standardized norms; if the means are different, it is advisable to use the normative group scores versus the standardized norms for comparisons because various studies have suggested that impairment rates may be underestimated if standardized norms are employed [22, 28].

Children at risk should not be included in standardization norms because this tends to inflate scores. Norms for specific clinical groups could be obtained for comparative purposes. Given that reconciliation efforts between the Bayley-II and Bayley-III have largely been unsuccessful, these should probably stop, particularly since the Bayley-III has been in use for the last decade. While it is possible that the Bayley-II underestimates development or the Bayley-III overestimates developmental scores (or both possibilities are true), this issue will most likely not have closure. Moreover, a Bayley-IV will be coming out in the future, and this will render such comparisons moot. With regard to categorical cutoff scores, there is no problem with Bayley-III cognitive scores of <70 being considered severe impairment. However, another category that includes children scoring in the 70–80 or 70–85 range should be considered. These children demonstrate compromised development and are categorically different than those in the normal range.

Controversial Topics Needing Resolution

A consensus is needed regarding several follow-up issues. This listing is by no means exhaustive, but rather it highlights three recurring concerns.

Correction for Prematurity

The question exists as to when children born preterm will catch up to peers. The answer depends on the specific outcome being considered and the actual degree of prematurity. Children born preterm arguably may never totally catch up [29]. Rather, differences between those born preterm and their peers become less obvious over time.

The general consensus is that correction is needed in the follow-up of preterm infants. Correction up through 2 years is perhaps the most popular convention, although correction to 3 years has also been endorsed. Some authors recommend correction into early childhood [30]. Recently, Wilson-Ching et al. [31] revisited the correction issue using Bayley-III normative data to compare corrected and uncorrected scores for 1, 2, 3, and 4 months of prematurity. They also compared these differences using baseline scores of 100, 85, and 70 (test mean, 1 and 2 standard deviations below the mean). These comparisons were made at 6, 12, 24, and 36 months. The authors reported that larger differences between scores were related to the degree of prematurity, higher baseline scores (85, 100), and younger age at time of assessment (particularly below 24 months).

With 1 and 2 months of prematurity, there was a 5-point difference at 2 years and no difference at 3 years, regardless of baseline score (70, 85, or 100). However, with 3 or 4 months of prematurity, differences ranged from 5 to 10 points at 2 years and were 5 points at the 3-year comparison. Greater differences were found at the upper two scores, suggesting a floor effect at the score of 70. The authors also report a several-point difference between corrected and uncorrected scores at later ages. Moreover, application of age bands on the Bayley-III [32] or later intelligence tests [30] has an impact. The difference between corrected and uncorrected scores increased when the child's assessment age was closer to the beginning cutoff of a normative age band [29].

It appears that correction should definitely be employed for all gestational ages or functional levels at 2 years for cognitive outcome. At 3 years there is evidence to suggest correction be used in children born at 28 weeks' gestation or earlier. Perhaps simply adding 5 points to the obtained score could be investigated further [31]. There still is the question as to whether similar findings would be obtained with regard to motor or language function. In addition, it is not established whether correction be made to 37, 40, or even 42 weeks [31, 33].

Autism Spectrum Disorders (ASD) and Prematurity

There are reports of an increased incidence of ASD in children born prematurely, particularly in those born <32 weeks' gestation or <1500 g, the OR being approximately 3. The incidence is also higher in males [34, 35]. The main concern here is that there is a confound between ASD and developmental disabilities in the preterm population [36]. There is a greater chance of receiving an ASD diagnosis in preterms who also have CP, hearing or visual impairment, or severe cognitive impairment. This makes it difficult to determine whether the ASD diagnosis is inappropriate, is comorbid with developmental disabilities, or is both. Therefore, caution is necessary because there may be a high false positive rate due to coexisting neurodevelopmental disabilities [24].

Definition of Neurodevelopmental Impairment

Cognitive and/or motor scores <70, cerebral palsy, bilateral blindness, and hearing loss requiring amplification are often combined when considering severe neurodevelopmental impairment. This practice produces a heterogeneous group because children may be included based on significantly different criteria or combinations of criteria. Inasmuch as the most frequent impairment is a developmental score <70, investigators have suggested not mixing these different conditions due to qualitative differences and variability in the rate of occurrence.

In summary, although progress continues to be made in our understanding and application of neurodevelopmental testing of children at risk, there still are areas that require further refinement and consensus as to best practice. These include, but are not restricted to, a more thoughtful approach in the developmental of the Bayley-IV that would enhance both diagnostic accuracy and predictive validity, a consensus on the best ages to test outcomes (i.e., 18–24 months and/or 3–4 years), agreement on how best to handle correction for prematurity (perhaps developing an algorithm that includes gestational age, age at testing, and area of function), and further investigation of the relationship between ASD and prematurity.

References

1. Flynn, J. R. (1999). Searching for justice. The discovery of IQ gains over time. *American Psychologist, 54*, 728–748.
2. Bayley, N. (2006). *Bayley scales of infant and toddler development* (3rd ed.). San Antonio, TX: The Psychological Corporation.
3. Aylward, G. P. (2009). Developmental screening and assessment: What are we thinking? *Journal of Developmental and Behavioral Pediatrics, 30*(2), 169–173.
4. McCall, R. B. (1983). A conceptual approach to early mental development. In M. Lewis (Ed.), *Origins of intelligence: Infancy and early childhood* (pp. 107–134). New York, NY: Plenum.
5. Volpe, J. J. (2009). Brain injury in premature infants: A complex amalgam of destructive and developmental disturbances. *Lancet Neurology, 40*, 110–124.
6. Taylor, H. G. (2015). Multiple risks for long-term cognitive impairments following preterm birth. *Acta Paediatrica, 104*, 218–220.
7. Dunn, L. M., & Dunn, D. M. (2007). *Peabody picture vocabulary test* (4th ed.). San Antonio, TX: Pearson.
8. Baron, I. S., & Rey-Casserly, C. (2010). Extremely preterm birth outcome: A review of four decades of cognitive research. *Neuropsychology Review, 20*, 430–452.
9. Sherman, M. P., Aylward, G. P., & Shoemaker, C.T. (2009). Follow-up of the NICU patient. *Pediatrics, eMedicine textbook* (Vol. 3, No. 1). Retrieved from www.eMedicine.com. Http://emedicine.medscape.com/article/977318-overview.
10. Vohr, B., Wright, L. L., Hack, M., Aylward, G. P., & Hirtz, D. (2004). Follow-up care high risk infants (NICHD/NINDS). *Pediatrics, 114*(Suppl), 1377–1397.
11. Bricker, D., & Squires, J. (2009). *The ages and stages questionnaire* (3rd ed.). Baltimore, MD: Brookes Publishing.
12. Aylward, G. P. (1995). *The Bayley infant neurodevelopmental screener.* San Antonio, TX: The Psychological Corporation.
13. Pittock, S. T., Juhn, Y. J., Adegbenro, A., & Voight, R. G. (2002). Ease of administration of the Cognitive Adaptive Test/Clinical Linguistic and Auditory Milestone Scale (CAT/CLAMS) and the Mental Developmental Index of the Bayley Scales of Infant Development. *Clinical Pediatrics, 41*, 397–403.
14. Aylward, G. P. (2005). Neurodevelopmental outcomes of infants born prematurely. *Journal of Developmental and Behavioral Pediatrics, 26*(6), 427–440.
15. Bayley, N. (1969). *Bayley scales of infant development.* New York, NY: The Psychological Corporation.

16. Bayley, N. (1993). *Bayley scales of infant development* (2nd ed.). San Antonio, TX: The Psychological Corporation.
17. Acton, B. V., Biggs, W. S. G., Creighton, D., et al. (2011). Overestimating neurodevelopment using the Bayley-III after early complex cardiac surgery. *Pediatrics, 128*, e794–e800.
18. Aylward, G. P. (2013). Continuing issues with the Bayley-III: Where to go from here. *Journal of Developmental and Behavioral Pediatrics, 34*(9), 697–701.
19. Pena, E. D., Spaulding, T. J., & Plante, E. (2006). The composition of normative groups and diagnostic decision making: Shooting ourselves in the foot. *American Journal of Speech and Language Pathology, 15*, 247–254.
20. Lowe, J. R., Erickson, S. J., Schrader, R., et al. (2012). Comparison of the Bayley II Mental Developmental Index and the Bayley III cognitive scale: Are we measuring the same thing? *Acta Paediatrica, 101*, e55–e58.
21. Hack, M. (2012). Dilemmas in the measurement of developmental outcomes of preterm children. *Journal of Pediatrics, 160*, 537–538.
22. Anderson, P. J., DeLuca, C. R., Hutchinson, E., Roberts, G., & Doyle, L. W. (2010). Underestimation of developmental delay by the new Bayley-III scale. *Archives of Pediatric and Adolescent Medicine, 164*, 332–336.
23. Moore, T., Johnson, S., Haider, S., et al. (2012). Relationship between test scores using the second and third editions of the Bayley Scales in extremely preterm children. *Journal of Pediatrics, 160*, 553–558.
24. Moore, T., Johnson, S., Hennessy, E., & Marlow, N. (2012). Screening for autism in extremely preterm infants: Problems in interpretation. *Developmental Medicine and Child Neurology, 54*, 514–520.
25. Milne, S., McDonald, J., & Comino, E. J. (2012). The use of the Bayley Scales of Infant and Toddler Development III with clinical populations: A preliminary exploration. *Physical and Occupational Therapy in Pediatrics, 32*, 24–33.
26. Vohr, B. R., Stephens, B. E., Higgins, R. D., et al. (2012). The NICHD Neonatal Research Network. Are outcomes of extremely preterm infants improving? Impact of Bayley assessment on outcomes. *Journal of Pediatrics, 161*, 222–228.
27. Aylward, G. P., & Aylward, B. S. (2011). The changing yardstick in measurement of cognitive abilities in infancy. *Journal of Developmental and Behavioral Pediatrics, 32*(6), 465–468.
28. Anderson, P. J. (2014). Neuropsychological outcomes of children born very preterm. *Seminars in Fetal and Neonatal Medicine, 19*, 90–96.
29. Doyle, L. W., & Anderson, P. J. (2016). Do we need to correct age for prematurity when assessing children? *Journal of Pediatrics, 173*, 11–12.
30. Van Veen, S., Aarnoudse-Moens, C. S., van Kaam, H., Oosterlaan, J., & van Wassenaer-Leemhuis, A. G. (2016). Consequence of correcting intelligence quotient for prematurity at age 5 years. *Journal of Pediatrics, 173*, 90–95.
31. Wilson-Ching, M., Pascoe, L., Doyle, L. W., & Anderson, P. J. (2014). Effects of correcting for prematurity on cognitive test scores in childhood. *Journal of Paediatrics and Child Health, 50*(3), 182–188.
32. Veldhuizen, S., Rodriguez, C., Wade, T. J., & Cairney, J. (2015). Misclassification due to age grouping in measures of child development. *Archives of Diseases in Children, 100*, 220–224.
33. Aylward, G. P. (2002). Cognitive and neuropsychological outcomes: More than IQ scores. *Mental Retardation and Developmental Disabilities Research Reviews, 8*, 234–240.
34. Johnson, S., Hollis, C., Kochar, P., Hennessy, E., Wolke, D., & Marlow, N. (2010). Autism spectrum disorders in extremely preterm children. *Journal of Pediatrics, 156*, 525–531.
35. Kuban, K. C., O'Shea, T. M., Allred, E. N., Tager-Flusberg, H., Goldstein, D. J., & Leviton, A. (2009). Positive screening on the Modified Checklist for Autism in Toddlers (M-CHAT) in extremely low gestational age newborns. *Journal of Pediatrics, 154*, 535–540.
36. Aylward, G. P. (2014). Update on neurodevelopmental outcomes of infants born prematurely. *Journal of Developmental and Behavioral Pediatrics, 35*(6), 392–393.

Best Practices in Test Construction for Developmental-Behavioral Measures: Quality Standards for Reviewers and Researchers

Frances Page Glascoe and John Cairney

Abstract Developmental-behavioral measurement is fundamental to clinical determinations about children's and families' needs. Such measurement consists of three main types (listed in order of complexity): (1) screening, (2) mid-level assessment, and (3) diagnostic tests. The three types of measures depend on the same psychometric precepts although screening tests, despite their inherent brevity, depend on an additional construct: proof of accuracy. In this chapter, we focus on standards in test construction with additional emphasis on the psychometry of screening tests because screens are deployed more frequently and serve as a fundamental decision point for whether more complex measures are needed. For example, screening test results help identify whether children require further vision, hearing, or lead screening referrals for further evaluation by special education services and/or to developmental-behavioral pediatricians or other subspecialists. The powerful role of screening tests in decisions that profoundly affect families' lives means that screens must be especially well-constructed. This review of methods in psychometry highlights how to research and review developmental-behavioral measures including screening tests. Crucial to test selection is an understanding of principles and policy in standardization, reliability, and validity, including accuracy computations in the case of screening tests, and utility—practical considerations in measurement.

F. P. Glascoe (✉)
Department of Pediatrics, Vanderbilt University, Nashville, TN, USA
e-mail: Frances.P.Glascoe@Vanderbilt.edu

J. Cairney
Department of Family Medicine, McMaster University, Hamilton, ON, Canada

Faculty of Kinesiology of Physical Education, University of Toronto, Toronto, ON, Canada
e-mail: john.cairney@utoronto.ca

H. Needelman, B. J. Jackson (eds.), *Follow-Up for NICU Graduates*,
https://doi.org/10.1007/978-3-319-73275-6_15

Developmental-behavioral tests have enormous impact of the lives of children and families. Tests help professionals decide whether: (a) there is a likelihood of problems requiring further evaluation; (b) children are progressing given specific conditions and treatments; (c) specific disabilities are present; (d) identified disabilities require unique medications and/or other treatment; (e) children are eligible for intervention programs, and if so which services are needed; and (f) parents need assistance (e.g., with mental health issues). Test results, when concerning, encourage nuanced clinical reasoning, including prompt explorations of potentially causal factors (e.g., whether receptive language deficits are due to hearing loss or whether motor delays are due to cerebral palsy) [1].

Given the crucial role of test results in professional decision-making, an understanding of psychometry is essential for (a) clinicians using established tools (but who perhaps wonder why results should be trusted or need a better understanding of the scores produced); (b) committees selecting tests for early detection initiatives or study protocols; (c) researchers constructing new questions or entire tests, measuring outcomes, or adapting tools for use in other nations; (d) reviewers of journal manuscripts involving experimental or established instruments; and (e) training in developmental-behavioral pediatrics including medical and nonmedical professionals. The goal is to summarize the precepts of test construction and the standards by which quality measures can be judged.

Methods

Reviewed were textbooks on psychometry including those created by professional societies establishing standards in test construction ($N = 14$), along with research studies on methodology ($N = 40$). Also scrutinized were technical manuals for various measures ($N = 132$).

Results

This review paper is prefaced by an explanation of measurement types, measurement methods, and types of test scores, followed by guidance on how to create and pilot new test items. Presented next are the five essential components of test construction along with a rationale for each component and the requisite supporting sub-studies: (a) standardization, (b) reliability, (c) validity, (d) accuracy (in the case of screening tests), and (e) utility.

Table 1 Types of developmental-behavioral tests, measurement methods, and test scores

Types of tests and measurement methods
1. *Screening tests* are brief tools that sort those with possible problems from those without [5]. Further evaluation is required to confirm and define possible problems. Screening tests are usually administered by parent self-report or interview, but sometimes via direct elicitation by clinicians, i.e., hands-on to children. Screening tests are considered *broad-band* if multiple developmental domains are measured versus *narrow-band* if only focused on a single domain (e.g., language or behavior) or a single condition (e.g., detection of autism spectrum disorder). Broad-band measures are administered first to facilitate identification of a range of possible problems. Narrow-band tools are then used to refine the type of possible problem
2. *Diagnostic tests* identify types of disabilities (or lack thereof). A multidisciplinary team is usually required (e.g., a psychologist, speech-language pathologist, developmental-behavioral pediatrician, social worker, special educator) with each professional administering tests within their sphere of expertise, adding clinical acumen and then collaborating in decisions about diagnostic assignations. Measurement methods with children are usually hands-on, in combination with skilled observation. For parent-report measures, skilled interviewing is used. Diagnostic tests are always narrow-band (e.g., measuring dimensions of intelligence, versus language, etc.), and so multiple diagnostic measures are required to render a diagnosis
3. *Mid-level assessment tests* are less complex than diagnostic measures but more detailed than screens. Jumping from screens to diagnostic evaluations is expensive, time-consuming, and wasteful of money better spent on actual intervention. Mid-level assessments offer an economical approach to measurement, a thoughtful view of strengths and weakness (e.g., scores across developmental domains), metrics useful for determining eligibility for services (e.g., under the US individuals with disabilities education act (IDEA)), and lessened needs for a costly multidisciplinary team. Although mid-level assessment measures do not render a final diagnosis, they tend to be broad-band in nature and produce age-equivalent and raw scores across domains. Such scores are responsive to effective treatment and are particularly useful for outcome studies and progress monitoring. Measurement methods may take many forms: Professional interview, parent self-report, direct elicitation of children's skills, or a combination of approaches.
Types of test scores
1. *Raw scores, standard deviations, standard scores.* Raw scores are a simple count of the responses to items (scaling could be binary, ordinal, or continuous). Changes in raw scores over time indicate differences in performance, but because the number of items within measures varies, raw scores neither enable a comparison across tests nor provide a uniform indicator of typical versus extreme performance. To add meaning to various test results, the z score transformation is used to convert raw scores for each age or grade group into standard deviation units [7, 8]. Z scores identify how much performance departs from average and usually have the following distribution (as shown in Fig. 1): More than 68% of children score within 1 standard deviation above or below the mean (i.e., $\pm 1\ \sigma$). Far fewer (<14%) score within the second standard deviation above or below the mean. Only ~3% score within the third standard deviations above or below the mean, thus signifying rare, asymptotic results. Because the z score range is restricted (from −3 to +3 with decimals in between), other standard score metrics are almost always applied to z scores
(a) *Quotients* are the most familiar standard score and assign a mean of 100 and standard deviation units of 15. So in viewing the Weschler scales of intelligence in Fig. 1, most children receive scores of 100 ± 15 points. Quotients generally have the following meanings but must be considered in terms of performance demands (e.g., for a child attending a rigorous private school, test performance in the average range may be below average in comparison with peers)

(continued)

Table 1 (continued)

<74	Very poor
77–84	Poor
85–90	Below average
91–114	Average
115–120	Above average
121–130	Superior
>130	Very superior
Quotients tend to remain the same over time, and when so, this means a child is continuing to learn skills at the same rate (but also see comments on Flynn effect and recommendations for adjustments when comparing various editions of measures below[a]) [2, 9, 10]. Dropping or rising quotients mean a child is learning more slowly or more quickly than prior testing predicted. For quotient changes to be considered significant, a difference of six or more points is needed [7, 11, 12]	
(b) *Scaled scores* are used to describe performance on subtests of diagnostic measures. For example, on the various Weschler scales of intelligence, each subtest produces a scaled score of 10 with a standard deviation of 3	
(c) *T*-scores express standard deviation scores but use a mean of 50 and a standard deviation of 10	
(d) *Percentile ranks* are another type of standard score into which grade placement, but preferably age is factored to produce rankings. Percentiles have a range between 0.1 and 99.9. A percentile of 20 means that a student performed in the lowest 20% of all children, while a percentile of 86 means that a child scored at the top 86%. Percentiles are helpful for expressing relatively fine differences in performance within the average range because they cluster heavily in the standard deviations just above and below the mean. Percentiles do less well at defining how high- and low-functioning children perform (e.g., all quotients <65 render the same score: the first percentile). The interpretation of percentiles varies according to the accepted meaning assigned to the variables measured and by the context of peer comparisons. For example, the 16th percentile through the 25th percentile are acceptable for height, weight, and head circumference but are always concerning when it comes to preschool and school skills: teachers tend to aim instruction toward the middle third of learners, and so children performing at the bottom of the class are rarely taught at their unique "proximal zone of development," i.e., beginning with lower level prerequisite skills [13, 14]. Difficulties with learning often occur when children perform at or below the 25th percentile. In addition, percentiles, as with any standard score, need to be evaluated in light of the performance of students' actual peers and immediate performance demands (e.g., scores at the 50th percentile are likely to be inadequate in a challenging private school if most learners achieve at the 80th percentile or higher)	
(e) *Stanines* (an abbreviation of "standard nine") divides the range of performance into 9 deviation units, with 1–3 indicating below average, 4–6 average, and 7–9 above average. Stanines are reported (along with percentiles) on group-administered academic achievement tests in elementary and secondary schools. Differences of two or more stanines are considered substantially discrepant and suggestive of possible learning disabilities. For clinicians reviewing school records, the stanine range offers clear information about performance levels	

(continued)

Table 1 (continued)

2.	*Age-equivalent scores/grade-equivalent scores.* As a consequence of the somewhat invariant nature of quotients or other standard scores, age-equivalents and raw scores offer a more responsive metric for outcome studies. Age/grade equivalents may be non-standardized and if so are created by viewing the average raw score for each age/grade group. But even if tied to standard scores, age-/grade-equivalent scores remain an uneven metric due to swift changes in skills especially in younger children (see Fig. 2). There is an enormous difference in skill sets between a child who is 1 year, 1 month of age and a child who is 1 year, 11 months of age. The latter is usually talking and walking, the former is often not. Similarly, a first grader performing at the first grade, 1-month level at the end of first grade (when most children are performing at the first grade, 9-month level) is a child who is surely failing with the tasks at hand
	Age-/grade-equivalents offer some practicality as an ever-changing continuous variable that is enormously responsive to developmental changes and interventions. Age-/grade-equivalents are also helpful for selecting instructional materials and age-appropriate toys. Nevertheless, the metric is often confusing to both parents and professionals. For example, a 10-year-old working at a 5-year level is not the same as a 5-year-old working at a 5-year-old level. The former child will have trouble with abstract thinking and problem-solving, while the latter child will not [7, 11]
	Age-equivalent scores are expressed as "year-month" (e.g., 2 years plus 3 months = 2-3). Grade equivalents are expressed as "grade-month" (e.g., second grade, third-month = 2.3). As with percentiles and other standard scores, when tests produce a range of age-/grade-equivalent scores, averaging these into a single result may be less than meaningful, especially if there are substantial strengths and weaknesses in skills [2, 7, 11, 12]
3.	*Cutoff scores* are used almost exclusively with screening tests to sort children with probable problems from those without. Cutoffs facilitate immediate clinical decisions (e.g., whether a referral for services is needed after a well-visit). Cutoffs are described with a range of terms although all have the same meaning (e.g., optimal versus suboptimal, milestones met versus unmet, high risk versus moderate or low risk). Issues in creating and using cutoff scores are described in section "Accuracy"

[a]Flynn effect refers to the finding that quotients, most particularly intelligence scores, tend to rise over time (e.g., 3–5 points per decade). Despite efforts of test authors and publishers to adjust for this when re-standardizing measures, differences in performance across test editions remain. Some studies of the Bayley-III scores found them to be almost one standard deviation higher than those rendered by the Bayley-II [8, 10]. This creates headaches for longitudinal studies using different versions of measures over time (e.g., NICU follow-up and for reevaluation of IDEA eligibility) (For an extended discussion of Flynn effect and how to compare the Bayley-II and –III, see [9, 10]). Even so, Flynn effect may not explain fully higher scores on the Bayley - III. Over-inclusion of children with psychosocial risk factors at some age levels is more likely - leading to an inflated indicator of "average" performance. For this reason, restandardization via the Bayley - IV is under-way and expected to be available by late 2019.

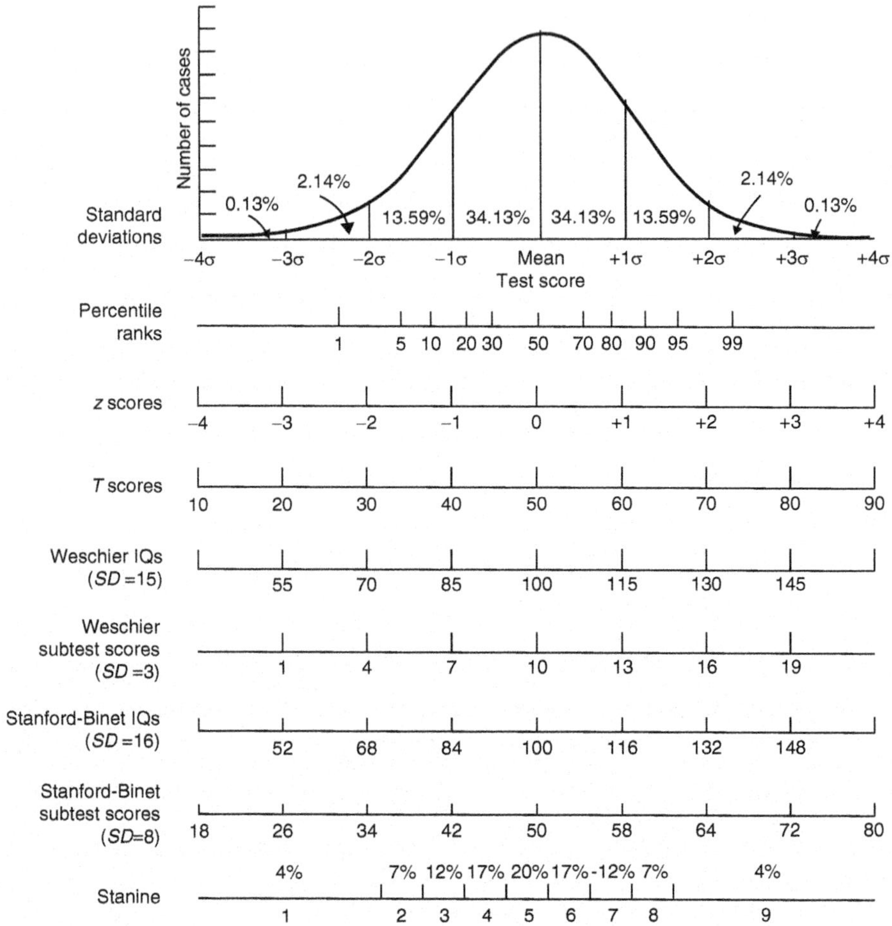

Fig. 1 Graph of performance distribution and types of standardized scores (© Glascoe FP, Marks KP, Poon JK, Macias MM. eds. *Detecting and Addressing Developmental and Behavioral Problems: A Practical Guide for Medical and Non-medical Professionals, Trainees, Researchers and Advocates*. Nolensville, Tennessee: PEDStest.com, LLC; 2013 *Used with permission*)

Types of Measures, Measurement Methods, and Test Scores

There are three essential types of developmental-behavioral measures varying by purpose, focus, complexity, measurement methods, metrics rendered, time demands, and requirements for professional skill [2–6]. Table 1 provides a description of the three types of tests, typical measurement methods, and types of test scores.

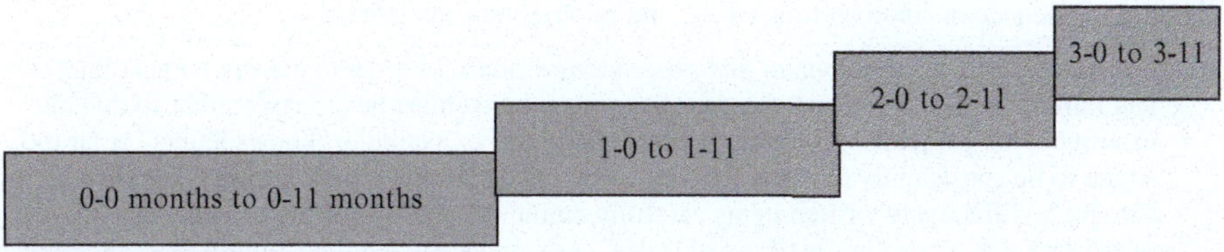

Fig. 2 Unevenness of equivalence scores (showing age-equivalents in years and months) (© Glascoe FP, Marks KP, Poon JK, Macias MM, eds. *Identifying and Addressing Developmental-Behavioral Problems: A Practical Guide for Medical and Non-medical Professionals, Trainees, Researchers and Advocates.* Nolensville, Tennessee: PEDStest.com, LLC; 2013, used with permission)

Creating and Piloting Items/Tests

After identifying theoretical constructs (e.g., domains of development, dimensions of mental health problems), thoroughly reviewing literature, and defining study goals, researchers may discover the need to create new questions or multiple items comprising a new test. Needed are skills in writing clear and understandable questions, ensuring quality translations, and piloting items to assess effectiveness. Table 2 lists considerations in creating, translating, and testing new questions.

Standardization: Purpose and Procedures

Although item development and piloting are usually conducted using local sites (e.g., nearby clinics), to ensure a new test is generalizable to a broad population (e.g., families across the USA), studies using national samples are required. Test standardization (sometimes referred to as "norming") involves multiple procedures that enable each examinees' performance to be compared to peers throughout the nation [2, 7, 8, 11, 12]. Table 3 lists guidelines for standardization and the various required studies.

Reliability

Reliability studies establish how well tests produce consistent results. Such studies reveal relationships among test items and whether measures render similar scores under differing conditions. Using the results, researchers can identify and remove redundant items or items that do not perform well. For example, a test item is less than helpful if all children of the same age are successful (meaning an item is too easy) or if all are unsuccessful (meaning an item is too difficult) [7, 8]. Table 4 describes the essential reliability studies in studies.

Table 2 Guidance in creating, translating, and piloting new items/tests

1. *Ensuring clarity of item content and proscribing stimuli.* In order to discern which children and families differ from peers for better or worse, items must be clearly written so that all examinees are presented with exactly the same task. For example, "knows letters" is far too vague to be consistently measureable (e.g., how many letters? Which letters? Uppercase? Cursive?). Effectively written items carefully define content (e.g., "names lowercase letters of the alphabet when presented out of order") as well as any required stimuli (e.g., pictures or objects) needed to elicit responses [15–17]. The Centers for Disease Control and Prevention has a particularly helpful publication on item writing and presentation of visual stimuli (www.cdc.gov/healthliteracy/pdf/Simply_Put.pdf accessed October 2016)
2. *Writing directions and prompts.* Directions to children or parents require careful stipulation. In the above example, "Point to the 'C'" would not elicit the skill of letter naming. Instead, examiners' questions should be, "What is this?" or "Tell me what this is?" Any needed follow-up prompts should be delineated (e.g., if a child answers, "that's a letter," examiners are prompted to ask, "Which letter? What's the name of this letter?"). Guidance is needed for what examiners are allowed to say after a task is attempted (e.g., non-specific feedback such as "Good trying." (whether or not the answer is correct) but not, "You remember, we talked about this letter yesterday.") [15–17]
3. *Ensuring readability and intelligibility.* Tests intended for youth or parent self-report should be written at or below the fourth grade level (average reading and receptive language skills of a 9-year-old) to ensure that 90% of parents can read and respond [18]. If test items are difficult to read or comprehend, problems occur (e.g., repeated requests for clarification, random answering, skipped items). The online *Plain Language Medical Dictionary* and its smartphone app can help researchers and clinicians identify commonly used, everyday terms (e.g., "tummy" not "abdomen"; "learning problem" not "developmental disability") (www.lib.umich.edu/plain-language-dictionary accessed October 2016)
Readability analyses are available on websites (e.g., https://readability-score.com/ accessed October 2016) and are useful for testing reading levels as well as demands on receptive language, i.e., intelligibility. Readability is influenced by polysyllabic words and sentence length. This means that short response options such as those on multiple-choice tests can falsely deflate reading indicators [19]. For example, first graders (average 6-year-old reading level) can often read words such as "yes," "no," and "sometimes" but not the complex sentences or questions that precede such answer options. Therefore readability/intelligibility indicators should be based on the stem questions with multiple-choice answers analyzed separately
4. *Translating items and vetting translations.* Although 337 languages are spoken (or signed) in the USA, the majority of the population speaks English (80%) followed by Spanish (13%) (www.census.gov accessed October 2016). In Canada, there are two official languages, English and French. Yet, 200 different languages are spoken across the country, with more than 20% (6.8 million people) of the population speaking a language in the home other than the two official languages (www12.statcan.gc.ca accessed October 2016). If current immigration patterns to Western nations persist, it is clear linguistic diversity will become the rule not the exception; this is significant challenge for test developers. Researchers creating test questions for national use in the USA will need to generate, at a minimum, both English and Spanish versions for standardization studies (Spanish is the de facto or an official language in four US states plus Puerto Rico). Spanish is a particular challenge because the USA has Spanish speakers from more than 20 national/territorial backgrounds. Idioms and colloquial expressions vary accordingly, and these may change the intended meaning of a word or phrase in unexpected ways. In Canada, English and French translations at a minimum are required

(continued)

Table 2 (continued)

	The history of advertising when translated from English to Spanish is rife with expensive debacles (e.g., the American Dairy Council's "Got Milk?" campaign rendered in Spanish, "Are you Lactating?"; Chevrolet's Nova did not sell well because Nova can be interpreted as "No Go") [20]. It is possible to create a single, simple Spanish translation that is commonly understood across regional and national language variants, but translation and back translation are only the beginning. Needed is vetting by multiple translators from a variety of national/regional backgrounds along with trials by clinicians and parents in a variety of geographic locations. Such an approach is recommended by the International Test Commission in their guidelines for translation and adaptation of tests [21]
5.	*Pilot testing*. After vetting test questions, response options, needed stimuli, and translations for clarity, researchers should pilot all items to see how well they work. "Cognitive interviewing" is often used whereby researchers ask a small group of subjects across the socioeconomic spectrum about their reactions to, and thinking about, test items [22]. This process may indicate that some questions are ineffective and that rewording is needed. Once satisfied with results, researchers often conduct additional pilot studies to determine whether any items are duplicative or if certain items fail to capture skills (e.g., at each age level)

Validity

Validity studies establish that the scores or results produced by instruments measure what they claim to measure and thus whether interpretation of results is supported by real-life evidence. For example, validation of a new test of school skills should reveal a close relationship with established diagnostic measures of academic achievement. There are many types of validity studies in quality test construction including concurrent, construct, convergent, discriminant, predictive, and criterion-related (described separately in section "Accuracy") [4, 5, 11, 15, 16]. Note that in contemporary psychometric theory, all forms of validation testing are part of construct validity. Nevertheless, we retain distinctions (e.g., content; convergent etc.,) to emphasize different methods of validation. Table 5 lists types of validity studies and analyses required.

Accuracy

In primary care, screening measures are used to make critical decisions about the need for further evaluation. In many nations, children are screened routinely (e.g., universally at specific ages, at targeted well-visits, upon enrollment into preschool programs, when referred to special education programs). Developmental-behavioral measures serve as a crucial conduit into intervention and treatment (e.g., by Head Start, IDEA, formal parent-training, developmental-behavioral/neurodevelopmental pediatricians, etc.). Accurate screens are known to improve detection and referral rates as well as enhance clinical acumen (e.g., whether hearing, vision, or lead screening is needed) [43–45]. Given that screens are applied to the

Table 3 The five steps in standardization of measures

1. *Identifying national demographics and administering measures to a nationally representative sample.* Authors need to identify the sociodemographic characteristics of the entire country in which norming will occur and then administer the new measure to a study sample that reflects the nation as a whole. In America, US Census Bureau data are used to determine population percentages for each variable essential in test norming (www.census.gov accessed October 2016). Variables include ethnicity, levels of parents' education, primary languages spoken at home, poverty levels, and locations (e.g., rural versus urban). Geographic regions (e.g., North, South, East, Midwest, etc.) should be thoroughly sampled because there are substantial differences across nations in the meaning of a "less than a high school education" (e.g., 4–6 completed grades is common in some areas, while in others 9–11 completed grades is more typical, http://factfinder2.census.gov accessed October 2016). A similar process could be conducted in countries where there is sufficient data on the demographic composition of the population. In Canada, for example, census data and routine national surveys both of which are conducted by Statistics Canada (a federal agency) and can be used to determine the composition of samples needed for norming
2. *Selecting sites, ensuring naturalistic sampling, and eliminating selection/spectrum bias.* If creating a test applicable to the general population of children and families, standardization sampling should occur in relevant settings (e.g., primary care, day care, preschool) ([3–5, 11, 12, 15, 16]. Such settings include families with a variety of demographic characteristics as well as children with disabilities and giftedness. While it may be surprising to include children with exceptionalities when standardizing a test, all are part of the overall population, and their presence in a sample ensures that test norms capture the full range of performance and thus inform the meaning of average. Simply adding in a group with known disabilities is not wise, because the continuum of performance may not be captured [3–5, 15, 16, 23]
Proportions of exceptional children in the study sample should be based on prevalence figures such as those provided by the Centers for Disease Control (www.cdc.gov accessed October 2016), World Health Organization (www.who.int, accessed October 2016), and statistics from the Individuals with Disabilities Act (IDEA) (www.ideadata.org accessed October 2016). For the birth through 18-year-old population, 16–18% is the accepted rate for disabilities (although there are differences in prevalence by age) and ~2% for giftedness
For most tests, it is best to avoid norming only on unique populations such as clinic-referred children, i.e., those with probable problems, because this skews the meaning of average (downward) and may, when it comes to accuracy studies, (described below) result in improbable sensitivity as well as under-referrals. Nevertheless, exceptions may be needed when tests are designed for, and used exclusively with, unique populations (e.g., to answer research questions such as, "How does this extremely low birth-weight child compare to others with this condition?")
Separate norms are not advised for subgroups (e.g., children with low socioeconomic status, various ethnicities) because all children are held to virtually the same curricular standards as they approach school entrance [24]. Tests need to reveal when mastery of critical skills, including prerequisites, has or has not occurred, regardless of psychosocial risk factors that often result in lower scores

(continued)

Table 3 (continued)

3. *Ensuring a sufficient sample size*. Diagnostic measures focused on children's skills require scoring tables for each month of age. For screening tests relying on milestones, scoring usually changes in 1-month increments in the first year of life, in 2-month increments in the second year of life, etc. A minimum of 100–200 children for each age group/scoring change are recommended. With the advent of web-based data collection, it is not uncommon to see tests standardized on thousands of children per age-range/scoring change. Large norming samples ensure that the range of performance is thoroughly captured for each age group and for each demographic variable [4, 16]
4. *Viewing collective performance* to determine what is *probably* average and what is not. This process enables authors/researchers to assign scoring criteria to items (e.g., how many numerals are named by most 5-year-olds versus 4-year-olds). For diagnostic tests, the determination of average performance by age/grade involves computing standard deviation units from which standardized scores (e.g., quotients or percentiles) are derived [7, 8, 15, 16, 21]. Establishing the cutoff scores needed for screening tests is described in section "Accuracy"
5. *Updating standardization on a regular basis*. Demographics change with time (e.g., in the USA, the Latino population is growing rapidly, languages other than English are increasingly common, etc.). School curricula often change—meaning that preschool readiness items must be updated (e.g., to include phonological awareness skills). Test stimuli can become dated. For example, many tests had to be revised because children, when shown a rotary dial telephone, could not name it correctly. The various societies focused on measurement standards and scholars in psychometry recommend re-norming tests at least every 10 years [4, 7, 8, 15–17, 21]

general population of children in order to sort into those with probable problems versus those unlikely to have problems, screens require additional research processes and standards not required for assessment or diagnostic tests, i.e., accuracy studies [3, 5, 6]. The steps toward establishing accuracy are shown in Table 6.

Utility

Utility, also called *Feasibility*, covers practical issues in testing and test selection focused on costs and time-related expenses. Table 7 provides a list of topics to consider.

Several studies produced scalable models for calculating testing expenses by first determining overhead (meaning hourly or per minute costs such as salaries, equipment, rent, insurance, materials, etc.) and then assigning time/costs spent on measurement compared to reimbursement or other funding [17, 58–60]. Quality improvement initiatives can thus focus on reducing the time-related costs of measurement in favor of the expense of far more fruitful activities, i.e., intervening. Scalable models also assist professionals selecting among measures that are comparable in scope but differ in terms of time, types of scores, training demands/costs, etc.

Table 4 Requisite reliability studies in test construction

1. *Test-retest reliability* reveals the consistency in performance over a short interval of time (usually 1–2 weeks) when a test is readministered to the same child by the same examiner. Dramatically inconsistent performance is often due to lack of clarity in directions or to stimuli that are ambiguous and easily misinterpreted. Nevertheless, minor performance differences are to be expected (due to fluctuations in memory, attention, fatigue, etc.). To express agreement, some authors use correlations, and some use the Kappa statistic (although standards for Kappa are vague because it accounts for chance agreement and lacks consensus among statisticians for rating significance) [25]. In fact, Kappa and the intraclass correlation coefficient have been shown to be statistically equivalent [26]. While most psychometrists prefer to express test-retest reliability as a percentage of agreement, caution is warranted because this does not take into account chance agreement. Standards for test-retest reliability are at least 85% agreement for brief screening tests but higher for diagnostic and assessment measures [7, 8, 11]. Although not commonly used, a statistic called relative improvement over chance (RIOC) was created to adjust for the problem of low prevalence conditions affecting agreement on tests such as Kappa [27]. The statistics is interpreted the same way as a correlation coefficient (RIOC = 1.0 representing perfect agreement)
2. *Inter-rater reliability* demonstrates how well two different examiners, when testing the same child across a short interval of time, can render virtually the same results. If a test can be administered in multiple ways (e.g., by interview versus parent self-administration versus hands-on by examiners), *inter-method reliability* studies should be conducted to test agreement across measurement approaches. Again, 85% agreement for screening tests, but higher for diagnostic and assessment tests, serves as a standard for both inter-rater and inter-method reliability [7, 8, 11]. Generalizability theory (G-theory) can be used to estimate sources of error/variability for different facets of reliability (e.g., test-retest, inter-rater, item level), all within a single study [26]. Practically, this avoids having to estimate different aspects reliability by conducting multiple studies. However, the design of so-called G-studies is more complex, more time-consuming, and therefore more expensive than typical inter-rater study designs. Once the largest sources of error to reliability have been identified using G-theory studies, D-theory or decision-theory (D-study) can be used to model changes in implementation of the test that will lead to increases in reliability. For example, if the largest source of error to reliability comes from low inter-rater agreement, increasing the number of raters (and/or providing better training and monitoring) may be advisable. D-studies allow the test developer to make decisions regarding test administration and/or item modifications for further development and testing of the scale
3. *Internal consistency* demonstrates that related items coalesce, i.e., items within the same domain cluster together and do not excessively relate to other domains. For example, if motor items are highly correlated with language items, that probably means verbal directions for motor tasks place too many demands on receptive language skills and thus penalize a child with good motor ability but poor language comprehension [7, 8, 11, 16, 17]
Measures of internal consistency also show which items are so highly intercorrelated as to be duplicative and thus safely removed. Techniques in item analysis and item discrimination are often used to identify and eliminate nonfunctioning items (e.g., those that are too difficult or too easy) with the goal of reducing test length to an optimal set of questions [7, 8, 11, 16, 17, 28]. Afterward, Guttman's Lamba, Cronbach's alpha, or similar statistical techniques are used in reporting internal consistency among the final set of items (including any subtests). Desirable values for Cronbach's alpha are usually set at >0.70; however, it should be noted that this statistic is very sensitive to the number of test items: the more items, the higher Cronbach's alpha will be, independent of whether the items truly coalesce. It is important to bear this in mind when interpreting results using this statistic

(continued)

Table 4 (continued)

Increasingly, a technique known as item response theory (IRT) is used for the purpose of item reduction (see Streiner [29] for a very accessible discussion of this technique). In brief, IRT allows the test developer to assess the discrimination ability (difficulty level) of individual items in a single test and thereby facilitate the identification of items with the greatest discriminative ability. It should be noted however that IRT was original developed in education for high-stakes testing involving hundreds of items (e.g., medical school admission or licensing examinations). In such situations, it is obviously desirable to reduce the items and therefore reduce respondent burden without a significant cost to either reliability or validity of the test results. For many screening scales used in child development, where the number of items is typically much smaller, it is not at all clear IRT offers much beyond what can be done with more conventional, psychometric approaches to item analysis
4. *Stability*, although not requisite in test construction, is helpful because stability indices indicate that scores are consistent over long intervals of time (e.g., across 3 months to 6 years). For example, if children perform in the above-average range at Time 1, stability coefficients will show (ideally) above-average performance at Time 2. Standards for stability indicators are not well established, which is understandable given the rapid changes in young children's skills (especially on repeated brief screening tests), and so a common way to show whether performance is broadly stable is by viewing agreement in performance as captured by standard deviations between Time 1 and Time 2 (e.g., if performing two standard deviations below the mean at Time 1, to what extent is performance within the same standard deviation at Time 2?). Identifying factors likely to promote or deter developmental progress (e.g., presence or absence of psychosocial risk and resilience factors, significant injuries or illnesses, early intervention) is much needed to interpret stability indices [30–32]
A particularly productive approach to viewing stability is use of change scores, sometimes called "growth indicators." By retesting children over time, rates of skill acquisition can be viewed; sorted into above average, average, or below average; and then used to identify children who learning more slowly than others

Discussion

Test construction is Euclidian in nature—there are postulates and axioms that must be followed for proving that measures work well. Although researchers summarize psychometric data in published studies, the in-depth information about the (gritty if somewhat soporific) details of item development, translations, standardization, reliability, validity, and accuracy are only to be found within test manuals. This means that panelists selecting measures for any programs need to obtain and read test manuals to best understand the strengths and weaknesses of specific tests.

Utility issues also deserve consideration: tests and testing incur costs to both parents and services. Given options among quality measures, choices should be as parsimonious as possible so that the bulk of professional time is spent not on testing but rather on advising families and facilitating access to needed services.

It is beyond the scope of this article to identify the many different instruments measuring child development and behavior. Although numerous reviews exist, new research and new editions of measures are published frequently. Thus only the most current reviews of tools are recommended and preferably reviews that clearly identify instruments meeting psychometric standards [60, 61].

Table 5 Types of validity studies

1. *Content/face validity* does not involve statistical analysis but focuses instead on gathering professional opinion that items are likely to tap the domains and skills of interest. Authors should vet new items prior to conducting standardization and reliability studies and evoke a range of professionals who are thoroughly grounded in the intended foci of the new tool (e.g., speech-language pathology researchers should be asked to evaluate language items for both scope and sequence) [4, 5, 11, 15, 16]. Often this is conducted is an unstructured, ad hoc way but it need not be. A formal process could involve selecting a sample of experts (which could include clinicians, psychometricians, and other content experts but also parents and/or other end users). Domain and corresponding items with detailed descriptions could then be provided to these respondents and each asked to rate all items using a simple ordinal scaling response (e.g., 1 = the item is not relevant to 4 = highly relevant). Statistics such as the content validity ratio (e.g., Lynn [33]) can then be used to assess each item based on expert opinion. Qualitative methods, such as respondent interviewing or focus groups to collect data on the perceived content validity of each item using expert input, could also be considered
2. *Concurrent validity* compares the test under study with reference-standard measures (informally referred to as "the gold standard," although in the field of child development, "bronze standard" is a more accurate referent), i.e., tests that are well-established, proven, effective, and diagnostic [2, 4, 5, 15–17]. The diagnostic battery should include measures tapping the same domains as the new test (e.g., if motor, language, and academic skills are assessed in the experimental test, then the criterion battery should include diagnostic measures of motor, language, and academics). Screening tests should not be used as the sole reference standard when validating a new measure—the error inherent in screening can compound rather than overlap. Thus "head-to-head" comparisons of new and existing screens should compare each screen to the diagnostic battery and not exclusively to other screens whenever possible
To conduct concurrent validity studies, a subset of randomly selected children from the standardization sample are administered the experimental measure along with the reference battery. To eliminate bias, examiners are usually blinded to results of the experimental test. Correlations between measures should show close and significant relationships between "like subtests" (e.g., language items on the experimental test correlate with language subtests on the reference battery) [11, 12, 16]. Concurrent validation research is highly expensive and thus inevitably dependent on funding whether from publishers or via grant support
3. *Construct/convergent validity* studies are much like internal consistency, but in this case, the focus is on whether items or subtests on the research tool have meaningful associations with established measurable theory about the phenomena studied. For example, if child development is considered to have verbal and nonverbal dimensions, then the expressive and receptive language items on an experimental measure should be shown to cluster with the verbal factors on the reference battery, while fine and gross motor items should correlate with nonverbal factors. Factor analysis is the most commonly used technique for establishing construct validity for new measures [11, 12, 16]. But more sophisticated statistical analyses, such as confirmatory factor analysis using structural equation modeling, should be considered when (1) the sample size is sufficiently large enough ($n > 200$ cases) and (2) there is a clearly hypothesized factor structure to be tested

(continued)

Table 5 (continued)

4. *Discriminant validity* is a subtype of construct/convergent validity but in this case determines appropriate lack of association between disparate constructs (e.g., verbal items on the tool under development should not have significant correlations with nonverbal factor scores on diagnostic measures). A more meaningful approach is the use of discriminant validity studies to search for distinct performance patterns according to the conditions of interest. To conduct these studies, types of disabilities are categorized via the reference standard battery and used as dependent variables via logistic regression, discriminant function analyses, latent class, or similar structural equation modeling [12, 16, 34]. Independent variables are the items or subtests on the new measure. An example of helpful results is that children whose only disability is motor impairment are found to perform differently than children for whom language impairment is the only disability. Such information is useful for determining whether a measure has limitations in detecting certain conditions and thus whether alternative tests should be used (See additional discussion of discriminant sensitivity in section "Accuracy".)
5. *Predictive validity* studies are not requisite in test construction but are desirable because these view how well a test measures enduring and meaningful dimensions of child development. Such research involves measuring children with both the experimental test and the diagnostic battery (Time 1) and then administering the same measures later in life (time 2). Time frames vary in studies and can be as little as 3 months or more than two decades [35–42]. It is sometimes also desirable to use other, clinically meaningful (or policy relevant) outcomes, such as performance on standardized educational tests or referrals to specialized medical or allied health services to examine predictive validity. This may be especially relevant in the context of population-level screening scales, where it is of interest to know if scoring in the delayed or atypical range on a test truly leads to negative developmental trajectories later on. In Australia, Canada, and most US States, standardized assessments of school readiness are administered to all children. Of much interest is determining whether scores on a screen administered prior to school entry can be used to predict which children are likely to have limited school success so that appropriate supports can be put in place as early as possible
Because child development is complex and ever-changing, there are no defined standards for predictive validity, but in general, a statistically significant correlation with outcome measures signifies a predictive relationship. Accuracy indicators such as sensitivity and specificity are sometimes used, but achieving the levels demanded for concurrent criterion-related validity (described below) should not be expected. It is worth noting that in predictive validity studies, it is not uncommon to find certain items or subtests that were not concurrent indicators but, instead, enjoy a long-term association with various outcomes or conditions diagnosed later in life (e.g., learning disabilities). Essential for longitudinal studies is creative exploration of results, careful tracking of families over time, and documenting intervening variables (e.g., participation in early intervention, changes in psychosocial risk factors) [38]. Loss to follow-up, whether by voluntary withdrawal or a failure in tracking, is of particular concern in longitudinal studies. Missing data can and often does introduce significant bias, particularly as it is often the most vulnerable families (and therefore those children with the greatest risk for developmental problems) who are lost to follow-up. Finally, as with all longitudinal research, ongoing funding is necessary to minimize subject (and researcher) attrition
6. *Criterion-related validity*, usually referred to as *Accuracy*, is the application of decision theory to screening test results [2, 5, 11, 16, 17]. Accuracy studies identify the likelihood of problems and thus facilitate clinical dispositions (e.g., whether or not to refer for further evaluation). Because a unique level of proof is needed to establish accuracy, the research procedures required are described in section "Accuracy"

Table 6 Standards for accuracy research with screening tests

1. *Identifying performance criteria* to be applied to the reference standard battery. A particularly practical approach is to use criteria for special education eligibility (e.g., for diagnosing learning disabilities, academic performance, that is, >1 standard deviation below IQ; for 0–3-year-old services, two >30% delays relative to chronological age) [15, 17, 46]. Although IDEA criteria vary across US States, there are more commonalities than differences in eligibility standards (http://ectacenter.org accessed October 2016). There will be some differences in standards across countries globally, but the logic remains the same
After identifying types of disabilities, these are then grouped into a single category, i.e., disabled, while children without diagnosed conditions are grouped into a single category, i.e., typically developing. But recognizing that developmental status is not binary (meaning there are children who perform below average and who are not eligible for special education but in need services such as Head Start or tutoring), researchers often create a third group from the nondisabled category to reflect children with delays (e.g., academic, language, and cognitive skills <25th percentile). Because tests of accuracy depend on binary groupings, the three groups (disabled, delayed, and typical) can then be clustered in more than one way to determine how well screens detect all three groups (e.g., disabled + delayed versus typical; disabled versus delayed + typical)
2. *Selecting a subsample that mirrors prevalence*. Test authors usually over-sample children with disabilities and families with psychosocial risk factors in order to view performance differences for validation studies. For accuracy studies of screens intended for use in general settings (e.g., community pediatric clinics, day care centers), it is essential to compute accuracy only on a group in which the incidence of risk factors, demographics, and exceptionalities mirrors prevalence (www.cdc.gov accessed October 2016). Statistical software offers sampling methods for constructing a representative population (e.g., www.spss.com accessed October 2016)
3. *Determining optimal cutoff scores on screens*. Having established one (and preferably more than one) way to group reference tests into binary results, the next step is to determine optimal cutoff scores for the screen. The goal for setting cutoffs is to detect the greatest number of children with problems while also correctly identifying the greatest number of children without problems. Receiver operating characteristic (ROC) is often used to plot scores on screens (e.g., number of items passed) against the binary categories created by grouping performance on reference standard measures (e.g., presence/absence of disability) [47]. ROC analyses help identify the optimal cut point for maximizing test accuracy, and some researchers spend vast amounts of journal space reporting "the area under the curve." But a simple comparison of continuous scores on screens against binary categorization (of problems versus no problems) on diagnostic measures more readily answers clinical questions such as, "My patient passed a screen but what is the chance that the screen missed a problem?" And conversely, when faced with a problematic result, a clinician may well wonder, "What is the chance my patient truly has a problem?" Figure 3 is an example of terms assigned to the intersection of screening test and reference standard results that are then used in the computation of screening accuracy
4. *Computation of screening test accuracy*. Using the subset of children administered both diagnostic and screening measures and ensuring that the subsample is nationally representative, there are a number of computations required and each has its own terminology

(continued)

Table 6 (continued)

(a) *Sensitivity* (*co-positivity*). This analysis answers the question, what percentage of children found to have problems on the diagnostic battery were detected by a screen? The computation involves dividing the numbers of true/co-positives by the sum of true positives/co-positives + false-negatives. In Fig. 4, 16 of 20 children diagnosed also performed poorly on screens, but 4 of the 20 diagnosable children passed the screen. So sensitivity is 16/20 = 75%
Ideally, all children with disabilities score below cutoffs on a screen and are thus identified as needing referrals for further evaluation. In reality, detection of disabilities is imperfect due to behavioral noncompliance, psychosocial malleability, age-related skill changes in development, and imperfections in reference tests (hence the reason for the term co-positivity), compounded by the necessary brevity of screens. So minimal standards for sensitivity are 75–80%. While these figures may seem low, many tests fail to attain this level of accuracy. Higher sensitivity may be found if using stringent performance criteria on the reference battery (e.g., Tenth percentile or lower), but such criteria are not appropriate if a screen is designed to identify not only children who probably need IDEA programs, but also those with milder delays who need services such as Head Start or quality day care (e.g., <16th percentile). Nevertheless, because screening tests should be readministered over time wherein information gleaned from surveillance activities can also be brought to bear especially on negative (meaning passed) screens, detection rates will improve given repeated measurement [43–45]
(b) *Specificity* (*co-negativity*). This analysis answers the question, what percentage of children found to have typical development on diagnostic measures also passed the screen? The computation involves dividing the number of true/co-negatives divided by the sum of true/co-negatives + false-positives. In this example, 80 children had typical development on diagnostic testing, and 70 passed the screen, but 10 did not. So specificity is 70/80 = 88%
Ideally, all typically developing children pass screening tests (100% specificity). But the realities of measuring children's development do not accord such accuracy. So 75–80% specificity is a minimum standard, but within that range, closer to 80% or higher is desirable because there are many more typically developing children than not: For each decrement in specificity, there is a geometric increase in potential over-referrals. Thus the balance of sensitivity and specificity needs to be tipped toward specificity, if at all possible, while still keeping sensitivity in the 75–80% ballpark for screening test standards [3, 5, 6, 46]
Although Fig. 4 provides an overall indicator of sensitivity and specificity, it is important to prove that a screen is accurate across age-ranges and unique age-associated cutoff scores. Figure 4 provides an example
(c) *Negative predictive value and under-referral rate* answer the question, if a child passes a screen, what is the chance that he or she does not have an actual problem? The computation involves dividing the number of true/co-negatives by the sum of true/co-negatives + false-negatives. In this case, 70 out of 74 children with a passed screen were not found to have problems on diagnostic testing. So the negative predictive value is 70/74 = 95%, meaning a 95% chance that a child who passed screens does not have a problem. *Under-referral rate* is a better way of explaining negative predictive value by answering the question: what percentage of children did a screening test fail to detect correctly? The computation divides the number of children without problems on screens by the total number without problems on diagnostic testing. In this case, four children who passed screening were found to have a problem on diagnostic evaluations, so 4/74 = 5%. This computation is the same as subtracting negative predictive value from 100%, and so in this example, 100% minus the 95% = 5% under-referral rate, i.e., a 5% chance that a problem was missed

(continued)

Table 6 (continued)

(d) *Positive predictive value and over-referral rate.* These analyses answer the question, if a child fails a screen, what is the chance that he or she truly has a problem, i.e., what percentage of problematic screening results are actually associated with problems on diagnostic measures? The computation involves dividing the number of true/co-positives by the sum of true/co-positives + false-positives. In this example, 16 of 26 children who failed the screen also had problems on diagnostic testing, but 10 of the 26 children who failed the screen did not have a problem on diagnostic testing. So positive predictive value is 16/26 = 62% but see the over-referral rate discussion below.
Over-referral rate answers a clinically relevant question related to positive predictive value: What percentage of children were over-referred for seemingly unnecessary evaluations? The computation involves dividing the false-positives (those who did poorly on screens but did not have a problem on diagnostic tests) by the total number of positive screening results. In this case, 26 children had positive results but 10 children were false-positives. So the over-referral rate is 10/26 = 38%. This computation is the same as subtracting positive predictive value from 100%. In this example, 100% minus the positive predictive value of 62% = 38% over-referral rate

It is important to note that the 38% over-referral rate does not reflect 38% of all patients, but rather only 10% (10 out of 100 patients). Even so, clinicians are often unnecessarily alarmed at over-referral rates and may be reluctant to refer as a consequence. Negating this assumption is a study (of four different screening tests compared to a reference standard battery) in which approximately 70% of over-referred children had numerous psychosocial risk factors and scored at or below the 25th percentile on diagnostic measures of intelligence, language, and academic achievement (the point below which regular classroom instruction is less than optimally effective) [49, 50]. Thus over-referrals often identify children who do not qualify for special education programs such as IDEA but still need other interventions (e.g., Head Start, Title I public school services, parent-training, quality daycare, social services, mental health programs, etc.). Although it is important to report positive predictive value/over-referral rates on screens, it is also important to parse the lower range of functioning into those likely eligible for special needs programs versus those not eligible but in need of other help, to enable clinicians to refer children to appropriate services [49]

(e) *Discriminant sensitivity (also known as discriminant accuracy)* is a relatively new but helpful technique in evaluating screening tests' strengths and weaknesses [17, 46, 51]. Discriminant sensitivity addresses the question: Is a screen equally sensitive and specific to below-average versus average/above-average performance across each developmental-behavioral domain (as measured by the reference battery)? Figure 5 provides an example

A related approach is to view a screen's sensitivity to each diagnosis rendered by the reference battery to answer such questions as: Is a screen as adept at identifying children with motor coordination difficulties versus language impairment or autism spectrum disorder? Limitations in sensitivity, i.e., <70% for a specific condition suggest the need to add a narrow-band screening tool to broad-band screening measures (e.g., a tool focused on detecting autism spectrum disorders) [17]. Nevertheless, narrow-band tools are not designed to detect the full range of disabilities and delays and should never serve as the sole developmental-behavioral screen

(continued)

Table 6 (continued)

(f)	*Longitudinal predictive accuracy* is another advance in screening test psychometry and answers such questions as: Do children who pass screens early in life, continue to perform well? Not all screening tests enjoy such studies and standards for longitudinal predictive accuracy are not well-established. Achieving levels expected for concurrent sensitivity and specificity (~≥75%) should not be expected (e.g., due to recovery from illness, changes in psychosocial risk, effects of intervention). Meanwhile, it is not uncommon to find certain screening test items or subtests that lack concurrent associations but instead have long-term associations with conditions diagnosed later in life. Essential for longitudinal studies is creative exploration of results, careful tracking of families over time, and documenting intervening variables [17, 36–42, 52–55]
(g)	*Hit-rates—an accuracy indicator to avoid*. Hit-rates are the total number of children for whom a screening test gave accurate information when compared to diagnostic measures. Co-positives and co-negatives are added together and then divided by the entire sample (co-positives + co-negatives + false-positives + false negatives). Like the Kappa statistic, hit rates are misleading because the preponderance of screening test results is co-negatives, i.e., most children are typically developing and perform well on both screening and on diagnostic measures. This means that specificity carries excessive weight. For example, a high hit rate (e.g., 91%) could mean that 99% of typically developing children were correctly detected (e.g., 89 out of 90) but only 20% of children with problems were correctly detected (e.g., 2 out of 10). So hit rates can mask serious flaws in accuracy (and most especially serious flaws in sensitivity), and reporting hit rates is actually outlawed in some countries [49]. Statistics such as RIOC can be used in situations where base prevalence is low to adjust for problems of skewed marginals (i.e., more non-cases than cases reflected in row and column tables in 2 × 2 tables)
(h)	*Other optional accuracy indices*. When healthcare providers serve a preponderance of low income or vulnerable patients, it is helpful to offer information on what to expect in terms of screening test failures. Computing *base rates* by clinic/population type is an effective way to alert clinicians that national prevalence figures for disabilities may be an underestimate of the frequency/incidence with which screening tests are failed in at-risk populations [46, 56, 57]. For example, Brixey et al. [56] found elevated rates of screening test failures in clinics serving mostly Medicaid patients. Roux et al. [57] showed that children screened through a non-emergent crisis call center (where parents with multiple psychosocial risk factors were calling about eviction, food instability, etc.) had even higher rates of screening test failures. Thus it is helpful if authors calculate base-rate performance differences (usually expressed as simple percentages of problematic results) according to various types of populations screened
Reporting *odds ratios* is a helpful adjunct to accuracy studies because they express the likelihood of diagnosable problems according to types of results rendered by screening tests (e.g., high risk = 11 times the chance of diagnosis as compared to low-risk children; moderate risk = 4 times the chance, etc.). Logistic regression is the statistical technique typically used to generate odds ratios/likelihood estimates	

For researchers studying existing measures, test manuals should be scrutinized carefully to ensure that research protocols deploy administration methods in accordance with each test's specific directions. In the absence of such adherence, it is not uncommon to find published studies in which quality instruments have been soundly criticized even while the researchers mis-administered the measures under study

Screening Test	Diagnosis: NO	Diagnosis: YES	
Pass	70 (co-negatives/ true-negatives)	4 (false-negatives)	74
Fail	10 (false positives)	16 (co-positives/ true positives)	26
	80	20	/100

Fig. 3 Terms used in accuracy indices for screening tests (Aylward et al., 2008; [3, 5, 6, 17, 48]). The figure should have a copyright notice below it reading: @Frances Page [46] (from Glascoe FP, Marks KP, Poon JK, Macias MM (eds). Identifying and Addressing Developmental and Behavioral Problems: A Practical Guide for Medical and Non-medical Professionals, Trainees, Researchers and Advocates. Nolensville, Tennessee: PEDStest.com, LLC, 2013). True/co-positives: the number of children with problematic screening scores who were also found to have problems on diagnostic testing ($N = 16$). True/co-negatives: the number of children who not only passed screening but also performed well on diagnostic testing ($N = 70$). False-positives: the number of children with problematic screening scores but who were not found to have problems on diagnostic testing ($N = 10$). False-negatives: the number of children who passed screening but were found to have problems on diagnostic testing ($N = 4$)

Age in months	Ns failing items/Ns below 16th %tile	Sensitivity	Ns passing items/Ns above 16th%tile	Specificity
0 – 2	41/59	70%	260/298	87%
3 – 4	61/65	94%	289/355	81%
5 – 7	75/87	86%	297/387	77%
8 – 10	58/76	76%	282/338	83%
11 – 13	70/82	85%	343/386	89%
14 – 16	43/51	84%	228/273	84%
17 – 19	68/80	85%	355/442	80%
20 – 22	41/49	84%	261/319	82%
23 – 25	67/72	93%	318/342	93%
26 – 28	24/30	80%	205/228	90%
29 – 33	67/77	87%	277/337	82%
34 – 28	58/67	86%	329/359	92%
39 – 43	93/111	84%	433/519	83%
44 – 48	60/68	88%	214/275	78%
49 – 53	60/79	76%	379/473	80%
54 – 58	24/34	71%	277/330	84%
59 – 65	44/56	79%	255/301	85%
66 – 72	24/30	80%	185/213	87%
73 – 83	32/41	78%	200/250	80%
84 – 95	67/82	82%	446/497	90%
TOTAL	**1077/1296**	**83%**	**5833/6922**	**84%**

Fig. 4 Example of age-focused accuracy studies showing sensitivity and specificity for a screening test in identifying total score performance on a diagnostic battery of developmental and behavioral/social-emotional skills (©From Glascoe FP, Robertshaw NS, Woods SK. *PEDS:Developmental Milestones Professional Manual.* Nolensville, TN: PEDStest.com, 2016. Used with permission)

Domain	Ns failing item/Ns below 16th %tile	Sensitivity	Ns passing items/Ns above 16th%tile	Specificity
Fine Motor	200/232	86%	911/1130	81%
Gross Motor	172/197	87%	678/828	82%
Expressive Language	146/176	83%	866/1003	86%
Self-Help	138/157	88%	922/1058	87%
Receptive Language	177/218	81%	938/1138	86%
Social-Emotional	133/177	75%	797/936	85%
Academic/Preacademic (for children 39 months and older)	111/ 139	80%	676/825	82%
TOTAL	**1077/1296**	**83%**	**5788/6918**	**84%**

Fig. 5 Example of a discriminant accuracy study identifying domain-specific performance on a reference-standard diagnostic battery (©From Glascoe FP, Robertshaw NS, Woods SK. *PEDS:Developmental Milestones Professional Manual.* Nolensville, TN: PEDStest.com, 2016. Used with permission)

Table 7 Considerations in utility/feasibility

(a)	*Material costs.* Although the price of a test kit is amortized across the many children to whom a measure is administered, ongoing costs include purchasing test protocols or photocopying (if allowed), as well as replenishing stimuli for hands-on tests because these are often lost or damaged over time (e.g., when needed blocks, dolls, scissors, crackers, cups)
(b)	*Administration and scoring costs.* The longer the measure and the more complicated the scoring, the more professional time is required. Hands-on measures cost more to administer than those relying on parent report. Even so, the more difficult the reading and thus intelligibility levels, the more lengthy interviews are required (along with re-explanations of item intent). For the same reasons, administration time/costs rise if translations are not carefully vetted (or if interpretation services such as language line (www.languageline.com) do not have copies of vetted translations)
(c)	*Report writing costs.* Time-related professional expense includes the creation of test reports such as a parent take-home summary and/or a referral letter. Absence of decision support within test manuals contributes to the post-service cognitive work required and thus to professional expense. Some tests enjoy web-based services that not only generate scores but also reports with recommendations based on test results (e.g., when to refer to special education services versus enrichment programs) such as Head Start versus parent-training. Although electronic applications tend to cost more (e.g., per encounter or per site), such expenses offset the costs of professional time
(d)	*Duplicative measurement costs.* When subspecialty clinics (e.g., cardiology and NICU follow-up) use measures rendering quotients but not the types of scores needed to determine early intervention eligibility (e.g., age-equivalents, percent of delay), early intervention professionals must administer alternative measures. Such unnecessary duplication creates expense and reduces monies available for intervention itself
(e)	*Training costs.* Some tests require professional training. Training may incur fees or may be freely available on publishers' websites. In any case, there are costs for professional time spent, with or without certification/training expenses

(e.g., substituting professional opinion for the proscribed method of parent/professional hands-on administration; failing to interview when test protocols were incomplete) [62–65].

Researchers asked to review manuscripts focused on new or existing tests, also need to be thoroughly familiar with each measure under study and with standards for test construction. Such familiarity helps discern careful research procedures from faulty ones (e.g., whether tests were administered correctly, whether validity studies used reference standard diagnostic batteries). Reviewers of journal articles and those making funding decisions about the development of new tools should be aware that authors may wax exuberantly about their emerging measures and may make undue recommendations for ubiquitous use, i.e., "Scientists cannot be too careful to avoid falling prey to their own enthusiasm" (p. 13) [66]. Avidity for deployment of tests not fully proven to work is a conflict of interest with the goal of providing optimal care for children and families—whether tools are freely or commercially available. This article should aid reviewers in identifying when studies of new tools remain in pilot stages and help authors report the limitations of their work.

For those creating new items or entirely new tests, the current review provides helpful highlights of essential standards in test construction. Although it is not possible to cover the wide-range of specific statistical methods applicable to item analysis, standardization, validation, etc., test authors are encouraged to explore titles in the reference section. All test authors must keep in mind that the goal of any instrument is to best detect or define children's difficulties so that professionals can, in turn, best assist families. This task cannot be accomplished confidently without adhering to standards in test psychometry.

References

1. Majnemer, A. (2012). *Selection and use of outcome measures.*
2. Aylward, G. P., Stancin, T., Wolraich, M. L., Drotar, D. D., Dworkin, P. H., & Perrin, E. C. (2008). Screening and assessment tools. In M. L. Wolraich (Ed.), *Developmental behavioral pediatrics evidence and practice* (pp. 123–130). Philadelphia, PA: Mosby Elsevier.
3. Barnes, K. E., & Charles, C. (1982). *Preschool screening: The measurement and prediction of children at risk.* Springfield, IL: Thomas.
4. Buros Center for Testing. (2014).
5. Frankenburg, W. K. (1974). Selection of diseases and tests in pediatric screening. *Pediatrics, 54*, 612–618.
6. Lichtenstein, R., & Ireton, H. (1984). *Preschool screening: Identifying young children with developmental and educational problems.* Orlando, FL: Grune & Stratton. Retrieved from www.amazon.com.
7. Anastasi, A., Urbina, S., & Cliffs, N. J. (2005). *Psychological testing*. Englewood Cliffs, NJ: Prentice-Hall.
8. Cronbach, L. J. (1990). *Essentials of psychological testing*. Harper & Row: New York.
9. Aylward, G. P. (2013). Continuing issues with the Bayley-III: Where to go from here. *Journal of Developmental and Behavioral Pediatrics, 34*, 697–701.

10. Aylward, G. P., & Aylward, B. S. (2011). The changing yardstick in measurement of cognitive abilities in infancy. *Journal of Developmental and Behavioral Pediatrics, 32*, 465–468.
11. Murphy, K. R., & Davidshofer, C. O. (2005). *Psychological testing: Principles and applications*. Upper Saddle River, NJ: Pearson Prentice Hall.
12. Miller, L. A., McIntire, S. A., & Lovler, R. L. (2010). *Foundations of psychological testing*. Thousand Oaks, CA: Sage.
13. Popham, W. J., & Alexandria, V. A. (2011). Transformative assessment in action: An inside look at applying the process. *Association for Supervision Curriculum Development.* Retrieved from www.ascd.org.
14. Vygotsky, L., & Harvard, P. (1978). *Mind in society: The development of higher psychological processes*. Cambridge: Harvard University Press.
15. American Educational Research Association. (2014). *American educational research on measurement in education*. Washington, DC: Author.
16. Downing, S. M., & Haladyna, T. M. (Eds.). (2006). *Handbook of test development.* Mahwah, NJ: Lawrence Erlbaum Associates.
17. Glascoe, F. P., Marks, K. P., Macias, M. M., Howard, B., Sturner, R., Aydin, M., et al. (2013). Test construction and other research in developmental-behavioral screening. In: *Identifying and addressing developmental behavioral problems guide for medical and nonmedical professionals trainees researchers and advocates* (pp. 423–51). Nolensville, TN: PEDStest.com, LLC. Retrieved from www.pedstest.org.
18. Epstein, A. S., & Jama. (2013). Not just words: Caring for the patient by caring about language. *Medicine, 173*(9), 727–728.
19. DuBay, W. H. (2006). *Smart language: Readers, readability, and the grading of text.* Costa Mesa, CA: Impact Information. Retrieved from www.impact-information.com.
20. Brooks, C. (2014). *Lost in translation: 9 International Marketing Fails* (Vol. 8).
21. ITC. (2014). *International test commission.*
22. Willis, G. B. (2005). *Cognitive interviewing: A tool for improving questionnaire design.* Thousand Oaks, CA: Sage.
23. Camp, B. W. (2007). Evaluating bias in validity studies of developmental/behavioral screening tests. *Journal of Developmental and Behavioral Pediatrics, 28*(3), 234–240.
24. Gottfredson, L. S. (1994). The science and politics of race-norming. *American Psychologist, 48*, 955–963.
25. Viera, A. J., & Garret, J. M. (2005). Understanding interobserver agreement: The kappa statistic. *Family Medicine, 37*(5), 360–363.
26. Streiner, D. L., Norman, G., & Cairney, J. (2015). *Health measurement scales: A practical guide to their development and use* (5th ed.). Oxford, UK: Oxford University Press.
27. Cairney, J., & Streiner, D. L. (2011). Using relative improvement over chance (RIOC) to examine agreement between tests: Three case examples using studies of developmental coordination disorder (DCD) in children. *Research in Developmental Disabilities, 32*(1), 87–92.
28. Baker, F. B., & Kim, S. (2004). *Item response theory: Parameter estimation techniques.* New York, NY: Marcel Dekker.
29. Streiner, D. L. (2010). Measure for measure: New developments in measurement and item response theory. *Canadian Journal of Psychiatry, 55*(3), 180–186.
30. Glascoe, F. P., & Leew, S. (2010). Parenting behaviors, perceptions, and psychosocial risk: Impacts on young children's development. *Pediatrics, 125*(2), 313–319.
31. Sameroff, A. J., Seifer, R., Barocas, R., Zax, M., & Greenspan, S. (1987). Intelligence quotient scores of 4-year-old children: Social-environmental risk factors. *Pediatrics, 79*(3), 343–350.
32. Taris, T. W., Bok, I. A., & Meijer, Z. Y. (1998). Assessing stability and change of psychometric properties of multi-item concepts across different situations: Approach. *The Journal of Psychology, 132*, 301–316.
33. Lynn, M. R. (1986). Determination and quantification of content validity. *Nursing Research, 35*(6), 382–385.
34. Campbell, D. T., & Fiske, D. W. (1959). Convergent and discriminant validation by the multitrait-multimethod matrix. *Psychological Bulletin, 56*, 81–105.

35. Anderson, L. M., Shinn, C., Fullilove, M. T., Scrimshaw, S. C., Fielding, J. E., & Normand, J. (2003). The effectiveness of early childhood development programs: A systematic review. *American Journal of Preventive Medicine, 24*(3), 32–46.
36. Bailey, D. B., Hebbeler, K., Spiker, D., Scarborough, A., Mallik, S., & Nelson, L. (2005). Thirty-six-month outcomes for families of children who have disabilities and participated in early intervention. *Pediatrics, 116*(6), 1346–1352.
37. Campbell, F. A., Ramey, C. T., Pungello, E., Sparling, J., & Miller-Johnson, S. (2002). Early childhood education: Young adult outcomes from the Abecedarian Project. *Applied Developmental Science, 6*(1), 42–57.
38. Marks, K. P., Shevell, M., Squires, J., Aylward, G., & Glascoe, F. P. (2008). The Thorny nature of predictive validity research on developmental screening tests. *Pediatrics, 122*(4), 866–868.
39. McCormick, M. C., Brooks-Gunn, J., Buka, S. L., Goldman, J., Yu, J., Salganik, M., et al. (2006). Early intervention in low birth weight premature infants: Results at 18 years of age for the infant health and development program. *Pediatrics, 117*, 771–780.
40. Muennig, P., Schweinhart, L., Montie, J., & Neidell, M. (2009). Effects of a prekindergarten education intervention on adult health: 37-year follow-up results of a randomized controlled trial. *American Journal of Public Health, 99*(8), 1431–1437.
41. Reynolds, A. J., Temple, J. A., Ou, S. R., Arteaga, I., & White, B. (2011). School-based early childhood education and age-28 well-being: Effects by timing, dosage, and subgroups. *Science, 333*(6040), 360–364.
42. Wake, M., Gerner, B., & Gallagher, S. (2005). Does parents' evaluation of developmental status at school entry predict language, achievement, and quality of life 2 years later? *Ambulatory Pediatrics, 5*(3), 143–149.
43. Guevara, J. P., Gerdes, M., Localio, R., Huang, Y. V., Pinto-Martin, J., & Minkovitz, C. S. (2013). Effectiveness of developmental screening in an urban setting. *Pediatrics, 131*(1), 30–37.
44. Hix-Small, H., Marks, K. P., Squires, J., & Nickel, R. (2007). Impact of implementing developmental screening at 12 and 24 months in a pediatric practice. *Pediatrics, 120*(2), 381–389.
45. Schonwald, A., Huntington, N., Chan, E., Risko, W., & Bridgemohan, C. (2009). Routine developmental screening implemented in urban primary care settings: More evidence of feasibility and effectiveness. *Pediatrics, 123*(2), 660–668.
46. Glascoe, F. P. (2013). *Collaborating with parents: Using parents' evaluation of developmental status in early detection and intervention.*
47. Brown, C., & Davis, H. (2006). Receiver operating characteristic curves and related decision measures: A tutorial. *Chemometrics and Intelligent Laboratory Systems, 80*, 24–38.
48. Aylward, E. M., Parrilo, P. A., & Slotine, J. J. E. (2008). Stability and robustness analysis of nonlinear systems via contraction metrics and SOS programming. *Automatica, 44*(8), 2163–2170.
49. Glascoe, F. P. (2001). Are over-referrals on developmental screening tests really a problem? *Archives of Pediatrics and Adolescent Medicine, 155*(1), 54–59.
50. Tarini, B. A., Clark, S. J., Pilli, S., Dombkowski, K. J., Korzeniewski, S. J., Gebremariam, A., et al. (2011). False-positive newborn screening result and future health care use in a state Medicaid cohort. *Pediatrics, 128*(4), 715–722.
51. Glascoe, F. P., Robertshaw, N. S., Woods, S. K., & Nolensville, T. N. (2016). *PEDS: Developmental milestones professional manual*. Retrieved from PEDStest.com.
52. Briggs-Gowan, M. J., & Carter, A. S. (2008). Social-emotional screening status in early childhood predicts elementary school outcomes. *Pediatrics, 121*, 957–962.
53. Guy, A., Seaton, S. E., Boyle, E. M., Draper, E. S., & Field, D. J. (2015). Infants born late/moderately preterm are at increased risk for a positive autism screen at 2 years of age. *Journal of Pediatrics, 166*, 269–275.
54. Ozonoff, S., Young, G. S., Steinfeld, M. B., Hill, M. M., Cook, I., & Hutman, T. (2009). How early do parent concerns predict later autism diagnosis? *Journal of Developmental and Behavioral Pediatrics, 30*(5), 367–373.

55. Sokolova, M., Japkowicz, N., & Szpakowicz, S. (2006). Beyond accuracy, F-score and ROC: A family of discriminant measures for performance evaluation. *Association for the advancement of artificial intelligence digital library proceedings and conference papers.* Retrieved from https://www.aaai.org/Papers/Workshops/2006/WS-06-06/WS06-06-006.pdf.
56. Brixey, S., Siddique, I., Cohn, J., Johnson, S., Hamilton, C., Li, S., et al. (2009, May). *Developmental screening in an urban pediatric clinic.* Presentation to the Pediatric Academic Societies Annual Meeting. Retrieved from www.pas-meeting.org.
57. Roux, A. M., Herrera, P., Wold, C. M., Dunkle, M. C., Glascoe, F. P., & Shattuck, P. T. (2012). Reaching underserved children with autism screening: The 211LA developmental screening project. *American Journal of Preventive Medicine, 43*, 514–530.
58. Dobrez, D., Sasso, A. L., Holl, J., Shalowitz, M., Leon, S., & Budetti, P. (2001). Estimating the cost of developmental and behavioral screening of preschool children in general pediatric practice. *Pediatrics, 108*(4), 913–922.
59. Glascoe, F. P., Foster, E. M., & Wolraich, M. L. (1997). An economic analysis of developmental detection methods. *Pediatrics, 99*(6), 830–837.
60. Glascoe, F. P., Marks, K. P., Bauer, N. S., Kliegman, R. M., Behrman, R. E., Jenson, H. B., et al. *Developmental screening and surveillance.* Philadelphia, PA
61. LaRosa, A. (2016). Developmental and behavioral screening tests in primary care. *PediatricsUpToDate.* Retrieved from www.uptodate.com.
62. Glascoe, F. P., & Squires, J. (2009). Questions about the ability of broad-band screens to detect children with ASD. *Journal of Developmental and Behavioral Pediatrics, 30*, 174.
63. Limbos, M. M., & Joyce, D. P. (2011). Comparison of the ASQ and PEDS in screening for developmental delay in children presenting for primary care. *Journal of Developmental and Behavioral Pediatrics, 32*(7), 499–511.
64. Marks, K. P. (2007). Should general pediatricians not select the Ages & Stages Questionnaire in light of the Rydz et al study? *Pediatrics, 120*(2), 457–458.
65. Pinto-Martin, J. A., Young, L. M., Mandell, D. S., Poghosyan, L., Giarelli, E., & Levy, S. E. (2008). Screening strategies for autism spectrum disorders in pediatric primary care. *Journal of Developmental Behavioral Pediatrics, 29*, 345–350.
66. Kiermer, V. (2014). Eureka once, eureka twice. *Scientific American, 310*(5), 13.

Part V
Relation to Early Intervention and the Medical Home

Primer on Special Education

Holly Roberts and Benjamin Kennert

Abstract Children who have required intervention in the Neonatal Intensive Care Unit (NICU) for prematurity, low birth weight, or medical conditions may require long-term services to optimize their success when they become school age. The specific type and level of intervention may vary across development. The purpose of this chapter is to provide a review of the types of services available to children throughout their school experience. Every child with a disability is entitled to a free and appropriate education (FAPE) under the Individuals with Disabilities Act (IDEA) from birth to age 21. Additionally, there are safeguards through Section 504 of the Rehabilitation Act (Section 504) of 1973, which is a civil rights law protecting the discrimination on the basis of a disability in organizations that benefit from federal funding (e.g., schools). Modifications under IDEA and Section 504 will be reviewed.

Children who are born premature or with complications that require a Neonatal Intensive Care Unit (NICU) stay may be at elevated risk for requiring specialized services in the school [1, 2]. Early intervention for these NICU graduates is optimal. Some of these NICU graduates may require ongoing accommodations in school that extends past preschool years. There are several factors that may predict the need for services in these children such as time spent in the hospital, illness, and days on mechanical ventilation. The purpose of this chapter is to describe special education law as indicated by the Individuals with Disabilities Act (IDEA) and Section 504 of the Rehabilitation Act of 1973 (Section 504).

A recent report from the National Center for Education Statistics indicated that 12.92% of all children between ages 3 and 21 in the United States receive special education services as a child with a disability, totaling 6.5 million children verified with a disability [3]. In 1975, Congress enacted Public Law 94-142, then titled the

H. Roberts (✉) · B. Kennert
University of Nebraska Medical Center, Munroe-Meyer Institute, Omaha, NE, USA
e-mail: hroberts@unmc.edu

H. Needelman, B. J. Jackson (eds.), *Follow-Up for NICU Graduates*,
https://doi.org/10.1007/978-3-319-73275-6_16

Education for All Handicapped Children Act. This was an extension of the previous Education of the Handicapped Act and was meant to ensure that children with a disability receive a free and appropriate education through providing financial assistance to schools. Prior to the enactment of Public Law 94-142, more than half of the over eight million children with disabilities in the United States were not receiving appropriate educational services, and more than one million identified children were excluded from the school system entirely [4].

Since it was first enacted, Public Law 94-142 has gone through several revisions, which began to emphasize the importance of also identifying children with a disability early. The most recent of these revisions is the Individuals with Disabilities Education Improvement Act [5]. Through these revisions, IDEA began to put more emphasis on inclusive parent roles and responsibilities in their child's special education, effective implementation of special education services, professional development for personnel who work with children with disabilities, providing incentives for whole-school services (i.e., a movement that focuses on increased integration of school and health services), and supporting the use of technology, such as assistive devices [5, 6]. Public schools must comply with IDEA in order to receive the funds provided, and noncompliance with IDEA can result in a removal of these funds from the schools.

There are four parts of IDEA: Parts A, B, C, and D. Part A includes general provisions, and Part D indicates national efforts for improving the education of children with disabilities. Parts B and C highlight the provision of services for children with disabilities, with Part C highlighting early intervention services for children from birth to 3 years old and Part B highlighting school-age services for children from ages 3 to 21 years.

Eligibility for Special Education

Under IDEA, a child with a disability must be identified in 1 of 13 eligibility categories: autism spectrum disorder, cognitive disability, deaf-blindness, developmental delay, emotional disability, hearing impairment, severe multiple disabilities, orthopedic impairment, other health impairment, specific learning disability, speech and language impairment, traumatic brain injury, and visual impairment. For more information on eligibility criteria for each category, see Appendix 1. Of particular interest to medical professionals is the category of other health impairment, which can include conditions such as attention-deficit/hyperactivity disorder (ADHD), diabetes, epilepsy, heart conditions, hemophilia, Tourette's syndrome, and sickle cell anemia. This category can also include any diagnosed medical condition which causes significant impairment in a child's ability to participate in learning and requires documentation from a medical professional (e.g., pediatrician) familiar with the child's condition.

Special education eligibility differs from a medical or clinical diagnosis of a disability in several ways. For a child to be verified as a child with a disability under

IDEA, there must be a documented need for special education services. In other words, a child may meet clinical criteria for a diagnosable disability, but if the child is able to succeed in an educational environment at an appropriate level without special education assistance, they are not a child with a disability under IDEA and therefore do not qualify for special education services. It must also be determined that the child's disability is not primarily a result of limited access to appropriate instruction, or the result of limited English proficiency. In these ways, a school-based evaluation can look much different from a medical or clinical evaluation, may draw from different sources of information, and may come to a different decision. This can be frustrating to many parents and professionals, as they are being told the child has a disability in one setting, but not in another setting. It is helpful for medical, clinical, and school professionals, as well as parents, to maintain open communication about the child's educational needs in order to coordinate care for the child most effectively between settings.

If a child does not meet eligibility criteria for special education services, it does not mean that they do not receive appropriate school-based interventions. IDEA reserves that 15% of funds may be used toward early intervening services, including school-wide and targeted interventions for children. Therefore, if a child is falling behind in school, rather than waiting for the child to fail, thus requiring special education services in the future, that child is best served by receiving immediate appropriate interventions, which are individualized to the child's level of need, regardless of the setting in which that child is in.

In addition, a school-based eligibility decision should be made by a multidisciplinary team of professionals, and information for eligibility should draw from a variety of sources, including aptitude and achievement tests, parent and teacher input, and information about the child's physical condition, social and cultural background, and adaptive behavior. It is important to recognize that a parent, as defined by IDEA, includes not only the child's biological parent but also adoptive parents, stepparents, guardians, or any other adult whom the child lives with and is responsible for their welfare [5]. No single assessment measure can be used to determine a child's eligibility, and the child must be assessed in all areas of suspected disability.

Part B

Under Part B, schools are required to provide a free and appropriate public education (FAPE). It is important to note here that FAPE guarantees only an "appropriate" education, not the "best possible" education. In a 1982 United States Supreme Court Case, a school district had denied an interpreter to a student with a hearing impairment who succeeded at expected levels in a general education classroom with the assistance of an FM hearing aid [7]. The Supreme Court determined that IDEA does not ensure a "potential-maximizing education" and that special education services must be "reasonably calculated to receive educational benefits [7]."

Additionally, Part B states that a child with a disability, to the maximum extent possible, be educated in the least restrictive environment possible. Therefore, children who are able to participate in a general education classroom with children without disabilities should be able to do so whenever possible, regardless of whether they need additional support or services in that environment.

Children who qualify with a disability under Part B are required to have an Individualized Education Program (IEP). An IEP states the child's current level of academic achievement and functional performance and states the child's goals for the year. The child's progress with IEP goals are documented within the IEP throughout the year, often on a quarterly basis. The goals should be clear and measurable. For example, consider the goal, "Michael will improve his ability to solve math problems as indicated by his teacher." This goal does not specify how the outcome will be measured in an objective way. Therefore, it would be difficult to determine whether Michael has met his goal in 3 months, 6 months, or over the course of the year. A better goal might be, "Michael will demonstrate an ability to solve 50 one-digit basic math facts during a 3-min timed test." This goal can be objectively measured and provides a clearer indication of whether Michael's goals need to be changed as progress is made. The IEP includes not only academic goals, but other functional goals as well. For example, depending on the child's need, an IEP may include goals for teaching social skills and adaptive skills or for addressing problematic behaviors.

A child's IEP also includes descriptions of the special education services the child will receive throughout the year. IDEA specifically states that these services be "based on peer-reviewed research, to the extent practicable ([5], p. 118)." Appropriate individual accommodations for the child to participate in learning, and in taking state- and district-wide assessments, are also provided within the child's IEP. This includes accommodations for problematic behaviors which interfere with learning, limited English proficiency of the child, sensory impairments (e.g., visual impairments, hearing impairments), communication needs, and environmental modifications. The IEP is reviewed at least on a yearly basis by the child's IEP team. Ideally, an IEP team consists of the child (when possible), parents of the child, a general education teacher, a special education teacher, a representative of the educational agency (i.e., a school administrator), a professional qualified to interpret assessment information (e.g., a psychologist), and anyone else the parent requests to participate on the team.

The Evaluation Process

To begin the initial evaluation process, either the child's parent, the state educational agency, a local educational agency, or another state agency submits a request for an initial evaluation [5]. Following the request, the child's parent must consent to the evaluation. Parents must be provided with a description of the evaluation procedures prior to giving their consent. IDEA requires that the time from the parent's consent to the completion of the evaluation be no more than 60 school days. However, states may employ a more stringent timeframe. For example, in Nebraska a special education evaluation must be completed within 45 school days from the time of parent consent.

Prior to beginning the testing process, a review of existing evaluation data (REED) is conducted. At this time, the MDT reviews any previous evaluation data, or evaluation data provided by the child's parent, and determines what additional information is necessary to gather. Children who have been determined to be eligible for special education services must be reevaluated every 3 years. However, the educational agency, or the parent of the child, can request an additional evaluation at any time. Parental consent must be obtained again to reevaluate the child, unless the school has documented reasonable attempts to obtain parent consent and the parent has not responded. If it is determined at the time of the REED that no new evaluation data are needed, the school may make an eligibility determination without testing, but must notify the parents that they have the right to request new testing be completed. Whenever a change in a child's special education provision is made by the educational agency, the parent is required to be informed of the change.

Part C

Part C of IDEA highlights early intervention services for children from birth through 3 years old. The lead agency for Part C is designated by the state and is often the state department of health, although this varies by state. In the state of Nebraska, for example, the lead for Part C services is shared by the state department of education and the state department of health and human services.

Under Part C, each state is mandated to have a method for identifying and evaluating all children with disabilities and determining need for special education services [5]. The evaluation process for Part C is similar, with parents giving consent prior to beginning the evaluation. However, under Part C only 45 days are allowed to complete the evaluation, compared to the 60 days allowed under Part B. There is only one eligibility category for Part C services, rather than the 13 eligibility categories under Part B. That category is developmental delay and requires one of the three following criteria for a child to be eligible [5]: (1) have a developmental delay in cognitive, physical, communication, social and emotional, or adaptive development; (2) have a diagnosed medical condition resulting in high likelihood of developmental delay; or (3) are considered at-risk infants or toddlers.

Under Part C, children are required to have an Individualized Family Service Plan (IFSP), rather than an IEP. There are three important differences to highlight between an IFSP and an IEP. The first is that an IFSP focuses on the needs and goals of the family, rather than the needs and goals of the child. Therefore, services can be provided to the family under an IFSP, but under an IEP services are specific to the child. Another difference is that the IFSP requires that services be provided within the child's natural environment. For this reason, Part C of IDEA does not guarantee

FAPE. A third difference is that the IFSP must be reviewed and revised every 6 months, as opposed to the IEP which is reviewed annually. Under Part C, a child receives year-round services, rather than services only during the academic year. A child's transition from Part C to Part B occurs before a child's third birthday.

Section 504

Some children with special needs do not require special education services covered under IDEA. Children with certain conditions may receive needed accommodations under Section 504 of the Rehabilitation Act of 1973 (Section 504). Section 504, a civil rights law, protects the right of students with a disability to receive FAPE without discrimination toward the student based on their disability [8]. Section 504 is meant to prevent three types of discrimination: (1) Exclusion from school programs and activities on the basis of disability, (2) harassment on the basis of disability, and (3) lack of accommodations are required which allow children to participate in school programs and activities equally [9]. Under Section 504, students with a disability receive a school-based plan which highlights accommodations the student will have available in order to receive FAPE. Accommodations are individualized based on the child's disability, but examples of accommodations may include preferential seating, additional time to take tests/exams, taking tests/exams in a room alone, adapted physical education curriculum, use of adaptive devices, fewer repetitions in assignments, rest periods, pretyped or prerecorded lessons or notes, designated time to complete homework, dietary accommodations, or peer support groups for the child.

Section 504 is monitored by the Office for Civil Rights (OCR), and parents who feel that their child is not receiving FAPE due to their disability have the right to file a complaint through the OCR. Under Section 504, the definition of a child with a disability is less stringent than for IDEA, so many children who do not meet eligibility criteria for IDEA, but have a documented disability, can still receive school-based accommodations through a Section 504 plan. Under Section 504, a student has to have a record of a physical or mental impairment which substantially limits one or more life activities (e.g., walking, speaking, breathing, learning; [8]). However, the OCR does not review special education decisions, such as a child's placement or the special education services they receive.

Conclusion

Given the unique needs of each NICU graduate, school programming may be a necessary consideration for these children. Whether a child would benefit from an evaluation and/or services during one phase of childhood or across development, special education laws exist providing a framework for how these services are delivered. Although there will be slight variations from state to state on how IDEA and 504 are interpreted, all children have a right to free and appropriate education in the United States. Advocating for the school-based needs of the NICU graduate should be considered by health-care professionals who work with this population.

Appendix 1: IDEA, 2004. Sec. 300.8 Child with a Disability

(a) General.

1. Child with a disability means a child evaluated in accordance with Sec. Sec. 300.304 through 300.311 as having mental retardation, a hearing impairment (including deafness), a speech or language impairment, a visual impairment (including blindness), a serious emotional disturbance (referred to in this part as "emotional disturbance"), an orthopedic impairment, autism, traumatic brain injury, another health impairment, a specific learning disability, deaf-blindness, or multiple disabilities, and who, by reason thereof, needs special education and related services.
2.
    (i) Subject to paragraph (a)(2)(ii) of this section, if it is determined, through an appropriate evaluation under Sec. Sec. 300.304 through 300.311, that a child has one of the disabilities identified in paragraph (a)(1) of this section, but only needs a related service and not special education, the child is not a child with a disability under this part.
    (ii) If, consistent with Sec. 300.39(a)(2), the related service required by the child is considered special education rather than a related service under State standards, the child would be determined to be a child with a disability under paragraph (a)(1) of this section.

(b) Children aged three through nine experiencing *developmental delays*. Child with a disability for children aged three through nine (or any subset of that age range, including ages three through five), may, subject to the conditions described in Sec. 300.111(b), include a child—

1. Who is experiencing developmental delays, as defined by the State and as measured by appropriate diagnostic instruments and procedures, in one or more of the following areas: physical development, cognitive development, communication development, social or emotional development, or adaptive development; and
2. Who, by reason thereof, needs special education and related services.

(c) Definitions of disability terms. The terms used in this definition of a child with a disability are defined as follows:

1.

(i) *Autism* means a developmental disability significantly affecting verbal and nonverbal communication and social interaction, generally evident before age three, that adversely affects a child's educational performance. Other characteristics often associated with autism are engagement in repetitive activities and stereotyped movements, resistance to environmental change or change in daily routines, and unusual responses to sensory experiences.

(ii) Autism does not apply if a child's educational performance is adversely affected primarily because the child has an emotional disturbance, as defined in paragraph (c)(4) of this section.

(iii) A child who manifests the characteristics of autism after age three could be identified as having autism if the criteria in paragraph (c)(1)(i) of this section are satisfied.

2. Deaf-blindness means concomitant hearing and visual impairments, the combination of which causes such severe communication and other developmental and educational needs that they cannot be accommodated in special education programs solely for children with deafness or children with blindness.
3. Deafness means a hearing impairment that is so severe that the child is impaired in processing linguistic information through hearing, with or without amplification that adversely affects a child's educational performance.
4.

(i) *Emotional disturbance* means a condition exhibiting one or more of the following characteristics over a long period of time and to a marked degree that adversely affects a child's educational performance:

A. An inability to learn that cannot be explained by intellectual, sensory, or health factors.
B. An inability to build or maintain satisfactory interpersonal relationships with peers and teachers.
C. Inappropriate types of behavior or feelings under normal circumstances.
D. A general pervasive mood of unhappiness or depression.
E. A tendency to develop physical symptoms or fears associated with personal or school problems.

(ii) Emotional disturbance includes schizophrenia. The term does not apply to children who are socially maladjusted, unless it is determined that they have an emotional disturbance under paragraph (c)(4)(i) of this section.

5. Hearing impairment means an impairment in hearing, whether permanent or fluctuating, that adversely affects a child's educational performance but that is not included under the definition of deafness in this section.
6. Cognitive disability means significantly subaverage general intellectual functioning, existing concurrently with deficits in adaptive behavior and

manifested during the developmental period, that adversely affects a child's educational performance.

7. Multiple disabilities means concomitant impairments (such as mental retardation-blindness or mental retardation-orthopedic impairment), the combination of which causes such severe educational needs that they cannot be accommodated in special education programs solely for one of the impairments. Multiple disabilities does not include deaf-blindness.
8. Orthopedic impairment means a severe orthopedic impairment that adversely affects a child's educational performance. The term includes impairments caused by a congenital anomaly, impairments caused by disease (e.g., poliomyelitis, bone tuberculosis), and impairments from other causes (e.g., cerebral palsy, amputations, and fractures or burns that cause contractures).
9. Other health impairment means having limited strength, vitality, or alertness, including a heightened alertness to environmental stimuli, that results in limited alertness with respect to the educational environment, that—

   (i) Is due to chronic or acute health problems such as asthma, attention deficit disorder or attention deficit hyperactivity disorder, diabetes, epilepsy, a heart condition, hemophilia, lead poisoning, leukemia, nephritis, rheumatic fever, sickle cell anemia, and Tourette syndrome; and
   (ii) Adversely affects a child's educational performance.

10. Specific learning disability

    (i) General. Specific learning disability means a disorder in one or more of the basic psychological processes involved in understanding or in using language, spoken or written, that may manifest itself in the imperfect ability to listen, think, speak, read, write, spell, or to do mathematical calculations, including conditions such as perceptual disabilities, brain injury, minimal brain dysfunction, dyslexia, and developmental aphasia.
    (ii) Disorders not included. Specific learning disability does not include learning problems that are primarily the result of visual, hearing, or motor disabilities, of mental retardation, of emotional disturbance, or of environmental, cultural, or economic disadvantage.

11. Speech or language impairment means a communication disorder, such as stuttering, impaired articulation, a language impairment, or a voice impairment, that adversely affects a child's educational performance.
12. Traumatic brain injury means an acquired injury to the brain caused by an external physical force, resulting in total or partial functional disability or psychosocial impairment, or both, that adversely affects a child's educational performance. Traumatic brain injury applies to open or closed head injuries resulting in impairments in one or more areas, such as cognition; language; memory; attention; reasoning; abstract thinking; judgment; problem-solving; sensory, perceptual, and motor abilities; psychosocial

behavior; physical functions; information processing; and speech. Traumatic brain injury does not apply to brain injuries that are congenital or degenerative, or to brain injuries induced by birth trauma.
13. Visual impairment including blindness means an impairment in vision that, even with correction, adversely affects a child's educational performance. The term includes both partial sight and blindness.

References

1. Vohr, B. R., & Msall, M. E. (1997). Neuropsychological and functional outcomes of very low birth weight infants. *Seminars in Perinatology, 21*, 202–220.
2. Vohr, B. R., Wright, L., Dusick, A., Mele, L., Verter, J., Steichen, J., et al. (2000). Neurodevelopmental and functional outcomes of extremely low birth weight infants in the National Institute of Child Health and Human Development Neonatal Research Network, 1993–1994. *Pediatrics, 105*(6), 1216–1226.
3. Kena, G., Hussar, W., McFarland, J., de Brey, C., Musu-Gillette, L., Wang, X., Zhang, J., Rathbun, A., WilkinsonFlicker, S., Diliberti, M., Barmer, A., Bullock Mann, F., & Dunlop Velez, E. (2016). *The condition of education 2016 (NCES 2016-144)*. Washington, DC: U.S. Department of Education, National Center for Education Statistics. Retrieved October 24, 2016, from http://nces.ed.gov/pubsearch.
4. Education of All Handicapped Children Act, 20 U.S.C. § 1401. (1975).
5. Individuals with Disabilities Education Act, 20 U.S.C. § 1400. (2004).
6. Lipkin, P. H., & Okamoto, J. (2015). The individuals with disabilities education act (IDEA) for children with special educational needs. *Pediatrics, 136*, 1650–1662.
7. Board of Education of Hendrick Hudson Central School District v. Rowley 458 U.S. 176. (1982).
8. Section 504, Rehabilitation Act of 1973, 29 U.S.C. § 701. (1973).
9. Jacob, S., Decker, D., & Timmerman Lugg, E. (2016). *Ethics and law for school psychologists* (7th ed.). Hoboken, NJ: Wiley.

Early Intervention for NICU Graduates

Kerry Miller

Abstract Neonatal intensive care unit (NICU) admissions are on the rise, and infants discharged from the NICU are at elevated risk for developmental delays. Early intervention programs, funded by Part C of the Individuals with Disabilities Education Act (IDEA, 2004), are one resource available to families and children to promote the growth and development of NICU graduates. This chapter provides a brief history of early intervention services provided by Part C of IDEA for infants and toddlers with disabilities and developmental delays or who may be at risk for delays, and it summarizes criteria for program enrollment and the key practices utilized with children and families in these early intervention programs. The efficacy of early intervention programs for children and families is discussed with a specific focus on the program outcomes for NICU graduates, including premature and low birth weight (LBW) infants. Early intervention enrollment and referral patterns for NICU graduates are presented. The implications of these patterns provide support for the need for partnerships between early intervention programs and NICU follow-up programs to maximize Part C-required Child Find efforts and ensure the best outcomes for NICU graduates and their families.

It is well known that the brain grows and develops rapidly during the prenatal period through the postnatal age of three [1]. From the moment a baby is born, the brain continues to grow at a rapid rate, and its development begins to be influenced by children's experiences and interactions with their parents and the environment around them. Parent-child interactions and relationships in the early years are a foundational component of children's future development. Specifically, the way parents interact with their children and the quality of their interactions influences cognition, language, and social-emotional skill development [2–6]; therefore, it is important for parents to feel confident in their interactions and experiences with their child during the critical early years.

K. Miller (✉)
University of Nebraska Medical Center, Munroe-Meyer Institute, Omaha, NE, USA
e-mail: kerry.miller@unmc.edu

H. Needelman, B. J. Jackson (eds.), *Follow-Up for NICU Graduates*,
https://doi.org/10.1007/978-3-319-73275-6_17

For infants born with a condition associated with developmental disabilities or a condition which places them at an elevated risk for developmental delays, it is essential to ensure that their parents have the necessary supports to foster optimal child growth and development during their daily interactions and activities. Early intervention programs are one resource available to families of children who are at risk for or demonstrate developmental delays. These programs aim to support parent confidence and competence in their daily interactions and activities with their children, promote child growth and development during the period of the most rapid brain development, and support and strengthen families.

What Is Early Intervention?

Numerous early intervention programs targeting children under the age of three and their families have emerged across the United States. Many of the programs, funded both privately and publicly, prioritize the enrollment of families and children with risk factors that may affect the children's development, such as low socioeconomic status (SES), single-parent household, parent mental health issues and substance abuse, children with an established disability or developmental delay, and children at risk for delays related to biological, medical, or environmental variables. Federal funds are often pooled with state and local funds to provide services for at-risk pregnant women and parents with young children. Home visitation programs for at-risk populations implement well-known evidence-based models such as Early Head Start, Healthy Families America, Nurse-Family Partnership, and Parents as Teachers with the goal of improving child health and development, preventing child abuse and neglect, improving school readiness, supporting positive parenting, and expanding family self-sufficiency. Programs administered under Part C of the Individuals with Disabilities Education Act share similar aims with programs intended for at-risk populations; however, the focus of Part C early intervention programs is on children with disabilities and children who are at risk for developmental delays and their families. Each program determines eligibility and enrollment based upon individually set program criteria variables. For the purposes of the population discussed in this chapter, neonatal intensive care unit (NICU) graduates, the definition of early intervention will align with the United States federal law in which early intervention is defined as special education services provided to children age birth to 36 months and their families (Individuals with Disabilities Education Act (IDEA), [7]).

The intent of the IDEA legislation in 1986 and its reauthorization in 2004 was to identify and support infants and toddlers with established disabilities or developmental delays or those who may be at risk for developmental delays and their families. By identifying children in the earliest years, early intervention supports children during the most critical developmental period with the goal of maximizing the benefits to children and families. Furthermore, early intervention under IDEA aims to enhance the capacity of families to meet the unique needs of identified infants and toddlers.

Early intervention does not strive to change the interactions and activities in which families engage with their children. The goal of early intervention is to affirm and enhance what is naturally occurring during daily activities to both encourage growth and development and to support children's participation in their families' routines and activities. In doing so, early intervention programs incorporate activities and approaches to support parents in their understanding of their children's early care needs and developmental skill progression, as well as their efforts to positively influence their children's developmental outcomes.

History of Early Intervention

Legislation related to early intervention has evolved over several decades. An important step in support of children with disabilities was the passage of the Education for All Handicapped Children Act (EHA) of 1975 (PL 94-142) [8]. The focus of this legislation was to provide free and appropriate public education to all school-age children; however, incentives were available for services provided to preschool-age children. Based on the knowledge gained from early service provision and lobbying by families and early intervention professionals, Public Law 94-142 was amended in 1986 into Public Law 99-457 [9]. This amendment mandated services for children 3–5 years with disabilities and incentivized states to voluntarily establish services for infants and toddlers with developmental delays and disabilities.

At the onset, Public Law 99-457 afforded each individual state the opportunity to develop a statewide plan for the provision of services for children with developmental delays/disabilities ages birth to 3 and their families. During program planning and development, each participating state designated a lead agency to administer early intervention services. Because of state-level program development, states developed and implemented programs with variables unique to their state. A diverse group of lead agencies (e.g., developmental services, education, rehabilitative services, health and social services) were chosen in each state to manage and guide the design of early intervention services. Additionally, each state independently established differing verification criteria for "developmental delay" to guide program eligibility, resulting in a diverse set of eligibility criteria that varied from state to state, ranging from including only those children with significant delays to inclusion of those at risk for delays.

IDEA legislation related to early intervention for infants and toddlers has undergone periodic revisions and reauthorizations since 1986 when it was considered Part H of the old Education for All Handicapped Children's Act (PL 94-142). It is now known as Part C of the IDEA [7]. These more recent amendments to the initial legislation did not change the requirements for the provision of service, and the original individual differences related to lead agency and verification criteria across states remain in place [10].

Although programs differ from state to state in who oversees the programs and who receives services, federal law specifies that state programs receiving Part C funds must meet specific guidelines. The lead agencies are responsible for providing early intervention evaluations, assessments, development of the Individualized Family Service Plan (IFSP), and service coordination to families at no cost. Some states provide the services outlined in the IFSP at no cost to families, while other states charge a sliding-scale fee. Programs must provide services to children and families, including but not limited to assistive technology devices/services; audiology services; family training, counseling, and home visits; nursing services; nutrition services; occupational therapy; physical therapy; psychological services; special instruction; speech-language pathology services; vision services; and service coordination [7]. The payment of actual services can be billed to Medicaid, private insurance, or Indian Health Services; however, the Part C agency must ask parent permission to access these funds to pay for services. It is important to recognize that family participation in early intervention programs is voluntary.

The foundation of IDEA's early intervention legislation lies in the recognition that children are part of a family and each family has individual values, needs, and priorities. Families and children eligible for early intervention services must have access to all relevant services and individual services to meet the individual needs of children and families. The provision of individual child and family services and the expected outcomes are determined during the development of the IFSP. The components of the plan, including outcomes, services, and frequency and duration of services, center on family and child needs and are collaboratively designated with the family and professionals. The service coordinator is responsible for the timely development of the IFSP, for the periodic review of progress toward outcomes, and for any necessary changes throughout the duration of the IFSP.

The role of a service coordinator is especially important to ensuring a system of coordinated care. The service coordinator on an early intervention team in a local community assumes the coordination, facilitation, and monitoring of the delivery of agreed-upon services, assists families in accessing additional identified services, and communicates and coordinates with medical providers, as needed. Service coordinators assume a neutral role on the early intervention team, as they ensure that families understand their parental rights and are aware of the availability of advocacy services.

Who Is Eligible for Early Intervention?

Part C of IDEA [7] mandates a statewide, comprehensive, multidisciplinary service system focused on addressing the needs of infants and toddlers who are experiencing developmental delays or have a diagnosed physical or mental condition with a high probability of an associated developmental disability in one or more of the following areas: cognitive development, physical development, language and/or speech development, social-emotional development, and self-help skills.

Every state now provides early intervention services for infants and toddlers with disabilities; however, as previously mentioned, eligibility criteria vary from state to state [11]. Each state is required to define developmental delay, including specific verification criteria. For a child to qualify for early intervention services, a comprehensive developmental evaluation must be completed within 45 calendar days of the date the referral is received and include an evaluation of the child's cognitive, speech and language, motor, adaptive behavior, and social-emotional functioning. Most states require a child to demonstrate a level of developmental delay as measured by a standardized, criterion-referenced assessment or have an established condition known to be associated with developmental delays, such as Down syndrome, hearing impairment, or myelomeningocele, to be eligible for early intervention services. Sometimes it is known prior to birth or shortly after birth that early intervention services will be essential in helping the child grow and develop. In some cases, parents are given a referral to their local early intervention program before the child is discharged from the hospital. Often it is the service coordinator who meets with the family while the child remains hospitalized to discuss needed supports and assist with planning the child's transition from hospital to home; however, under Part C guidelines, the child is eligible to receive needed early intervention services prior to discharge. Receipt of early intervention services as an inpatient is less common as the child's needs are often met by their medical team.

As part of the eligibility criteria outlined by each individual state, there is discretion to define at-risk children and subsequently provide services to children identified as at risk based on outlined factors. Common variables that place an infant or toddler at risk for developmental delays include prematurity, exposure to toxins through maternal substance abuse, low birth weight (LBW), intraventricular hemorrhage, seizures, and congenital anomalies associated with developmental issues. Given the option for provision of services for at-risk children, only a limited number (10%) of states consider risk criteria related to biological, medical, or environmental risk factors as determinants of children's eligibility for services [12]. The other states require the child to demonstrate a level of delay in at least one developmental area as indicated by standardized testing. Nationally, approximately 3% of infants and toddlers receive early intervention services [13].

To ensure all children who may be eligible for early intervention services have access to needed services, federal legislation mandates that early intervention programs funded by Part C of IDEA engage in Child Find activities. To meet the set Child Find guidelines, states must develop and implement a comprehensive system to locate, identify, and refer children and families to early intervention services as early as possible [7]. Child Find activities are essential to ensuring identification of children with developmental delays or those at risk for delays in their development and ensure they receive services as early as possible.

Early Intervention Practices

As early intervention programs have evolved over the years, research has investigated practices related to service provision. To link research to practice, the Division for Early Childhood of the Council for Exceptional Children (DEC) developed *Recommended Practices in Early Intervention/Early Childhood Special Education* [10] as a resource to guide the work of practitioners and families as they partner together "to improve the learning outcomes and promote the development of young children birth through 5 years of age, who have or are at risk for developmental delays or disabilities" (p. 1). The recommendations encompass evidence-based practices in early intervention believed to have the greatest impact on child and family outcomes. The following sections provide an overview of commonly implemented early intervention practices guided by the recommendations outlined by DEC. These include focus on family, supports in natural environments, instructional practices, the role of interactions, teaming and collaboration, and transition.

Focus on Family. Grounded in an ecological model of human development [14], early intervention recognizes the child as just one element in the complex relationship among families. Subsequently, the goals of early intervention programs should not focus solely on the child, but rather to enhance family-professional and parent-child interactions and address both child and family needs. With the focus on the family, the service delivery model often utilized in early intervention is the family-centered model.

The family-centered model recognizes the family as an integral constant in the child's life, and the model relies on active involvement of family members in the support and care of their children. Professionals view families as capable of making informed choices and, when provided needed supports, feel they can act on their choices [15]. Implementation of the family-centered model requires the professional to build a trusting and respectful relationship with the family with the focus on strengthening existing skills and promoting acquisition of new skills that will result in positive family and child outcomes.

Supports in Natural Environments. Infants and toddlers learn through their everyday activities, routines, and interactions. Under the direction of Part C of IDEA [7], infants and toddlers with disabilities must "receive needed early intervention services in natural environments to the maximum extent appropriate" (sec. 612(a)(5)). A natural environment is a setting in which infants and toddlers without disabilities spend their time. For most children, the bulk of their daily activities and interactions occur in their home or child care setting; therefore, the environment in which infants and toddlers with disabilities receive supports and services should be consistent with the environment in which their typically developing peers spend their time.

Most infants and toddlers with disabilities receive supports and services in their home, with a smaller number of children receiving supports and services in other settings, such as hospitals, clinics, service provider offices, and community-based child care settings [16]. Given the home is the most common location for service provision for infants and toddlers with disabilities, it is important to understand the benefits of home visitation.

Home visits allow the professional and family to incorporate learning opportunities into naturally occurring daily routines and activities. These learning opportunities are not elaborate; instead, they involve extending and expanding what the family is already doing in their daily routines and activities. Providing home visits encourages the involvement of all family members, as well as care providers, such as nursing staff or respite providers working with the child, in activities, goal identification, and participation in the development of a plan to address the family's concerns and priorities.

Instructional Practices. DEC [17] defines instructional practices as "intentional and systematic strategies to maximize learning" (p. 11) and suggests interventions be embedded in naturally occurring daily activities. With home visitation as the most common service delivery model, it is important to consider how to best implement recommended instructional practices. There is a common misconception that early intervention services are one-on-one, direct interactions between the professional and the child; however, the family-centered model seeks to build family capacity with the core belief that parents play a critical role in the development of their young children [15]. Instead of direct instruction with the child, the professional seeks to influence parent competence and confidence during interactions with their child with the subsequent benefit of maximizing their child's learning. In this model, the parent becomes the learner; therefore, professionals must be knowledgeable and capable of implementing adult learning strategies during their interactions with parents.

Early intervention professionals utilize various adult learning methods during their interactions with parents, including collaborative consultation [18], coaching [19], and/or partnership and collaboration [20]. Regardless of the method, the professionals' focus is to strengthen parents' existing skills and promote varied interactions that will result in positive parent and child outcomes [15]. By strengthening and promoting parental skills, the professional in turn supports the child's learning and participation in everyday, naturally occurring activities; thus, the intervention becomes embedded in these activities.

Family-professional relationships form the basis for the successful use of these instructional practices. These relationships are a key component of family engagement in home visitation programs and promote trust and responsive parenting [21, 22]. Further, quality professional-parent relationships are an important factor in parent's ability to benefit from home visitation services [23, 24]. It is essential to recognize that within the home visitor-parent dyad, each person influences the other as they coordinate their interactions with one another with mutually positive feelings and individual strengths that can contribute to the desired outcomes of the working relationship [25]. Much as each individual family brings various traits to the relationship, each home visitor has different knowledge, skills, values, and personal qualities which he/she contributes to the relationship.

The Role of Interactions. DEC [17] recommendations convey that "sensitive and responsive interactional practices are the foundation for promoting the development of a child's language and cognitive and emotional competence" (p. 13). To build on this notion, it is important to recognize that parents have numerous opportunities to interact with their young children each day through their daily activities. During these interactions, parents and their children develop their relationship through their actions, responses, sensitivity, and joint attention. These relationships are foundational to children's future development [2, 3, 5, 26, 27]. Because of the integral role parent-child interactions play in children's development, many early intervention programs have begun to focus on the quality of these interactions. Accordingly, professionals employ adult learning principles to influence parent confidence and competence during their interactions with their children and support parents in their understanding of their children's early care needs and developmental progression. Early intervention can positively influence parent self-efficacy, which has positive effects on children's developmental outcomes and participation in daily activities and routines [28].

Teaming and Collaboration. In thinking about the journey families and professionals take together in early intervention programs, teamwork and collaboration are essential components. In early intervention, teaming and collaboration are not only critical to the family-professional relationships but also to professional-professional relationships, as each group has much to learn from one another. Many services are available to families and children found eligible for early intervention programs. As an example, one family may work with a physical therapist, an occupational therapist, a speech-language pathologist, a vision specialist, and a service coordinator to meet their child and family's needs. If all the specialists work autonomously and provide individual home visits and subsequently provide separate suggestions and supports to the family, the family would more than likely become overwhelmed. This scenario is recognized and consequently DEC advises professionals and families to collaborate to "plan and implement supports and services to meet the unique needs of each child and family and to systematically and regularly exchange expertise, knowledge, and information to build team capacity and jointly solve problems, plan, and implement interventions" (p. 14).

Early intervention teams vary in how they implement this recommended practice and historically function as a multidisciplinary, an interdisciplinary, or a transdisciplinary team [29]. In the multidisciplinary approach, practitioners provide early intervention services to the child and family independently, and the professionals do not discuss the child's development or the progress toward meeting the child and family goals. An interdisciplinary approach is similar to the multidisciplinary approach in that the practitioners provide services independently; however, with this approach practitioners meet to discuss the child and family on a regular basis. The transdisciplinary approach is unique to the field of early intervention [30]. Guided by practice and research, the transdisciplinary model is most frequently utilized by early intervention teams and, when it is implemented correctly, it is frequently acknowledged as the most effective early intervention model [31–33]. In a transdisciplinary team model, professionals collaborate extensively, and they understand individual team roles, understand the disciplines outside of their scope of practice, and demonstrate a willingness to work together. This model most closely aligns with family-centered practices and strives to coordinate and integrate services to meet the complex needs of children with disabilities and their families. The transdisciplinary model is also referred to as the primary service provider (PSP) model [18, 34]. In this model, the PSP is typically chosen based upon a match between the expertise of the provider and the family's major identified area of concern. The PSP is responsible for the consistent provision of services and supports for a family on an ongoing basis. The primary provider collaborates and consults with team members from other disciplines who provide expert consultation and who may accompany the primary provider on visits with the family. The aim of these joint home visits is for the team member to support the PSP by conducting assessment, demonstrate specific interventions, and provide parent education in an effort to extend and expand what the parent and PSP are working on together [18].

Transition. Key changes that occur when a child enters or exits early intervention or changes that occur during the time the child is enrolled in early intervention are called transitions. Examples of transitions during the infant and toddler years include transitioning from hospital to home, transitioning into early intervention programs, and transitioning out of early intervention to community early childhood programs or special education programs for 3- to 5-year-olds.

Key transitions in a child's life can be difficult. In an effort to foster greater parent satisfaction, better adjustment for both child and parents, and better child outcomes, DEC [17] recommends early intervention programs develop a transition plan for these events. As a part of the plan, professionals of both the sending and receiving programs exchange information about supportive practices and employ identified strategies before, during, and after the transition to ensure the child and family's successful adjustment and positive outcomes. As teams develop transition plans, they should gather information from parents and professionals including specific family and child needs and parent's vision for their child. The plan should include specific steps to facilitate an efficient and smooth transition.

Effectiveness of Early Intervention

Widespread implementation of early intervention programs and evidence-based practices over the past several decades prompted research on the effectiveness of early intervention for children with disabilities who require special education or related services or those who are at risk for developmental delays due to genetic and/or environmental risk factors. Confounding factors related to variability among program service delivery models include duration and frequency of services, provider qualifications and experiences, child and family characteristics, and family demographics which influence the generalizability of the findings across programs. Although variables differ between studies, there are reports of similar findings of positive program efficacy across settings and populations.

Many of the initial investigations examined the effectiveness of early intervention program factors and benefits for children identified with a disability or at risk for developmental delay. The results showed that early intervention programs can be an effective agent of change in children's early learning and development ([35–37]). With the establishment that early intervention can be effective, researchers were encouraged to further examine the effects of specialized interventions, the response of various populations to intervention, and the outcomes of early intervention for not only children but also their families [29].

Because of the call for further research, several studies have explored the effects of specialized interventions targeting specific populations, such as children with autism, Down syndrome, and other medical conditions [38, 39], as well as the effects of early intervention on family outcomes [40, 41]. One specific population targeted for research due to its increased risk for neurodevelopmental delays are children with a history of hospitalization in the NICU. Whereas multiple factors related to NICU admissions place children at high risk for developmental delay [42], the population most extensively studied in relation to early intervention efficacy are premature infants and those with secondary risk factors often associated with premature birth, such as low birth weight (LBW).

Unlike their full-term peers, children born prematurely are at an increased risk for delays in their early childhood years [43–45], and their prematurity can affect future school success. Upon reaching school-age, children born prematurely have greater probability of cognitive and language delays, learning disabilities, behavior and attentional difficulties, and social-emotional difficulties [46–49]. Additional concern lies in family outcomes, as parents of preterm infants are at risk for increased stress, anxiety, and depression [50, 51] as well as decreased quality of parent-child interactions and diminished reports of parent self-efficacy [52, 53].

To identify efficacious early intervention practices for preterm infants and their families, multiple explorations of implemented practices and resulting child and family outcomes are reported. Although few studies have examined programs explicitly administered under Part C of IDEA, the findings relate to both child and parent outcomes; therefore, it is beneficial to generalize the findings to overall early intervention practices. Studies on early intervention programs used varying models and measures, making it difficult to apply the findings across programs and populations. To account for existing variability, several meta-analyses were conducted to explore relationships between early intervention programs and outcomes for children born prematurely or with low birth weight and their families.

Park, Maitra, Achon, Loyola, and Rincón [54] reviewed the literature published between 2000 and 2013 and targeted studies examining the effects of early intervention on cognitive or neuromusculoskeletal and movement-related functions. The 16 included studies each compared an intervention group comprised of children born preterm or with low birth weight with a comparison group with most of the studies reported as randomized control studies. The studies included interventions provided after discharge from the NICU, and several included a home visitation component. The results generally showed the intervention groups demonstrated significant improvement in cognitive functions, such as executive functioning, verbal performance, and IQ scores, when compared with the nonintervention groups. Further, the intervention groups generally demonstrated improved neurobehavioral and movement-related functional outcomes compared to their peers not receiving intervention.

Benzies, Magill-Evans, Hayden, and Ballantyne [55] conducted a meta-analysis of the literature published between 1990 and 2011 with focus on the relationship between key components of early intervention programs for preterm infants and maternal outcomes of stress, anxiety, depression, self-efficacy, and responsiveness.

Eighteen studies were included in the analysis and included varied program models. Most of the programs included parent support and education components. The programs provided services to infants born prematurely and/or with low birth weight and their families. The studies explored the interrelatedness between programs emphasizing parental support and involvement and improved child and parental outcomes. They measured parent stress, anxiety, depression, and parent reports of self-efficacy and parental responsivity. The analyses found evidence demonstrating positive and meaningful effects of interventions. Parent education programs provided to mothers of preterm infants decreased levels of maternal anxiety and depression and positively influenced parent self-efficacy resulting in improved child outcomes.

In support of these meta-analyses, an independent study conducted by McManus, Carle, and Poehlmann [56] found that strengthening the supports for mothers of preterm and low birth weight infants via Part C early intervention programs improved cognitive function trajectories measured at 16 months, 24 months, and 36 months adjusted age. The most benefit was noted at 24 months of age. The short-term effects of early intervention programs for preterm and low birth weight infants are established; however, less is known about the long-term effects of these programs. Since early intervention programs for other populations have shown positive effects on children as they move into school and adolescence [57, 58], one might surmise that the benefits of early intervention for preterm and low birth weight infants would persist beyond the early childhood years. The longitudinal study by McCormick et al. [59] supports this theory. In a randomized control study, researchers followed children born with low birth weight until the age of 18. Participants in the intervention group received early intervention home visits from birth to 3 years of age, a daily center-based program from 1 to 3 years of age, and included a parent support component. The comparison group did not receive early intervention. At the age of 18, intervention group participants' academic achievement, behavior, cognitive abilities, and physical health were measured. The results indicate that the intervention group obtained higher IQ scores and improved vocabulary, mathematics, and reading skills compared to their comparison group peers. These findings are encouraging; however, further longitudinal investigation about the long-term efficacy of early intervention programs for this population is needed.

The mentioned studies emphasize the importance of early intervention for high-risk and preterm infants, as well as their parents. Whereas the short- and long-term benefits of early intervention programs exist, there are multiple gaps and barriers identified for Part C early intervention programs. Identified gaps and barriers such as inadequate funding; variable evaluation practices; poor communication between referral source, early intervention programs, and families; lack of family receptivity and understanding of early intervention programs; and staffing issues often affect referral, eligibility, and enrollment of many children who are at risk for developmental delays [60]. The impact of these gaps and barriers ultimately affects child outcomes; therefore, early intervention programs and NICU follow-up programs must ensure children are referred, evaluated for eligibility, and enrolled in early intervention programs when deemed appropriate and necessary.

NICU Follow-Up Programs and Early Intervention: A Key Partnership

There are a small number of children referred to and enrolled in early intervention programs when they transition from the NICU to their home due to conditions that are known to be associated with developmental delays, such as some syndromes, congenital anomalies, and complex medical issues [61]; however, most children are not referred at discharge and remain at risk for delays. One major barrier in connecting at-risk NICU graduates with early intervention programs is the variability in program eligibility for this population. Recall that each state has the discretion to define at-risk children and eligibility criteria for at-risk populations. Given this discretion, only a small percentage of states automatically deem children eligible based on biological, medical, or environmental risk factors [12]. Nevertheless, the risk of adverse developmental effects for these children is high, specifically for preterm and low birth weight infants [45, 62, 63].

The American Academy of Pediatrics (AAP) recommends that high-risk infants, including preterm infants, infants born with low birth weight, and infants with special health-care needs, enroll in a NICU follow-up program to receive periodic evaluation to monitor their neurodevelopmental progress and facilitate referral to early intervention programs as soon as developmental concerns arise [64]. In support of this recommendation, many NICU follow-up programs are in place across the country. The model, focus, and staffing of these programs vary from site to site; however, the results from a small survey of follow-up programs across the country show that most of the responding programs complete neurodevelopmental assessment as one component of their program model [65]. Many programs report conducting outcomes research utilizing the collected developmental assessment data, yet there is less knowledge about the percent of children referred to early intervention and how the NICU follow-up programs employ assessment data to inform referral to early intervention.

Recent studies have explored the prevalence of referrals made to early intervention by follow-up programs. The results indicate that, despite an increased awareness among the medical community of the developmental risks that many NICU graduates face, primary physicians rely heavily on follow-up programs to play a key role in monitoring infant and toddler development and subsequently act as a referral agent to statewide early intervention programs [66, 67]. Due to the discretionary interpretation of early intervention eligibility across the country and the increased risk for developmental delays for this population of children, follow-up programs must be viewed as a key partner in federally mandated Child Find efforts to ensure identification of children with developmental delays or those at risk for developmental delays as early as possible.

Facilitation of referrals to early intervention programs is just one important role follow-up programs can assume in the Part C Child Find process. Not all children referred to early intervention will meet eligibility criteria at the time of referral, as delays associated with prematurity may not present until closer to the preschool years [68] or delays may not be significant enough to meet state-specific eligibility criteria. Furthermore, disparities exist between the number of children eligible for early intervention programs and the number of children enrolled in early intervention programs. Rosenberg, Robinson, Shaw, and Ellison suggest the eligibility criteria in some states may be too rigorous, thus requiring a child to demonstrate a significant delay to be deemed eligible for early intervention services [11]. Despite the acknowledged risks, retrospective analyses of early intervention eligible ELBW children in one state found that only about one half of the children with ELBW were enrolled in early intervention services. Of the enrolled children, those with more severe medical needs, longer length of NICU stay, and lower birth weight were more likely to be enrolled than their healthy, heavier peers [69]. These factors precipitate the need for close developmental monitoring of NICU graduates during their most critical early years and subsequent facilitation of referrals to and enrollment in early intervention as warranted. To inform referral practices, it is important for follow-up program professionals to understand the patterns of referral to early intervention programs in the state and the common predictors associated with early intervention program enrollment for NICU graduates.

The NICU Graduate: Predictors of Early Intervention Referral and Enrollment

Many infants are admitted to NICUs, and these children often benefit from early intervention services during their early years. Admission to an NICU alone identifies children at an increased risk for developmental delays and disabilities. The length of stay and conditions warranting NICU admissions clearly play key roles in children's long-term needs and outcomes. Despite the acknowledged benefits of early intervention programs, limited investigations are available regarding the early intervention referral and enrollment processes for NICU graduates. Explorations have begun to examine the factors influencing the timing of referrals to early intervention programs and program enrollment to inform referral practices for this population. An examination of specific variables related to referral to early intervention programs indicates the referral rate for extremely low birth weight (ELBW) (<1000 g) infants is nearly double than in the first year of their peers who weighed over 1000 g at birth [67]. At 20 months of age, children diagnosed with chronic lung disease or born to a multiple birth have a higher likelihood of referral to early intervention programs [69]. Additional variables can affect NICU graduate's eligibility for and enrollment in early intervention programs. Both medical and sociodemographic factors predict who is likely to receive early intervention services. Children born at VLBW (<1500 g) or ELBW have higher and more consistent early intervention enrollment rates when compared to their peers [67]. Program eligibility and enrollment increases for males [70] and for children whose families have low SES; however, eligible children from low SES are less likely to enroll and begin services [71].

Delays in development are often more apparent as children age and developmental expectations increase. To ensure developmental surveillance for an appropriate duration and identification of potential delays as early as possible, the length of NICU follow-up is an important program consideration. The average age span targeted for program follow-up varies, with most follow-ups occurring between 12 and 36 months [65]. One population benefitting from a longer duration of follow-up is late preterm infants (34–36 weeks' gestation). Previous studies on the outcomes of late preterm infants at 24 months found that these children had worse developmental outcomes compared to their full-term counterparts [72, 73]. However, in a more recent study, which controlled for covariates, the differences in outcomes did not emerge until the children reached preschool age [68]. Monitoring of late preterm infant development is a critical consideration for follow-up programs due to the potential for future developmental issues as this population moves into preschool and school-age years. Late preterm infants are at risk for delays in neurocognitive development affecting their school readiness and performance in areas such as reading, mathematics, language, social, and motor skills [74–76].

With the potential emergence of delays after the first 2 years of life, it is imperative that NICU follow-up programs consider the effect of continuity of developmental surveillance through the first 3 years. Early identification of delays and referral to early intervention programs is the first step to ensuring the connection of children and families with appropriate early intervention supports during the developmentally important early childhood years.

Considerations

NICU admission rates are on the rise [77], and many infants discharged from the NICU are at elevated risk for developmental delays at some point in their first years of life. Early intervention programs, funded by Part C of IDEA, are well positioned to provide supports to promote the growth and development of NICU graduates referred to their programs.

To maximize Part C Child Find efforts, NICU follow-up programs and early intervention programs must collaborate to monitor children for delays in development and ensure that children and their families are connected to appropriate resources to meet their needs and address developmental concerns. As with any successful partnership, ongoing communication, collaboration, and training are essential to success. Professionals from early intervention and follow-up programs need background and training on the roles and responsibilities of each program. Follow-up professionals must have a comprehensive understanding of early intervention models and practices including DEC recommended practices, the referral process in their state, and state-specific eligibility criteria. In turn, early intervention programs must have a good understanding of the medical and intervention needs of NICU graduates and their families during the critical early childhood years. Furthermore, it is essential that professionals working in both follow-up and early intervention programs understand how the eligibility criteria in their state relate specifically to NICU graduates.

Education about the needs of NICU graduates and their families must begin in professional training. Preservice programs providing training for professionals who may work in early intervention programs postgraduation (e.g., physical, occupational, and speech-language therapy, early childhood education, special education) should consider incorporating coursework to increase knowledge about the unique needs of this population, and medical professions should consider including a basic overview of early intervention practices and efficacy in their course work and residency programs. With follow-up programs predominantly ending once a child turns 3 years old, both follow-up and early intervention programs must ensure that parents and physicians are knowledgeable about the continued possibility for developmental delays as this population of children transition into the preschool and school-age years. Additionally, they also need to know how to access appropriate assessments for children after the age 3 years through the public school early childhood special education programs.

Studies on early intervention effectiveness for NICU graduates and program referral and enrollment patterns provide a platform to inform professional policy and practice for both follow-up and early intervention programs. Additional examination of referral and enrollment patterns from follow-up programs, at both the local, state, and national level, will further inform Child Find efforts and aid in recognizing barriers that affect the identification and enrollment of NICU graduates in early intervention programs. Specific consideration should be made to examine enrollment and utilization patterns for children from lower SES. The paired risk associated with low SES and a NICU stay call for efforts to increase parent awareness of the important benefits of early intervention programs for NICU graduates from lower SES.

Conclusion

The partnership between early intervention programs and NICU follow-up programs is an integral component in the care of the growing population of NICU graduates. Effective partnership between these two programs has immense potential given their shared missions to identify children with developmental delays or those at risk for delays as early as possible. Successful collaboration between the programs can lead to maximized Child Find efforts and the connection of children and families to early intervention programs with the goal to ensure the best outcomes for NICU graduates and their families.

References

1. Shonkoff, J., & Phillips, D. (2000). *From neurons to neighborhoods: The science of early childhood development*. Washington, DC: National Academy Press.
2. Clark, K., & Ladd, G. (2000). Connectedness and autonomy support in parent–child relationships: Links to children's socioemotional orientation and peer relationships. *Developmental Psychology, 36*(4), 485.

3. Hart, B., & Risley, T. (1995). *Meaningful differences in the everyday experience of young American children*. Baltimore, MA: Brookes Publishing.
4. Ipsa, J., Fine, M., Halgunseth, L., Harper, S., & Robinson, J. (2004). Maternal intrusiveness, maternal warmth, and mother–toddler relationship outcomes: Variations across low-income ethnic and acculturation groups. *Child Development, 75*, 1613–1631.
5. Landry, S., Smith, K., Swank, P., & Guttentag, C. (2008). A responsive parenting intervention: The optimal timing across early childhood for impacting maternal behaviors and child outcomes. *Developmental Psychology, 44*(5), 1335.
6. Lee Raby, K., Roisman, G., Fraley, R., & Simpson, J. (2015). The enduring predictive significance of early maternal sensitivity: Social and academic competence through age 32 years. *Child Development, 86*(3), 695–708.
7. Individuals with Disabilities Education Act. (2004). 20 U.S.C. § 1400. Washington, DC.
8. Education for All Handicapped Children Act of 1975, PL 94-142. 20 U.S.C. § 1400. Washington, DC.
9. Education of the Handicapped Act Amendments of 1986, PL 99-457. 20 U.S.C. § 1400 Washington, DC.
10. Early Childhood Technical Assistance Center. (2014). *Part C lead agencies*. NC: Chapel Hill. Retrieved from http://ectacenter.org/partc/ptclead.asp.
11. Rosenberg, S. A., Robinson, C. C., Shaw, E. F., & Ellison, M. C. (2013). Part C early intervention for infants and toddlers: Percentage eligible versus served. *Pediatrics, 131*(1), 38–46.
12. Ringwalt, S. (2015). *Summary table of states' and territories' definitions of/criteria for IDEA Part C eligibility*. Early Childhood Technical Assistance Center website: Retrieved from http://ectacenter.org/~pdfs/topics/earlyid/partc_elig_table.pdf.
13. U.S. Department of Education: Office of Special Education Programs. (2016). *Annual Federal Appropriations and Number of Children Served*. Retrieved from http://ectacenter.org/partc/partcdata.asp.
14. Bronfenbrenner, U. (1992). *Ecological systems theory*. London: England: Jessica Kingsley.
15. Dunst, C., Boyd, K., Trivette, C., & Hamby, D. (2002). Family-oriented program models and professional help giving practices. *Family Relations, 51*(3), 221–229.
16. U.S. Department of Education. (2014). *EMAPS: IDEA Part C Child Count and Settings Survey*. Retrieved from http://www.ed.gov/edfacts.
17. Division for Early Childhood. (2014). *DEC recommended practices in early intervention/early childhood special education 2014*. Retrieved from http://www.dec-sped.org/dec-recommended-practices.
18. McWilliam, R. A. (2010). *Working with families of young children with special needs*. New York, NY: Guilford Press.
19. Rush, D., & Shelden, M. (2011). *The early childhood coaching handbook*. Baltimore, MA: Brookes Publishing.
20. Dunst, C., Trivette, C., & Deal, A. (1994). *Supporting and strengthening families*. Cambridge, MA: Brookline Books.
21. Bernstein, V., Hans, S., & Percansky, C. (1991). Advocating for the young child in need through strengthening the parent-child relationship. *Journal of Clinical Child and Adolescent Psychology, 20*(1), 28–41.
22. Roggman, L., Boyce, L., Cook, G., & Jump, V. (2001). Inside home visits: A collaborative look at process and quality. *Early Childhood Research Quarterly, 16*(1), 53–71.
23. Korfmacher, J., Green, B., Spellmann, M., & Thornburg, K. (2007). The helping relationship and program participation in early childhood home visiting. *Infant Mental Health Journal, 28*(5), 459–480.
24. Spieker, S., Solchany, J., McKenna, M., DeKlyen, M., & Barnard, K. (2000). The story of mothers who are difficult to engage. In J. D. Osofsky & H. I. Fitzgerald (Eds.), *World association of infant mental health handbook of infant mental health, Vol 3: Parenting and child care* (pp. 172–209). New York, NY: Wiley.
25. Bronfenbrenner, U. (1979). *The ecology of human development*. Cambridge, MA: Harvard University Press.

26. Hirsh-Pasek, K., Adamson, L., Bakeman, R., Owen, M., Golinkoff, R., Pace, A., Yust, P., & Suma, K. (2015). The contribution of early communication quality to low-income children's language success. *Psychological Science, 26*(7), 1071–1083.
27. Knoche, L. L., Edwards, C. P., Sheridan, S. M., Kupzyk, K. A., Marvin, C. A., Cline, K. D., & Clarke, B. L. (2012). Getting ready: Results of a randomized trial of a relationship-focused intervention on the parent–infant relationship in rural early head start. *Infant Mental Health Journal, 33*(5), 439–458.
28. Salisbury, C. L., & Copeland, C. G. (2013). Progress of infants/toddlers with severe disabilities: Perceived and measured change. *Topics in Early Childhood Special Education, 33*(2), 68–77.
29. Bruder, M. (2010). Early childhood intervention: A promise to children and families for their future. *Exceptional Children, 76*(3), 339–355.
30. Briggs, M. (1997). *Building early intervention teams. Working together for children and families*. Gaithersburg: Aspen Publishers, Inc..
31. Bruder, M. (2000). Family-centered early intervention clarifying our values for the new millennium. *Topics in Early Childhood Special Education, 20*(2), 105–115.
32. Guralnick, M. (2001). A developmental systems model for early intervention. *Infants & Young Children, 14*(2), 1–18.
33. King, G., Strachan, D., Tucker, M., Duwyn, B., Desserud, S., & Shillington, M. (2009). The application of a transdisciplinary model for early intervention services. *Infants & Young Children, 22*(3), 211–223.
34. Shelden, M., & Rush, D. (2013). *The early intervention teaming handbook: The primary service provider approach*. Baltimore, MA: Brookes Publishing.
35. Blair, C., Ramey, C., & Hardin, M. (1995). Early intervention for low birth weight premature infants: Participation and intellectual development. *American Journal of Mental Retardation, 99*, 542–554.
36. Guralnick, M., & Bricker, D. (1987). *The effectiveness of early intervention for children with cognitive and general developmental delays*. San Diego, CA: Academic Press.
37. Shonkoff, J., & Hauser-Cram, P. (1987). Early intervention for disabled infants and their families: A quantitative analysis. *Pediatrics, 80*(5), 650–658.
38. Bradshaw, J., Steiner, A., Gengoux, G., & Koegel, L. (2015). Feasibility and effectiveness of very early intervention for infants at-risk for autism spectrum disorder: A systematic review. *Journal of Autism and Developmental Disorders, 45*(3), 778–794.
39. Brown, J., & Woods, J. (2015). Effects of a triadic parent-implemented home-based communication intervention for toddlers. *Journal of Early Intervention, 37*(1), 44–68.
40. Corr, C., Santos, R., & Fowler, S. (2015). The components of early intervention services for families living in poverty: A review of the literature. *Topics in Early Childhood Special Education, 36*(1), 55–64.
41. Dunst, C., Bruder, M., & Espe-Sherwindt, M. (2014). Family capacity-building in early childhood intervention: Do context and setting matter? *School Community Journal, 24*(1), 37.
42. Spiker, D., Sangeeta, X., Elebbeler, K., Scarborough, A., Simeonsson, R., & Baily, D. (2004). Birth history and health status of children entering early intervention. *NEILS Data Report, 5*.
43. Aarnoudse-Moens, C., Weisglas-Kuperus, N., van Goudoever, J., & Oosterlaan, J. (2009). Meta-analysis of neurobehavioral outcomes in very preterm and/or very low birth weight children. *Pediatrics, 124*(2), 717–728.
44. Beck, S., Wojdyla, D., Say, L., Betran, A., Merialdi, M., Requejo, J., et al. (2010). The worldwide incidence of preterm birth: A systematic review of maternal mortality and morbidity. *Bulletin of the World Health Organization, 88*(1), 31–38.
45. Petrini, J., Dias, T., McCormick, M., Massolo, M., Green, N., & Escobar, G. (2009). Increased risk of adverse neurological development for late preterm infants. *The Journal of Pediatrics, 154*(2), 169–176.
46. Bhutta, A., Cleves, M., Casey, P., Cradock, M., & Anand, K. (2002). Cognitive and behavioral outcomes of school-aged children who were born preterm: A meta-analysis. *JAMA, 288*(6), 728–737.

47. Gidley Larson, J., Baron, I., Erickson, K., Ahronovich, M., Baker, R., & Litman, F. (2011). Neuromotor outcomes at school age after extremely low birth weight: Early detection of subtle signs. *Neuropsychology, 25*(1), 66.
48. Jennische, M., & Sedin, G. (2001). Spontaneous speech at 6½ years of age in children who required neonatal intensive care in 1986–1989. *Acta Paediatrica, 90*(1), 22–33.
49. Moster, D., Lie, R., & Markestad, T. (2008). Long-term medical and social consequences of preterm birth. *New England Journal of Medicine, 359*(3), 262–273.
50. Miles, M., Holditch-Davis, D., Schwartz, T., & Scher, M. (2007). Depressive symptoms in mothers of prematurely born infants. *Journal of Developmental & Behavioral Pediatrics, 28*(1), 36–44.
51. Singer, L., Salvator, A., Guo, S., Collin, M., Lilien, L., & Baley, J. (1999). Maternal psychological distress and parenting stress after the birth of a very low-birth-weight infant. *JAMA Pediatrics, 281*(9), 799–805.
52. Field, T. (2010). Postpartum depression effects on early interactions, parenting, and safety practices: A review. *Infant Behavior & Development, 33*(1), 1–6.
53. Forcada-Guex, M., Pierrehumbert, B., Borghini, A., Moessinger, A., & Muller-Nix, C. (2006). Early dyadic patterns of mother–infant interactions and outcomes of prematurity at 18 months. *Pediatrics, 118*(1), e107–e114.
54. Park, H., Maitra, K., Achon, J., Loyola, E., & Rincón, M. (2014). Effects of early intervention on mental or neuromusculoskeletal and movement-related functions in children born low birth weight or preterm: A meta-analysis. *American Journal of Occupational Therapy, 68*(3), 268–276.
55. Benzies, K., Magill-Evans, J., Hayden, K., & Ballantyne, M. (2013). Key components of early intervention programs for preterm infants and their parents: A systematic review and meta-analysis. *BMC Pregnancy and Childbirth, 13*(1), 1.
56. McManus, B., Carle, A., & Poehlmann, J. (2012). Effectiveness of Part C early intervention physical, occupational, and speech therapy services for preterm or low birth weight infants in Wisconsin, United States. *Academic Pediatrics, 12*(2), 96–103.
57. Barnett, W. S. (1995). Long-term effects of early childhood programs on cognitive and school outcomes. *The Future of Children, 5*(3), 25–50.
58. Ramey, C. T., & Ramey, S. L. (2004). Early learning and school readiness: Can early intervention make a difference? *Merrill-Palmer Quarterly, 50*(4), 471–491.
59. McCormick, M., Brooks-Gunn, J., Buka, S., Goldman, J., Yu, J., Salganik, M., et al. (2006). Early intervention in low birth weight premature infants: Results at 18 years of age for the Infant Health and Development Program. *Pediatrics, 117*(3), 771–780.
60. Little, A., Kamholz, K., Corwin, B., Barrero-Castillero, A., & Wang, C. (2015). Understanding barriers to early intervention services for preterm infants: Lessons from two states. *Academic Pediatrics, 15*(4), 430–438.
61. Jackson, B., & Needelman, H. (2007). Building a system of child find through a 3-tiered model of follow-up. *Infants & Young Children, 20*(3), 255–265.
62. Beaino, G., Khoshnood, B., Kaminski, M., Marret, S., Pierrat, V., Vieux, R., et al. (2011). Predictors of the risk of cognitive deficiency in very preterm infants: The EPIPAGE prospective cohort. *Acta Paediatrica, 100*(3), 370–378.
63. Wilson-Costello, D., Friedman, H., Minich, N., Fanaroff, A., & Hack, M. (2005). Improved survival rates with increased neurodevelopmental disability for extremely low birth weight infants in the 1990s. *Pediatrics, 115*(4), 997–1003.
64. American Academy of Pediatrics Committee on Fetus and Newborn. (2008). Hospital discharge of the high-risk neonate. *Pediatrics, 122*(5), 1119–1126.
65. Kuppala, V., Tabangin, M., Haberman, B., Steichen, J., & Yolton, K. (2012). Current state of high-risk infant follow-up care in the United States: Results of a national survey of academic follow-up programs. *Journal of Perinatology, 32*(4), 293–298.
66. Barfield, W., Clements, K., Lee, K., Kotelchuck, M., Wilber, N., & Wise, P. (2008). Using linked data to assess patterns of early intervention (EI) referral among very low birth weight infants. *Maternal and Child Health Journal, 12*(1), 24–33.

67. Greene, M., & Patra, K. (2016). Part C early intervention utilization in preterm infants: Opportunity for referral from a NICU follow-up clinic. *Research in Developmental Disabilities, 53*, 287–295.
68. Shah, P., Kaciroti, N., Richards, B., & Lumeng, J. (2016). Gestational age and kindergarten school readiness in a national sample of preterm infants. *The Journal of Pediatrics, 178*, 61–67. https://doi.org/10.1016/j.jpeds.2016.06.062.
69. Wang, C., Elliott, M., Rogowski, J., Lim, N., Ratner, J., & Schuster, M. (2009). Factors influencing the enrollment of eligible extremely-low-birth-weight children in the Part C early intervention program. *Academic Pediatrics, 9*(4), 283–287.
70. Vohr, B., Wright, L., Dusick, A., Mele, L., Verter, J., Steichen, J., et al. (2000). Neurodevelopmental and functional outcomes of extremely low birth weight infants in the National Institute of Child Health and Human Development Neonatal Research Network, 1993–1994. *Pediatrics, 105*(6), 1216–1226.
71. McManus, B., Robert, S., Albanese, A., Sadek-Badawi, M., & Palta, M. (2013). Predictors of receiving therapy among very low birth weight 2-year olds eligible for Part C early intervention in Wisconsin. *BioMed Central Pediatrics, 13*(1), 1.
72. Nepomnyaschy, L., Hegyi, T., Ostfeld, B., & Reichman, N. (2012). Developmental outcomes of late-preterm infants at 2 and 4 years. *Maternal and Child Health Journal, 16*(8), 1612–1624.
73. Woythaler, M. A., McCormick, M. C., & Smith, V. C. (2011). Late preterm infants have worse 24-month neurodevelopmental outcomes than term infants. *Pediatrics, 127*(3), 622–629.
74. Baron, I., Litman, F., Ahronovich, M., & Baker, R. (2012). Late preterm birth: A review of medical and neuropsychological childhood outcomes. *Neuropsychology Review, 22*(4), 438–450.
75. Chan, E., & Quigley, M. (2014). School performance at age 7 years in late preterm and early term birth: A cohort study. *Archives of Disease in Childhood. Fetal and Neonatal Edition, 99*(6), 451–457.
76. Kirkegaard, I., Obel, C., Hedegaard, M., & Henriksen, T. (2006). Gestational age and birth weight in relation to school performance of 10-year-old children: A follow-up study of children born after 32 completed weeks. *Pediatrics, 118*(4), 1600–1606.
77. Harrison, W., & Goodman, D. (2015). Epidemiologic trends in neonatal intensive care, 2007–2012. *JAMA Pediatrics, 169*(9), 855–862.

Quality Coordinated Health Care for the High-Risk Infant: The Evolving Concept of the Medical Home

Linda Copeland and Shalin Parekh

Abstract Sick neonates that are treated in the neonatal intensive care unit are at high risk of health, developmental, and social problems. The complexities of managing such high-risk infants can be overwhelming for a single primary care provider. These infants require input from multiple specialties and agencies. Delivering quality care includes coordination of these multiple agencies and professionals that is best provided by means of a patient-centered medical home. There is growing evidence that such a model improves patient outcomes and reduces readmissions to the hospital. Team communication and coordination is at the heart of the medical home model. In an ideal setting, patient interaction and provider communication with the child's medical home should begin before discharge from the NICU and should be a part of the NICU discharge process. This is especially important as survival of extremely sick and fragile infants is improving and a number of infants are reliant on medical technologies for a prolonged period of time following discharge for their survival. A medical home should not only deal with the medical problems of the individual child but should also address the psychosocial impact of caring for a sick child and should keep the child and the family at the center of its care. The child's medical home can maintain close coordination with the high-risk infant follow-up clinic and can help the primary care provider in navigating subspecialty services. There are numerous barriers to the medical home model. Key among these are logistic difficulties and a lack of resources in coordinating services. The American Academy of Pediatrics (AAP) has set out a number of guidelines to overcome these issues. The medical home model is expected to be more efficient while at the same time decreasing health-care costs by reducing unnecessary duplication of services. Medical homes should function as a team with each member having formal responsibilities. Such a model will not only improve patient care but will also result in greater satisfaction among patients and their families. Increasingly, funding sources are recognizing the benefits of the medical home model through payments given specifically for care coordination services.

L. Copeland (✉) · S. Parekh
UCSF Fresno Center for Medical Education and Research, San Francisco, CA, USA
e-mail: lecopeland372@gmail.com

H. Needelman, B. J. Jackson (eds.), *Follow-Up for NICU Graduates*,
https://doi.org/10.1007/978-3-319-73275-6_18

Introduction

The annual societal costs of preterm birth are estimated to be $26 billion in the United States. These include medical costs, costs of early intervention services, costs of special education, and costs associated with child care and consequent loss of labor and productivity [1]. Infants graduating from the neonatal intensive care unit (NICU) or at high risk for any reason can have multiple health, developmental, and psychosocial needs. About 25% of extremely preterm infants have some form of visual impairment [2], 5–10% of extremely preterm infants have hearing deficits [3], and up to 60% of extremely preterm infants have chronic lung disease [4, 5]. About 11% of these infants have cerebral palsy [6], and up to 20% of infants born less than 25 weeks can have severe neurological disability [7]. The multiple care needs of the infant from the impact of prematurity can be overwhelming for the family. Studies have shown that NICU graduates are at significantly higher risk of hospital admissions including PICU admissions and even death in the first year of life [8–12]. It has been shown that an enhanced medical home that provides comprehensive care reduces the rate of PICU admissions by 40% and reduces total hospital and clinic costs by about $10,000 overall for each high-risk child with chronic illness [13, 14]. It is thus imperative that high-quality coordinated comprehensive care be delivered to these infants for the best possible outcomes.

However, delivery of such high-quality medical care for these infants can be overwhelming to a primary care health provider due to the range and complexity of health, developmental, family, and systems issues that must be dealt with. Factors that impact the ongoing needs and long-term outcome of high-risk infants are best addressed in a comprehensive and coordinated manner by multidisciplinary partnerships. Stakeholders include the infant; family; NICU; primary care practitioner; medical subspecialists; medical equipment suppliers; health-care funding sources (federal, state, and private); social, behavioral, and public-health professionals; early intervention professionals (such as speech, occupational, and physical therapists as well as infant program specialists); the educational system; advocacy groups; and ultimately society at large [15, 16]. The high-risk infant follow-up clinic (also referred to as the NICU follow-up clinic or the neurodevelopmental clinic) plays an important supportive role for the family and for the health-care practitioner in determining the needs of the infant and family for medical and developmental services as well as for psychosocial supports.

Achieving the desired level of effective, efficient, and compassionate care coordination for medically fragile infants is challenging. The medical home model has been developed with these challenges in mind. The ideal source of the medical home in the community is the child's primary care provider. It cannot be overemphasized that appreciating their respective roles and ensuring close communication between high-risk infant follow-up clinics, the primary care provider, and all members of the child's medical and early intervention teams are essential.

History of the Medical Home

The concept of the medical home has several decades of history and evolution as a philosophy and as operationalized practice applying to all patients. From inception, the medical home has had care coordination as an essential element [17]. There is growing momentum strongly focused on the patient/family-centered medical home as one of the most promising models for cost-effective, optimal health outcomes [18]. In particular, the American Academy of Pediatrics (AAP) describes the *pediatric* medical home as a cultivated partnership between the patient, family, and primary care provider in close cooperation with any required medical specialists and community partners [16]. Coordination of care across settings promotes an integration of services that is centered on the comprehensive needs of the patient and family. Distinguishing features of the pediatric versus the adult medical home include a focus on the child's developmental trajectory, their dependency on adults, differential epidemiology of neonatal chronic disease, demographic patterns of diversity and poverty, and total dollars spent on children as compared to adults [19]. The patient and family are the focal point of the AAP envisioned model. Crucial characteristics of the medical home are defined as medical care that is accessible, patient and family centered, continuous, comprehensive, coordinated, compassionate, and culturally effective [16, 20].

In 2007 the American Academy of Pediatrics (AAP) joined with the American Academy of Family Physicians (AAFP), the American College of Physicians (ACP), and the American Osteopathic Association (AOA) to agree upon the Joint Principles of the Patient-Centered Medical Home [21]. There are seven general principles which are briefly paraphrased here:

1. Every individual/child will have a personal physician.
2. For the physician-directed medical practice, the personal physician leads the medical team.
3. There is a whole person orientation in the delivery of medical care.
4. Medical care is coordinated and integrated.
5. Quality and safety of medical care is a priority within the medical home.
6. There is appropriate and timely access to care ("enhanced access"), including to any needed medical subspecialties and procedures.
7. Payment should appropriately recognize the added value of coordinated care that goes beyond the work of the face-to-face visit (e.g., there is pay for services associated with medical home care coordination).

The Maternal and Child Health Bureau (MCHB) at the US Health Resources and Services Administration (HRSA) embraces the principles outlined above and identifies specific criteria when a child's health care meets the definition of a medical home [22, 23]. This definition stipulates that every child should have at least one personal doctor, nurse practitioner, or nurse who knows him or her well and is a usual source of medical care. The child should have no problems getting referrals to specialty care and access to any needed therapies, services, and health-care equipment. A goal is to have a family who is very satisfied with the level of communication and care delivery among their child's doctors, therapists, and service programs. The family should get sufficient help coordinating their child's health care when needed, for effective care coordination. The child's doctors should spend enough time with the family, listen thoughtfully to concerns, and be sensitive to the values and customs of the family. Health-care providers should provide any care-related information the family needs. The family should feel like a partner in their child's care. To optimize communication an interpreter should always be available when needed. Additionally, the infant's health care includes behavioral and oral health needs. Thus, the medical home is a comprehensive health-care strategy aimed to improve and integrate all aspects of care to increase family satisfaction and the child's health outcome. Too often such family satisfaction and quality health care of the high-risk infant is not achieved, particularly when the medical home model is not followed, increasing family stress as well as deleterious infant outcomes. Without quality integrated medical care, the high-risk infant is more likely to suffer an array of medical issues including failure to thrive, respiratory complications, serious infection, increased use of the emergency room (with health-care providers who may lack specific expertise with high-risk infants), and rehospitalization [11]. Developmental issues may go unrecognized [16]. There has been a great deal of discussion at the state and national level about the need to improve coordinated care to optimize the medical home. Coordinated care involves the identification and involvement of many support services. There needs to be a determination of all the various types of follow-up care that the high-risk infant will need. This also includes recognizing an infant's developmental needs and initiating developmental services through early intervention where indicated. The providers or sources of care need to be clearly designated. The family and primary care provider should be empowered to use community resources, both public and private. Home-nursing visits may be an important care component, with experienced nurses who are qualified in their tasks and who can train family members where appropriate. The goal is to keep the focus on providing coordinated care and family support with efficient teamwork by all the health-care professionals and community service providers involved. A greater need has been found for family care coordination when communication between health-care providers was inadequate [24]. In contrast, care coordination within primary care pediatric practices is associated with increased family satisfaction, decreased barriers to care, decreased unnecessary office visits and emergency department visits, increased family satisfaction, and reduced unplanned hospitalizations [25–28].

The high-risk infant follow-up clinic plays vital roles both in identifying when an infant requires medical subspecialty referral and when further targeting of developmental issues must occur [16].

The NICU Graduate and the High-Risk Infant

Medical homes are deemed to be of particular benefit to high-risk infants who may have a host of factors such as prematurity, congenital birth defects, genetic syndromes, intrauterine drug exposure, technology dependency, combinations of medical and social impairments, and other medical issues [29, 30]. Families with at-risk factors such as a single-parent home, poverty, substance abuse, or other family stresses likewise benefit from the medical home model in caring for their medically fragile child [20]. The American Academy of Pediatrics and the Institute of Medicine note the importance of generalist-specialist communication for chronic care of children with special health-care needs, abbreviated as CSHCN [31]. For instance, in the 2005–2006 National Survey of Children with Special Healthcare Needs, almost half the children were identified by their families as requiring assistive technologies in their care [32]. It has been demonstrated that medically fragile children are most likely to get their medical technology needs met if they receive care in a qualified medical home [33, 34]. The National Survey of Children with Special Healthcare Needs data also revealed positive associations between care coordination, family-provider relations, and family/child outcomes. Family satisfaction with services was increased, and referrals were achieved with greater ease. Additionally, families reported that care coordination provided them with lower out-of-pocket expenses, less financial burden, and fewer required hours per week spent coordinating care, thus minimizing impact on parental employment. A final benefit was fewer school absences and emergency room visits [35].

Discharge Planning and Challenges

It has been postulated that hospital care in the NICU should be conducted according to medical home principles even prior to discharge, as these principles foster a cohesive team with the infant and family at the center [36, 37]. Medical home-oriented NICU care recognizes right from the outset the unique needs of the infant and family and allows for a smooth transition from hospital NICU care to the many challenges of going home and obtaining quality community care. Parents encounter multiple health-care professionals from a variety of disciplines during their infant's NICU stay. This mirrors the later variety of disciplines the family will encounter upon transitioning to community care. Preterm birth and prolonged hospitalization are family stressors that create risk factors for later family dysfunction and even a higher rate of child abuse [38]. Recognizing emotional stressors on parents related to having a sick infant can help provide support to parents, promote bonding, and increase parenting skills for caring for their infant from admission through discharge and into home care. Family-centered care in the NICU involves timely, sensitive reciprocal communication between the parents and the many multidisciplinary NICU team members, especially for a coordinated care transition. This includes giving parents emotional and educational support and considering home visitor support for these families after discharge [39].

A big issue in transitioning care of an infant from the NICU to the community involves when to discharge the baby from the hospital. Many factors must be considered as an infant graduates from neonatal intensive care. High-risk infants have been classified into four broad categories. These include the preterm infant, any infant with special health-care needs, infants whose special needs depend on technology, the infant at risk due to family issues, and the infant with anticipated early death, as noted in the American Academy of Pediatrics (AAP) guidelines for hospital discharge of the high-risk neonate [12]. The AAP guidelines specify that discharge decisions involve the infant medical status, family readiness, home care needs, and financial considerations both for families and for hospital costs. Additionally, there are four broad infant variables to be addressed, including the infant's physical health, mental health, cognitive development, and quality of life [40]. Individualized planning, physician judgment, and a coordinated care approach are needed to optimize long-term health outcome for every child and family. In weighing the risks and benefits of NICU discharge, the infant should have achieved a healthy degree of physiologic stability. An advantage of discharge is that it can decrease the risks associated with hospital-acquired morbidity and it can shorten the separation between infants and parents that can interfere with the process of attachment.

Societal and economic forces bear upon the timing and process of discharge and quality of home care. It takes time for the family of a high-risk infant to prepare to care for their infant in a home setting. There often must be mobilization of an extensive variety of community resources. It has been found that children with combinations of impairments such as seen in various genetic syndromes may particularly benefit from quality medical homes upon discharge [29, 30]. Families connected to a quality medical home are more likely to receive respite care, transportation, and rehabilitation services [41]. Supportive services can assist families in navigating insurance issues, accessibility to care, and financial burdens.

NICU Graduates Requiring Medical Technologies for Survival

With the increased survival of very preterm and ill infants, many are now discharged with unresolved medical issues that complicate ongoing medical care and put added stress on the family [42]. For instance, through coordinating home oxygen therapy for infants with bronchopulmonary dysplasia, these infants can achieve earlier hospital discharge while still avoiding risks of marginal oxygenation such as growth failure and cor pulmonale [43]. Tracheostomy may be required for neonates with upper airway abnormalities who have not yet been weaned from assisted ventilation [44]. Home ventilation clearly requires qualified personnel such as home nursing support to coordinate with the family to provide bedside care in the home at least part time. Most often, the technology includes nutritional and respiratory support. Other technological support that is less common but not rare includes such things as intravenous medications or nutrition, bladder catheterization, and renal replacement therapy. Parenteral nutrition in the home requires careful assessment of the caregivers and home environment, as well as the support of a well-qualified home care company [45]. When choosing a home care company or agency for technology-dependent infants, a key medical home task is to review previous company performance and existing quality control programs of the home care company [44]. Good parental teaching and coordinated multidisciplinary follow-up care is obviously essential for infants requiring such a high level of medical support. The primary care provider ideally will have the training and professional comfort to act as the medical coordinator and as the first-line resource for the family with regard to integrating all this care and empowering the family in their role.

In circumstances in which the primary care provider does not have the medical expertise to deal with the severity of the high-risk infant's medical problems, models have been studied for what has been termed transitional medical home care. The transitional medical home model involves substantial support from neonatologists and their tertiary care center staff for the first year or so of the medically fragile infant's life. One demonstration of a transitional medical home program came from a Duke University study [46]. The study examined a cohort of 172 neonatal infants hospitalized in the Duke NICU over a 2-year period. Study participants were enrolled in the transitional medical home intervention if they were less than 27 days of age and had a history of extremely low birth weight (less than 1000 g), a chronic illness requiring multiple medications, and/or technology dependence. The enrolled infants had follow-up from trained NICU professionals within 1 to 2 weeks after discharge along with care coordination with their primary care provider. There was also ongoing surveillance and treatment of any acute and chronic medical issues by highly experienced physicians in the transitional care program, who were available on pager. The transition medical home service engaged with the families with a weekly call for the first month. There was 24 h per day access to the medical intervention team. Among other services, the program resulted in over 200 pages from families made directly to the intervention team, potentially reducing the number of needless emergency room visits. Through communication with the primary care pediatricians, the program furthered their education on future care of complex medical problems in young babies. More research is needed on such models which hopefully will reduce emergency room visits and rehospitalizations, minimize subspecialty visits, and reduce costs without sacrificing quality.

The defining components of the pediatric medical home as discussed thus far make sense when considering the broad range of health-care issues the NICU graduating infant and their family will face at discharge. Families climb a steep learning curve as they adjust to technical and detailed medical care. After their infant's hospital discharge, parents often spend long hours per week in the overall management and provision of care for their child with special needs. This complex set of responsibilities and stresses can impact parental mental, physical, and financial health. Parents are often dealing with their own issues while struggling with ongoing serious medical issues in their infants [39]. Infants born preterm with low birth weight who require NICU care have much higher rates of hospital readmission and death in the first year of life compared with healthy term infants [9, 10]. Careful preparation for discharge, good follow-up, and family support can reduce the risk of morbidity and mortality for such infants. But it is not enough just to address specific medical concerns without addressing how the high-risk infant is progressing in overall development.

The Follow-Up Clinic and Developmental Assessments

As discussed in preceding chapters in this book, a likely majority of, though not all, NICU graduates will qualify for enrollment with a neurodevelopmental (high-risk) follow-up clinic. This high-risk follow-up clinic is multidisciplinary in nature, assessing the medical, nutritional, neurologic, developmental, and psychosocial needs of the high-risk infant after discharge. It is standard for the clinic to measure growth parameters, conduct physical and neurological exams, give nutritional counseling, conduct developmental assessments, and offer parent support. Sometimes the developmental assessment is integrated with the infant's medical follow-up visit with their neonatologist. Most NICU follow-up clinics are associated with academic institutions and large tertiary care centers [20]. Kuppala et al. (2012) conducted a national survey of NICUs associated with pediatric residency programs [47]. The survey found that 93% of such NICUs reported having an associated high-risk infant follow-up program. Of the 7% of NICUs with no associated follow-up program, lack of financial resources and trained personnel were cited as the most common barriers. With regard to existing follow-up clinics, the survey found that they generally strive to be multipurpose. These clinics often extend and facilitate continuity of clinical care; provide anticipatory guidance; make referrals to early intervention programs; offer education for health-care trainees such as pediatric residents, medical students, and neonatology fellows; and facilitate outcome data research. Sixty percent of the follow-up programs were predominately clinical care oriented, 37% noted both clinical and research roles, and only one follow-up program identified itself as primarily research oriented. For those follow-up programs that conduct research, they serve as a measure of quality control of NICU innovations. Such innovations require outcome research based on morbidity and mortality data as well as infant developmental assessment data to determine if the innovations actually positively impact the short- and long-term outcomes for these special needs infants. Contributing to a comprehensive medical home approach, the national survey found that the follow-up programs often had nutritionists, case managers, and research coordinators available, especially if the follow-up clinic had an association with a neonatology fellowship program. Greater than 90% of the programs noted that growth measurements, infant neurological status, and neurodevelopmental progress of the infants were regularly monitored. Other additional services offered frequently included assessments by occupational, physical, and speech therapy along with feeding and behavioral assessments and social worker support as needed. The majority of follow-up clinics conducted infant monitoring up to at least 24 months of age and many to 31–36 months of age. However, a few clinics followed NICU graduates even up to 60 months of age. Patient insurance was by far the most common funding source for high-risk follow-up services (for 81% of visits), but the majority of programs reported multiple sources of funding.

For areas with limited resources, the American Academy of Pediatrics (AAP) recommends a partnership between NICU programs and community physicians to conduct developmental screenings and/or assessments [20], a practice certainly in keeping with a medical home model. As has been noted earlier in this book, standardized developmental assessments are required at specific ages through early childhood and are performed by the follow-up clinic [48]. The purpose of these developmental assessments is not only to get a data-driven overview of the infant's evolving developmental status but also to determine if referrals to any medical subspecialties are required and/or whether further developmental evaluation is necessary to see if the infant qualifies for early intervention services [49, 50].

The follow-up clinic shares their findings with the parents and generates a medical report with the results of developmental testing and the overall evaluation. Recommendations based on the findings are typically included in the report. Often the follow-up clinic also writes a brief summary letter to the primary care provider with a synopsis of the developmental test results and the medical and developmental recommendations. As highlighted in the introduction of this chapter, it is vital that there be close communication between the follow-up clinic and the primary care provider acting as the medical home. This is because it is typically the primary care provider who needs to then initiate any needed medical specialty referrals through the child's health-care coverage plan. The follow-up clinic plays a vital role in helping families navigate the subspecialty clinic system, but this process is often dependent on the primary care provider to ultimately ensure that needed services and referrals happen the way they should.

In addition to the infant's physical health issues, the primary care provider needs to be aware of any developmental delays the infant exhibits and assist the family in connecting with early intervention services. It is not uncommon to have breakdowns in communication, especially between those clinics located at large tertiary centers, the community primary care provider, and early intervention agencies. Such communication breakdowns can delay the infant receiving early intervention services as well as needed medical subspecialty care.

Since not all NICU graduates are eligible for high-risk infant follow-up, and since follow-up clinic appointments are scheduled only at a few specific ages, high-risk infants also benefit from periodic developmental surveillance and screening by their primary care physician [48]. For example, even late-preterm infants born between 34 and 37 weeks' gestation who do not qualify for high-risk infant clinics are at an increased risk for both medical and developmental problems. These include hyperbilirubinemia and feeding problems after discharge. There is also mounting evidence that these infants are at risk of developmental delay [51]. Thus, close follow-up and coordinated care is needed even for mildly premature infants who have a relatively unremarkable hospital course. The general pediatrician is in a unique position to assess and support families with children with special health-care needs on many levels, medical and developmental, even when the infant and family are not initially eligible for other agency services.

For clinics with high volumes of patients qualifying for early intervention services, an argument can be made for integration of early intervention services directly into the medical home model. When this can be achieved, it streamlines the number of visits parents have to make for their children. Partnership and close communication between early intervention services and the child's medical home is certainly beneficial given that the child may need to access a variety of other services from, for instance, neurology, psychology, speech therapy, occupational therapy, and a dietician in order to maximize the benefits gained from early intervention. The child's medical home is likely the best place to allow such coordination of services [52].

Barriers

Many pediatricians in the United States do not currently have the resources in information technology, laboratory facilities, or quality-of-care reporting needed to qualify as a certified medical home [53]. Pediatricians may feel unable to care for medically complex infants because of limited funding and inadequate medical staff or because of perceived lack of training in developmental pediatrics and care coordination [47]. In this same vein, a survey of a large group of pediatricians 1–5 years post-residency training found that 18% reported a need for additional training in coordination of care for complex illnesses [54]. General pediatricians found that they did not adequately integrate medical care with the plans of other health-care providers or agencies or with families' needs for nonmedical services (e.g., transportation or developmental services) [31]. General pediatricians also note communication barriers such as delays in receiving consultants' notes, inability to speak directly with consultants, and difficulty coordinating communication among multiple specialists [31]. One study found that for only one of four regions of the United States did children with special health-care needs consistently receive all their needed preventive care in a previous 12-month period [55]. Low family income and inadequate health insurance are consistent barriers to neurodevelopmental follow-up [34] and being enrolled in a primary care medical home [52]. Parents find such lack of care coordination to be a significant burden [56].

Overcoming Barriers

To help with the complexities of care for medically fragile infants, the American Academy of Pediatrics (AAP) has issued a number of important guides to assist primary care physicians. This includes clinical reports with guidance on the hospital-to-home transition for children, including those with dependence on technologies [57, 58]. If the pediatrician's office has a staff member who is dedicated to care coordination, this can increase access to services for families [59]. The NICU follow-up clinic can also assist the primary care provider and family to navigate systems such as medical subspecialty referrals and early intervention systems. Well-resourced tertiary care centers or integrated clinics that specifically engineer their clinic practice to function as a medical home often have the resources to improve care coordination and communication and thereby decrease family burden [60]. Without such care coordination, families may feel isolated and experience diminished ability to function [48, 61–63].

Pediatricians and other primary care providers can offer families an identified central source of medical care and support so that they need not feel isolated and overwhelmed by their infant's care. As a specific example, it has been found that social supports and adequately trained health caregivers can be provided to assist family with complex issues like ventilators, feeding pumps, and other technical devices. Such in-home health-care support can positively impact parents in their overall physical and mental health, in their level of broader social participation, and in obtaining adequate sleep [63, 64]. The treating physician should be aware that families also face financial pressures in raising a high-risk infant. In this regard, one study found that direct medical costs for low birth weight infants were more than nine times higher than caring for full-term infants within the first year after discharge. These same families also faced a sevenfold increase in child care costs [65]. Referrals to social work agencies can connect families with services that may offer at least some financial relief.

Medical care for high-risk infants must not only be continuous and compassionate, but must come from a well-trained primary care provider with expertise at the required level of scope and practice. The specific training pediatricians and other primary care providers have on medical issues that commonly affect infants has great impact on ultimate health and developmental outcomes. Pediatricians as specialists typically have extensive clinical training in such issues as infant feeding problems, adequate nutrition and physical growth, the special nutritional needs of premature infants, an infant's head size and shape (for issues such as plagiocephaly and craniosynostosis), eye muscle tracking and strabismus, abnormalities in an infant's motor tone and motor development (for possible evolving cerebral palsy or other neurological issues), hearing or vision impairment, and issues of general developmental delays or deficits. Pediatric nurse practitioners who specialize in infant care have similar competencies. By providing access to quality primary care, such health-care providers are fulfilling at least one major requirement of a medical home model. Similarly, other health-care providers such as family practice physicians and non-pediatric nurse practitioners can do the same. However, depending on the health-care provider's clinical experience, they may or may not feel comfortable with the range of common medical problems that can occur in any infant, but especially in NICU graduates. It behooves the family practice physician or nurse practitioner who has a good comfort level with typically developing infants but may have a lack of experience with high-risk infants to consider their pediatric colleagues as specialists to rely on for consultation at least early in the first year or two of life for infants at high risk. This is akin to the pediatrician relying on a transitional medical home offered by a tertiary care facility when available. Through such medical collaboration and consultation, evolving medical and developmental problems of the infant can be promptly addressed before development is negatively impacted or before complications occur that could even become life-threatening.

The concept of care through a medical home model involves but goes beyond specific disease case management, which primarily focuses on certain medical diagnoses. In a disease or condition case management model, case managers offer disease-related services that are intrinsic to their specific agency, often within the constraints of eligibility criteria. In contrast, medical home care coordinators should work with and guide the whole team process, which is driven by the needs of patients and families for comprehensive health care that goes beyond a diagnostic category and reaches across community agencies. It is critical that medical home team members focus on their competencies, job descriptions, and functions in a physician-led team caring for the patient and family. Team members should have formal responsibilities. The team process should build on existing strengths of patients and their family support systems. Communication (often involving signed releases of information) is needed across all systems (medical and nonmedical) that are involved in the child's care while adhering to Health Insurances Portability and Accountability Act (HIPAA) rules and Family Educational Rights and Privacy Act regulations. Consent should be driven by the wishes of the patient and, for an infant, by the wishes of the family. Transitions between agencies and other parties are best facilitated by a medical home care coordinator to maximize outcomes. This includes transitions between community partners such as the high-risk infant follow-up clinic, public health nursing, social work agencies, urgent/emergency care facilities, the hospital, specialists, and the primary care office. Such collaboration and coordination needs to be sustained across time. The coordinated care team should assess needs and establish clear goals developed jointly with the patient, family, health-care team, and system. Part of this process involves the creation and implementation of a formal written plan of care. Family and patient input should be the bedrock of the care plan. This formal written plan needs to be periodically updated. The plan of care should be sensitive to the patient's and family's language, values, and culture. Examples of written plans of care can be found at the National Center for Medical Home Implementation web site [66] and the Center for Medical Home Improvement web site [67].

The patient-centered medical home has some similarity to managed care and health maintenance organizations (HMOs) but some important functional differences as well. The medical home model asks providers to focus on improving care as the primary focus rather than managing costs. Cost savings are expected as a result of providing better care, not from withholding needed care. The primary care physician leads a team that serves as a patient advocate and guide to navigate the health-care system rather than serving as a gatekeeper. Networking allows for assistance from the NICU follow-up clinic in the case of a high-risk infant. Data has shown that patients who have a primary care provider (where the medical home model should start) incur about a third less in health-care expenditures, validating the potential for cost savings [21].

Since 2012, at least 41 states have promoted medical home policies within their Medicaid programs [68]. To qualify as a patient-centered medical home, standards and criteria must be met to participate in a state's medical home program [69]. To determine the effectiveness of the medical homes, states have also developed benchmarks of performance and measures that medical homes are required to track and report on [69]. The AAP follows state *pediatric* medical home activities specifically, and this includes the Medicaid medical home initiatives. Some states have especially worked to improve delivery of care for low-income children and other specific pediatric populations, such as children with special health-care needs (CSHCN), including high-risk infants [70].

In 2009 the Commonwealth Fund gave financial backing for the National Academy of State Health Policy (NASHP) [71] to launch collaborative learning models among states for implementing projects aimed at maximizing child health and development through better pediatric care coordination. One such model, the Assuring Better Child Health and Development (ABCD) III learning collaborative, involves the states of Arkansas, Illinois, Minnesota, Oklahoma, and Oregon and has a strong focus on the pediatric medical home model [72]. Four medical home strategies that are used by this learning collaborative are:

1. Use of pediatric-specific criteria in the qualification standards for medical homes
2. Creation of *pediatric* learning collaboratives for medical home providers
3. Educating nonmedical providers on the value of patient-centered medical homes
4. Use of child health and development data to help medical home providers track progress and work on quality improvement efforts

Federal legislation that reauthorized the Children's Health Insurance Program (CHIP) created ongoing opportunities for states to implement pediatric medical homes. For example, after the CHIP Reauthorization Act or "CHIPRA" of 2009, the following year, the federal government awarded 10 CHIPRA grants in 18 states for projects enhancing medical home initiatives. Rhode Island was a recipient of this grant program and in 2011 approved a medical home project with the CEDARR (Comprehensive Evaluation Diagnosis Assessment Referral and Reevaluation) family center as the health home provider serving children with special health-care needs [73]. A 2012 analysis of state medical home initiatives by the National Academy for State Health Policy found that fully 24 state Medicaid programs offered enhanced payment specifically for pediatric medical home services [72]. The number of patient-centered medical home initiatives is clearly growing. Such expansion requires methods to evaluate the impact of medical homes. The Patient-Centered Medical Home Evaluators' Collaborative has created a core set of standardized measures to assess the impact of the medical home on clinical utilization, quality, and cost [74]. Some states require a recertification process after a Medicaid practice or provider has been initially qualified by the state as a medical home for quality control and to promote ongoing innovation. For instance, in the state of Minnesota, recertification requires that the medical home practice demonstrate an ongoing partnership with at least one community resource to which they refer patients [75]. The states of Oregon and Illinois have improved pediatric care coordination by promoting data sharing between Medicaid and early intervention programs. Providers can access an online provider portal through a Medicaid claims-based data exchange system called the Medical Electronic Data Exchange (MEDI). Illinois is integrating the MEDI system with a previously separate early intervention electronic data system that tracks point of entry. This integration allows for information sharing for care coordination and follow-up on early intervention referrals [72].

Additionally, regions and states have expanded medical home access to some degree through private insurers via pilot initiatives [15]. Even a small pediatric practice can aim to function as a medical home through such state grants and programs, though this typically requires some redesign of the pediatric practice, with a key component being the hiring of a patient care coordinator [76]. The Community Access to Child Health (CATCH) grant program of the AAP supports development of pediatric medical homes as well [77]. Tools have been developed to evaluate a practice as a pediatric medical home [78]. One tool in particular, the Primary Care Assessment Tool [79], has shown acceptable reliability and validity and is designed to assess both the structure and process of medical home features within a primary care setting [80]. Innovative and promising practices within the category of "small pediatric practices" have been identified by an expert workgroup through the National Center for Medical Home Implementation (NCMHI) in such states as North Carolina, Arkansas, and Florida [81].

An AAP policy statement in 2014 [16] listed the NCMHI as one of the foremost resources for medical home tools, along with other organizations that support medical home implementation. For example, the NCMHI provides tools and resources for care coordination with specific supports, templates, and guides for pediatricians [82].

The patient-centered medical home (PCMH) recognition program [83] was developed by the National Committee for Quality Assurance (NCQA) and is one of two processes for certification as a patient-centered medical home. The NCQA PCMH certification process supports local practices in development of medical home models [15]. The PCMH shares information on how practices can organize care in a manner that makes the patient's needs the center of focus, develop effective team processes, and improve the delivery and tracking of care coordination over time.

The second process for certification as a medical home is through the Joint Commission on Accreditation of Healthcare Organizations (JCAHO) [84]. Another resource for practices aspiring for medical home certification has been developed by the Agency for Healthcare Research and Quality (ARHQ) and reviews a number of measures from the Care Coordination Measures Atlas that are tailored for primary care practice [85].

Engineering a medical practice or health-care system to function as a medical home often requires that new staff positions be developed for the time-consuming work of care coordination [76, 86]. For instance, the Shriners Hospital for Children system uses social workers and/or nurses individually or in teams to work as care coordinators [87]. Suggested disciplines and minimal educational requirements for care coordinators vary across practices, but commonly the position calls for a social worker, nurse, or the equivalent [88, 89]. In some cases, a bachelor's degree-prepared nurse is suggested, while other programs have used bachelor's degree-prepared early childhood experts with at least 3 years of related experience [90, 91]. For care coordination of medically fragile, high-risk infants, a higher level of medical background is beneficial. Pediatric nurse practitioners are used for care coordination by the Pediatric Alliance for Coordinated Care [25]. Published job descriptions for a pediatric care coordinator are available [76]. The US Maternal and Child Health Bureau offers medical home grants and learning collaborative opportunities which support systematic implementation of care coordination activities as crucial to the medical home process [92]. Care coordination evolves through a quality improvement process by allocating time for dedicated staff members to function as the practice-based care coordinator, with protected time to implement, document, and assess the efficacy of care coordination services [76].

A Model of the Medical Home

Our institution [93] has long functioned as a medical home and is currently undergoing a formal process for certification as a qualified medical home through the JCAHO certification process. At our institution, pediatric residents, under attending supervision, work as primary care providers (PCPs) and are thus central in the management of the medical home model. The clinic is located in an inner city area in California and serves a large proportion of patients from lower socioeconomic backgrounds as well as from a variety of different ethnicities.

In addition to having pediatric subspecialty clinics in the same facility as the outpatient children's health center, we have a social worker, dietician, developmental pediatrician, mental health case managers, and the services of a child psychiatrist. Laboratory and radiology services are shared with the adult outpatient services but are also located within the same building. There is a weekly high-risk infant clinic led by an experienced nurse practitioner that operates within the same premises. The resident primary care provider (PCP) makes referrals to the necessary services and maintains close coordination between these different services. This tight-knit coordination is readily facilitated by the co-location of so many services in one area. The high-risk infant clinic is thus able to focus on the neurodevelopmental outcomes of the high-risk infant and quickly advise team members on any other subspecialty input that the child may require in a collaborative fashion. This allows the resident PCP to fulfil the primary care needs of the patient and maintain close contact with all the other agencies and services that have input into the child's care. As far as possible, well-child check appointments are coordinated with subspecialty appointments so as to reduce the number of visits the family has to make to the office. Social worker and dietician input can be obtained within the same well-child visit along with obtaining "curbside" opinions from subspecialists for any urgent or semi-urgent issues that emerge during the well-child visit. This system has been found to be remarkably beneficial in improving efficiency and utilization of resources in addition to providing better overall care for the patient and family.

Of note, child development regional centers under California law operate to determine eligibility for federal Part C early intervention services. Our clinic works closely with the local regional center office located in the same town by making referrals for early intervention eligibility determination. Our clinic receives written reports from the regional center which describe the findings from the infant eligibility assessments. We also work closely with the early intervention service providers that contract with the regional center. Thus, resident PCPs have the opportunity to

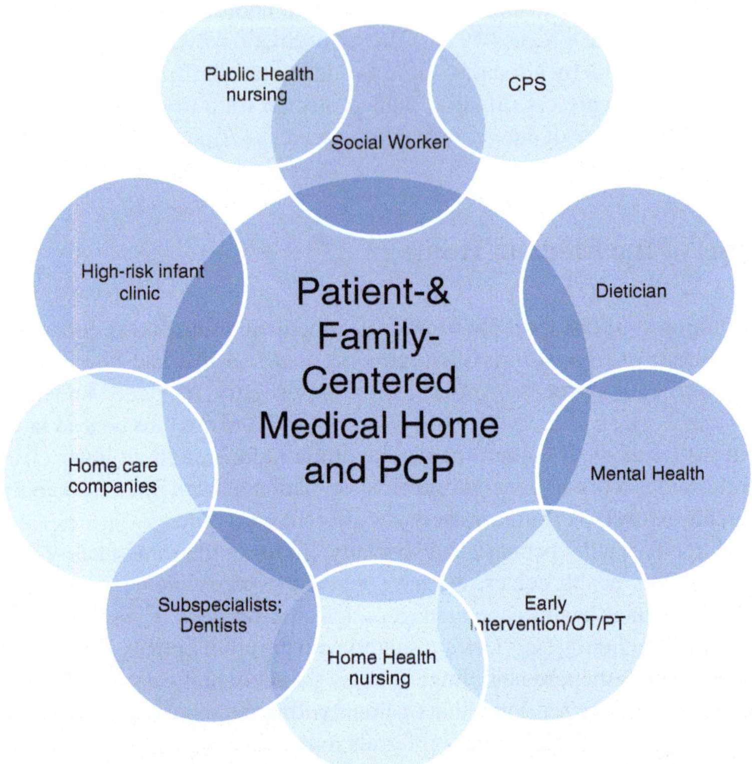

Fig. 1 Our model of the medical home. Certain services are provided by external agencies with whom we work closely. These services are represented by the lighter circles. Subspecialties offered within our clinic include cardiology, pulmonology, endocrine, genetics, adolescent medicine, gastroenterology, neurology, and dermatology. Referrals are made to external providers for all other subspecialties and to dentists

spend dedicated time with early intervention service providers and the local regional center to enhance their learning in the area of child development. Where funding allows, we are able to make direct referrals to and receive reports from the various early intervention service providers as well. We also receive reports from home health nursing, and through our clinic social workers are able to maintain close contact with public health nurses and Child Protection Services (CPS). The figure below summarizes our clinic as a medical home model (Fig. 1):

The Importance of Promoting Medical Home Strategies

It is incumbent on the US health-care system to reduce costs of care while improving quality outcomes [16]. Recent legislative and regulatory efforts to achieve these savings include the redesign of systems of care. Promising redesigns include the development of efficient medical health-care homes for both children and adults. Fragmentation of care leads to inefficiency within and across health systems. Reducing fragmentation through systematic care coordination enhances the care-giving capabilities of families [26]. Care coordination is essential to transform the health-care delivery system in this country. Through good coordination it is possible to achieve optimal quality and cost outcomes in the context of culturally sensitive and patient- and family-centered care. As repeatedly emphasized, this requires partnerships across the full range of health-care delivery settings and community resources. The delivery system should have provisions for services that support the coordination of care across settings and professionals. The foundation for efficient and effective systems of care delivery resides in the patient-/family-centered medical home that is engineered for action and accountability, not lip service. Optimal outcomes for all children but especially children with special health-care needs, such as infants graduating from the NICU, require interfacing among multiple care systems and individuals.

There is a body of growing evidence that care coordination can lead to improved outcomes but at the same time decreased health-care costs, reduction in fragmented care, and improvement in the patient and family experience of care. Related to this, there is a growing national consensus that the medical home model has strong supporting evidence of efficacy for achieving the triple aims of "better care, better health and lower cost" [94]. There are other promising models for redesigning US health-care systems, such as Accountable Care Organizations (ACOs), but the ACO model focuses on service to a defined population of patients, not a philosophy of care [95]. Hence, the medical home is increasingly viewed as philosophically setting a standard of care for all children and adults [96]. This is especially true for high-risk infants.

Payment Issues

Over the last 10 years, payment for care coordination, which is essential to build quality medical homes, has had some success. For instance, the American Medical Association developed care coordination codes 99487–99489 to add to the Current Procedural Terminology (CPT) manual. These codes deal with care coordination for patients with complicated, ongoing health issues within a medical home, accountable care organization, or similar model [16]. Use of these codes permits health-care providers to document and bill for coordinating care between community service agencies, link patients to resources, support the transition of patients from inpatient to other settings, and to work to limit potentially preventable readmissions [97]. As practices obtain formal recognition or certification as a patient-centered, family-centered medical home through the PCMH recognition program and/or through JCAHO certification, this facilitates billing and payment for care coordination services. Ongoing advocacy is still needed by primary care providers so that third-party payers will recognize and fund such CPT codes in their regions [16]. Payment reform and innovations will be a critical driver to expand medical home models [15, 98].

Conclusion

The patient- and family-centered medical home is positioned to provide coordinated, compassionate family-centered health care by forming strong links between the primary care provider team, the infant and family, specialist teams, nurses, social workers, educators, hospitals, other utilized health-care facilities, as well as other appropriate community agencies. Care coordination should be a crosscutting system intervention with thoughtful and deliberative organization among all these participants. Organized care marshals personnel and other resources needed to carry out all required patient care activities in a manner that is patient and family centered. Successful care coordination considers the continuum of health, education, early child care, early intervention, nutrition, mental/behavioral/emotional health, community partnerships, and social services, is competent with regard to culture and languages, and is delivered with compassion. Additionally, payment for all these services is essential to improve the quality of care for children with special health-care needs.

References

1. Behrman, R., & Butler, A. (2007). *Institute of Medicine (US) Committee on Understanding Premature Birth and Assuring Healthy Outcomes. Societal Costs of Preterm Birth.* Washington, DC: National Academies Press.
2. O'Connor, A., Wilson, C., & Fielder, A. (2007). Ophthalmological problems associated with preterm birth. *Eye (London, England), 21*(10), 1254–1260.
3. Marlow, N., Wolke, D., Bracewell, M., et al. (2005). Neurological and developmental disability at six years of age after extremely preterm birth. *NEJM: New England Journal of Medicine, 352*(1), 9–19.
4. Stoll, B., Hansen, N., Bell, E., et al. (2010). Neonatal outcomes of extremely preterm infants from the NICHD Neonatal Research Network. *Pediatrics, 125*(3), 443–456.
5. Greenough, A. (2012). Long term respiratory outcomes of very premature birth (<32 weeks). *Seminars in Fetal & Neonatal Medicine, 17*(2), 73–76.
6. Vohr, B., Wright, L., Poole, W., et al. (2005). Neurodevelopmental outcomes of extremely low birth weight infants <32 weeks' gestation between 1993 and 1998. *Pediatrics, 116*(3), 635–643.
7. Moore, T., Hennessy, E., Myles, J., et al. (2012). Neurological and developmental outcome in extremely preterm children born in England in 1995 and 2006: The EPICure studies. *BMJ [British Medical Journal], 345*, e7961.
8. Hulsey, T., Hudson, M., & Pittard, W., 3rd. (1994). Predictors of hospital postdischarge infant mortality: Implications for high-risk infant follow-up efforts. *Journal of Perinatology, 14*(3), 219–225.
9. Lamarche-Vadel, A., Blondel, B., Truffer, P., et al. (2004). Re-hospitalization in infants younger than 29 weeks' gestation in the EPIPAGE cohort. *Acta Paediatrica, 93*(10), 1340–1345.
10. Smith, V., Zupancic, A., McCormick, M., et al. (2004). Rehospitalization in the first year of life among infants with bronchopulmonary dysplasia. *The Journal of Pediatrics, 144*(6), 799–803.
11. Resch, B., Pasnocht, A., Gusenleitner, W., et al. (2005). Rehospitalisations for respiratory disease and respiratory syncytial virus infection in preterm infants of 29–36 weeks gestational age. *The Journal of Infection, 50*(5), 397–403.

12. American Academy of Pediatrics. (2008). Hospital discharge of the high-risk neonate. *Pediatrics, 122*(5), 1119–1126.
13. Mosquera, R., Avritscher, E., Samuels, C., et al. (2014). Effect of an enhanced medical home on serious illness and cost of care among high-risk children with chronic illness: A randomized clinical trial. *JAMA, 312*(24), 2640–2648.
14. Broyles, R., Tyson, J., Heyne, E., et al. (2000). Comprehensive follow-up care and life-threatening illnesses among high-risk infants: A randomized controlled trial. *JAMA, 284*(16), 2070–2076.
15. Bachrach, A., Isakson, E., Seith, D., et al. (2011). *Pediatric Medical Homes: Laying the Foundation of a Promising Model of Care*. New York, NY: Columbia University. National Center for Children in Poverty, Mailman School of Public Health, Columbia University.
16. Council on Children with Disabilities and Medical Home Implementation Project Advisory Committee. (2014). Patient- and family-centered care coordination: A framework for integrating care for children and youth across multiple systems. *Pediatrics, 133*(5), e1451–e1460.
17. Horst, L., Werner, R., & Werner, C. (2000). Case management for children and families. *Journal of Child and Family Nursing, 3*, 5–14.
18. Lorenzo, S., Bertness, J., DeFrancis, B. et al. (2014, November). *Medical Home. National Center for Education in Maternal and Child Health. Georgetown University*. Retrieved July 9, 2016, from http://ncemch.org/evidence/NPM-11-medical-home.php.
19. Stille, C., Turchi, R., Antonelli, R., et al. (2010). Academic Pediatric Association Task Force on Family-Centered Medical Home. The family-centered medical home: Specific considerations for child health research and policy. *Academic Pediatrics, 10*(4), 211–217.
20. Boss, R., Hobbs, J. (2013, October 1). Continuity of care for NICU graduates. *Contemporary Pediatrics*. Retrieved July 9, 2016, from http://contemporarypediatrics.modernmedicine.com/contemporary-pediatrics/content/tags/american-academypediatrics/continuity-care-nicu-graduates
21. Franks, P., & Fiscella, K. (1998). Primary care physicians and specialists as personal physicians. Health care expenditures and mortality experience. *The Journal of Family Practice, 47*(2), 105–109.
22. U.S. Department of Health and Human Services. (2016). *Implementation of the Medical Home Program*. Retrieved July 9, 2016, from http://mchb.hrsa.gov/programs/medicalhome/.
23. Health Resources and Service Agency (HRSA; U.S. Department of Health and Human Services). *What is a medical home? Why is it important?* Retrieved July 9, 2016, from http://www.hrsa.gov/healthit/toolbox/Childrenstoolbox/BuildingMedicalHome/whyimportant.html.
24. Rosenberg, D., Onufer, C., Clark, G., et al. (2005). The need for care coordination among children with special health care needs in Illinois. *Maternal and Child Health Journal, 9*(2 Suppl), S41–S47.
25. Palfrey, J., Sofis, L., Davidson, E., et al. (2004). The pediatric alliance for coordinated care: Evaluation of a medical home model. *Pediatrics, 113*(5 Suppl), 1507–1516.
26. Antonelli, R., Stille, C., & Antonelli, D. (2008). Care coordination for children and youth with special health care needs: A descriptive, multisite study of activities, personnel costs, and outcomes. *Pediatrics, 122*(1), e209–e216.
27. Cooley, W., McAllister, J., Sherrieb, K., et al. (2009). Improved outcomes associated with medical home implementation in pediatric primary care. *Pediatrics, 124*(1), 358–364.
28. Wood, D., McCaskill, Q., Winterbauer, N., et al. (2009). A multi-method assessment of satisfaction with services in the medical home by parents of children and youth with special health care needs (CYSHCN). *Maternal and Child Health Journal, 12*(1), 5–17.
29. Kogan, M., Strickland, B., Blumberg, S., et al. (2008). A national profile of health care experiences and family impact of autism spectrum disorder among children in the United States, 2005-2006. *Pediatrics, 122*(6), e1149–e1158.
30. Phelps, R., Pinter, J., Lollar, D., et al. (2012). Health care needs of children with Down syndrome and impact of health system performance on children and their families. *Journal of Developmental and Behavioral Pediatrics, 3*(3), 214–220.

31. Stille, C., Primack, W., & Savageau, J. (2003). Generalist-subspecialist communication for children with chronic conditions: A regional physician survey. *Pediatrics, 112*(6 PT 1), 1314–1320.
32. Child and Adolescent Health Measurement Initiative. (2005/2006). *National Survey of Children with Special Health Care Needs, Data Resource Center for Child and Adolescent Health website*. Retrieved from www.cshcndata.org.
33. Benedict, R., & Baumgardner, A. (2009). A population approach to understanding children's access to assistive technology. *Disability and Rehabilitation, 31*(7), 582–592.
34. Benedict, R. (2008). Quality medical homes: Meeting children's needs for therapeutic and supportive services. *Pediatrics, 121*(1), e127034.
35. Turchi, R., Berhane, Z., Bethell, C., et al. (2009). Care coordination for CSHCN: Associations with family-provider relations and family/child outcomes. *Pediatrics, 124*(suppl 4), S428–S434.
36. Thomson, G., Moran, V., Axelin, A., et al. (2013). Integrating a sense of coherence into the neonatal environment. *BMC Pediatrics, 84*.
37. Klassen, A., Lee, S., Raina, P., et al. (2004). Psychological health of family caregivers of children admitted at birth to a NICU and healthy children: A population-based cross-sectional survey. *BMC Pediatrics, 4*(1), 24.
38. Spencer, N., Wallace, A., Sundrum, R., et al. (2006). Child abuse registration, fetal growth, and preterm birth: A population based study. *Journal of Epidemiology and Community Health, 60*(4), 337–340.
39. Purdy, I., Craig, J., & Zeanah, P. (2015). NICU discharge planning and beyond: Recommendations for parent psychosocial support. *Journal of Perinatology, 35*(Suppl 1), S24–S28.
40. Doyle, L., Anderson, P., Battin, M., et al. (2014). Long term follow up of high risk children: Who, why and how? *BMC Pediatr England, 279*.
41. Hamilton, L., Lerner, C., Presson, A., et al. (2013). Effects of a medical home program for children with special health care needs on parental perceptions of care in an ethnically diverse patient population. *Maternal and Child Health Journal, 17*(3), 463–469.
42. Puddu, M. (2010). Home care for the high-risk newborn infant. *Minerva Pediatrica, 62*(3 Suppl 1), 11–14.
43. Moyer-Mileur, L., Nielson, D., Pfeffer, K., et al. (1996). Eliminating sleep-associated hypoxemia improves growth in infants with bronchopulmonary dysplasia. *Pediatrics, 98*(4 Pt 1), 779–783.
44. Edwards, E., Hsiao, K., & Nixon, G. (2005). Paediatric home ventilator support: The Auckland experience. *Journal of Paediatrics and Child Health, 41*(12), 652–658.
45. George, D. (2002). American Academy of Pediatrics, Section on Home Health Care. Home parenteral and enteral nutrition. In M. McConnel & S. Imaizum (Eds.), *Guidelines for Pediatric Home Health Care* (pp. 113–123). Elk Grove Village, IL: American Academy of Pediatrics.
46. Goldstein, R. (2015, October 2), Abstract 3059. *Presented at: 2015 American Academy of Pediatrics National Conference and Exhibition*.
47. Kuppala, V., Tabangin, M., Haberman, B., et al. (2012). Current state of high-risk infant follow-up care in the United States: Results of a national survey of academic follow-up programs. *Journal of Perinatology, 32*(4), 293–298.
48. Vohr, B., Wright, L., Hack, M., et al. (2004). Follow-up care of high-risk infants. *Pediatrics, 114*(5 Suppl), 1377–1397.
49. Newman, K., & Parrott, J. (2013). The NICU graduate: Implications for pediatric primary care. *Newborn and Infant Nursing Reviews, 13*(2), 94–100.
50. Gong, A., Johnson, Y., Livingston, J., et al. (2015). Newborn intensive care survivors: A review and a plan for collaboration in Texas. *Maternal Health, Neonatology and Perinatology, 1*, 24. https://doi.org/10.1186/s40748-015-0025.2.
51. Parekh, S., Boyle, E., Guy, A., et al. (2016). Correcting for prematurity affects developmental test scores in infants born late and moderately preterm. *Early Human Development, 94*, 1–6.

52. Strickland, B., Jones, J., Ghandour, R., et al. (2011). The medical home: Health care access and impact for children and youth in the United States. *Pediatrics, 127*(4), 604–611.
53. Zickafoose, J., Clark, S., Sakshaug, J., et al. (2013). Readiness of primary care practices for medical home certification. *Pediatrics, 131*, 473–482.
54. Freed, G., Dunham, K., Switalski, K., et al. (2009). Recently trained general pediatricians: Perspectives on residency training and scope of practice. *Pediatrics, 123*(Suppl 1), S38–S43. https://doi.org/10.1542/peds.2008-1578J.
55. Fulda, K., Johnson, K., Hahn, K., et al. (2013). Do unmet needs differ geographically for children with special health care needs? *Maternal and Child Health Journal, 17*(3), 505–511.
56. Gulmans, J., Vollenbroek-Hutten, M., Van Gemert-Pijnen, J., et al. (2009). Evaluating patient care communication in integrated care settings: Application of a mixed method approach in cerebral palsy programs. *International Journal for Quality in Health Care, 21*(1), 58–65.
57. American Academy of Pediatrics, Section on Home Health Care. (2002). Guidelines for pediatric home health care. In M. Mconnel & S. Imaizum (Eds.), *Guidelines for pediatric home health care* (pp. 113–123). Elk Grove Village, IL: American Academy of Pediatrics.
58. Elias, E., & Murphy, N. (2012). Home care of children and youth with complex health care needs and technology dependencies. *Pediatrics, 129*(5), 996–1005.
59. Tien, C.-L., Peterson, C. A., & Shelley, M. C., II. (2002). Postdischarge service use by families of neonatal intensive care unit graduates. *Journal of Early Intervention, 25*(1), 42–57.
60. Cohen, E., Freidman, J., Nicholas, D., et al. (2012). A home for medically complex children: The role of hospital programs. *Journal for Healthcare Quality, 30*, 7–15.
61. Gordon, J. (2009). An evidence-based approach for supporting parents experiencing chronic sorrow. *Pediatric Nursing, 35*(2), 115–119.
62. Kuo, D., Cohen, E., Agrawal, R., et al. (2011). A national profile of caregiver challenges of more-complex children with special health care needs. *Archives of Pediatrics & Adolescent Medicine, 165*(11), 1020–1026.
63. Gallagher, S., Phillips, A., & Carroll, D. (2010). Parental stress is associated with poor sleep quality in parents caring for children with developmental disabilities. *Journal of Pediatric Psychology, 35*(7), 728–737.
64. Heaton, J., Noyes, J., Sloper, P., et al. (2005). Families' experiences caring for technology-dependent children: A temporal perspective. *Health & Social Care in the Community, 13*(5), 441–450.
65. McCormick, M., Bernbaum, J., Eisenberg, J., et al. (1991). Costs incurred by parents of very low birth weight infants after the initial neonatal hospitalization. *Pediatrics, 88*(3), 533–541.
66. National Center for Medical Home Implementation Web Site. Retrieved September 4, 2016, from www.medicalhomeinfo.org.
67. Center for Medical Home Improvement Web Site. Retrieved September 4, 2016, from www.medicalhomeimprovement.org.
68. National Academy for State Health Policy, "*Medical Home and Patient-Centered Care*." Retrieved July 9, 2016, from http://www.nashp.org/med-home-map.
69. Kaye, N., Buxbaum, J., Takach, M. (2011, December). *National Academy for State Health Policy, The Commonwealth Fund. Building Medical Homes: Lessons from eight states with emerging programs*. Retrieved September 4, 2016, from http://nashp.org/sites/default/files/building.medical.homes_.emerging.states.pdf.
70. American Academy of Pediatrics, National Center for Medical Home Implementation in "*Medical Home Initiatives and Resources by State*." Retrieved July 9, 2016, from http://www.medicalhomeinfo.org/state_pages/.
71. National Academy for State Health Policy, "*Medical Home and Patient-Centered Care*." Retrieved from http://www.nashp.org/med-home-map.
72. Hanlon, C. (2012, November). *In briefing: A publication for the National Academy for State Health Policy*. Retrieved July 9, 2016, from http://www.nashp.org/sites/default/files/child.development.medical.home_.abcd_.III_.pdf.

73. Florio, D. (2011, October 4). Rhode Island Health Home Initiative. *Presentation at NASHAP's 24th Annual State Health Policy Conference*. Retrieved July 9, 2016, from https://www.statereforum.org/system/files/ri_health_homes_-_cedarr_family_centers.ppt.
74. Rosenthal, M., Abrams, M., Bitton, A., et al. (2012, May). *Recommended core measures for evaluating the patient-centered medical home: Cost, utilization, and clinical quality*. New York, NY: The Commonwealth Fund. Retrieved July 9, 2016, from http://www.commonwealthfund.org//media/Files/Publications/Data%20Brief/2012/1601_Rosenthal_recommended_core_measures_PCMH_v2.pdf.
75. Minnesota Department of Health. (2011, September). *Health Care Home Reassessment Tool*. Retrieved from http://www.health.state.mn.us/healthreform/homes/certification/recertifyassesstool.pdf.
76. McAllister, J., Presler, E., & Cooley, W. (2007). Practice-based care coordination: A medical home essential. *Pediatrics, 120*(3), e723–e733.
77. Sia, C., Tonniges, T., Osterhus, E., et al. (2004). History of the medical home concept. *Pediatrics, 113*, 1473–1478.
78. Malouin, R., Merten, S. Measuring medical homes: Tools to evaluate the pediatric patient- and family-centered medical home. *American Academy of Pediatrics. National Center for Medical Home Implementation*. Retrieved July 2, 2016, from https://childhealthdata.org/docs/medical-home/monograph_final_march29-pdf.pdf?sfvrsn=2.
79. Primary Care Assessment Tools (PCAT). (2009). Baltimore, MD: The Johns Hopkins Primary Care Policy Center.
80. Malouin, R., Starfield, B., & Sepulveda, M. (2009). Evaluating the tools used to assess the medical home. *Managed Care, 18*(6), 44–48.
81. National Center for Medical Home Implementation (NCMHI). Retrieved July 9, 2016, from https://medicalhomeinfor.aap.org/practices/Pages/Promising-Practices-Archives.aspx.
82. National Center for Medical Home Implementation (NCMHI). Retrieved July 9, 2016, from http://www.medicalhomeinfo.org.
83. Patient-Centered Medical Home (PCMH) Recognition Program Website. Retrieved July 9, 2016, from http://www.ncqa.org/Programs/Recognition/PatientCenteredMedicalHomePCMH.aspx.
84. Joint Commission on Accreditation of Healthcare Organizations (JCAHO) Website. Retrieved July 9, 2016, from https://www.jointcommission.org/accreditation/pchi.aspx.
85. Agency for healthcare Research and Quality (ARHQ): Care Coordination Accountability Measures for Primary Care Practice Website. Retrieved July 9, 2016, from http://www.ahrq.gov/qual/pcpaccountability/pcpaccountability.pdf.
86. The Commonwealth Fund; Fryer, A., Doty, M., Audet, A. (2011, March 29). *Sharing resources: Opportunities for smaller primary care practices to increase their capacity for patient care*. Retrieved September 5, 2016, from http://www.commonwealthfund.org/publications/issue-briefs/2011/mar/sharing-resources.
87. Shriners Hospital for Children. (1999). *Children's healthcare options improved through collaborative efforts and service core team. Recommendations of the task force on care coordination*. Lexington, HY: Shriners Hospital for Children.
88. Lindeke, L., Leonard, B., Presler, B., et al. (2002). A family-centered care coordination for children with special health care needs across settings. *Journal of Pediatric Health Care, 16*, 290–297.
89. Cooley, W., & McAllister, J. (2004). Building Medical Homes: Improvement strategies in primary care for children with special health care needs. *Pediatrics, 113*(5 Suppl), 1499–1506.
90. Jackson, P., & Vessy, V. (2000). *Primary care of the child with a chronic condition*. St Louis, MO: Mosby.
91. Desmond, K., Johnson, W. (2006). Effective approaches to ensuring preventive health care and developmental screening for children. Presented at: *Centers for Disease Control and Prevention National Health Promotion Conference*; September 12–14, 2006; Atlanta, GA.
92. Homer, C., & Baron, R. (2010). How to scale up primary care transformation: What we know and what we need to know? *Journal of General Internal Medicine, 25*(6), 625–629.
93. University of California San Francisco-Fresno Pediatric Residency Program. Fresno, CA.

94. Berwick, D., Nolan, T., & Whittington, J. (2008). The triple aim: Care, health, and cost. *Health Aff (Milwood), 27*(3), 759–769.
95. Joint Principles for Accountable Care Organizations. (2010, November 19). Retrieved June 26, 2016, from www.aafp.org/media-center/releases-statements/all/2010/aco-jointprinciples.html.
96. Turchi, R., & Mann, M. (2013). Building a medical home for children and youth with special health care needs. In D. Hollar (Ed.), *Handbook of children with special health care needs* (pp. 399–418). New York, NY: Springer.
97. Elliott, V. (2012, August 16). Revised CPT book includes new codes for care coordination. *AmedNews*. Retrieved July 9, 2016, from www.amednews.com/article/20120816/business/308169997/.
98. Bailit, M., Hughes, C. (2011). Bailit Health Purchasing, LLC.: Key Design Elements of Shared-Savings Payment Arrangements. *The Commonwealth Fund. Issue Brief August 2011*. Retrieved September 5, 2016, from http://www.commonwealthfund.org/~/media/Files/Publications/Issue%20Brief/2011/Aug/1539_Bailit_key_design_elements_sharedsavings_ib_v2.pdf.

The State of Neonatal Follow-Up Programs

Brian Tang

Abstract Neonatal follow-up (NFU) programs have existed for several decades and have evolved as mortality rates of high-risk infants have declined due to technological advances in neonatal intensive care unit (NICU). Initially developed as a way to audit these interventions following NICU discharge, NFU programs now serve a variety of roles that include research, teaching, surveillance, and neurodevelopmental assessments. Several surveys suggest wide variation in services provided/available, eligibility criteria, funding sources, and clinical structure. Standards of care are lacking. No-show rates to NFU are high and are associated with several sociodemographic and child factors. The successful future of NFU will depend on the role of the child's medical home, developing meaningful outcome measures, and the needs of the region and community for which the NFU program is situated.

The proliferation of neonatal follow-up (NFU) programs can be directly tied to the increasing survival rates of infants born preterm. Outcome research studies dating back to the mid-twentieth century documented alarming disability rates, thus establishing the need to develop outpatient programs to audit the impact of neonatal intensive care unit (NICU) events and interventions [1]. Since the 1970s, NFU programs have become an essential component in the continuity of care and management of high-risk infants discharged from NICUs. Their evolution today has come to incorporate a variety of important functions beyond outcome research. Many programs provide educational opportunities for neonatal-perinatal medicine fellows, pediatric residents, psychology interns, and nurse practitioners. NFU programs can also provide specialized and comprehensive care. Some serve as the medical home for these high-risk infants.

An American Academy of Pediatrics policy statement published in 1998, and reaffirmed in 2008, describes a variety of discharge planning concerns and recommendations for outpatient medical assessments that include primary care and

B. Tang (✉)
Stanford University, Stanford, CA, USA
e-mail: bgtang@stanford.edu

H. Needelman, B. J. Jackson (eds.), *Follow-Up for NICU Graduates*,
https://doi.org/10.1007/978-3-319-73275-6_19

neurodevelopmental evaluations [2]. A medical home should be identified well before discharge and facilitate the coordination of care for the child. An interdisciplinary model is ideal for children discharged with any combination of special healthcare needs (e.g., lung disease), psychosocial challenges (e.g., young single parent), and developmental delay. The statement recommended that high-risk infants be enrolled in a NFU program that conducts neurodevelopmental assessment using standardized tools.

A workshop on the follow-up care of high-risk infants in 2002 at the National Institutes of Health made several recommendations for follow-up care [3]. Surveillance and research were considered the central responsibilities of NFU according to this report. Optimal methods to assess high-risk infants following NICU were proposed. In spite of these recommendations, standardized guidelines for provision of follow-up services do not exist to this day. Although many NICUs are associated with regional or national networks that require follow-up, a national network of NFU programs to set standards for practice or allow opportunities for multi-site research studies to identify best practices in follow-up care is lacking, though some regional programs exist [4].

While the value of NFU is widely accepted, their purposes remain varied and at times competing. Poor attendance and the lack of standards of care are ongoing issues of debate and discussion. Despite a call over 10 years ago for more standardized practice in NFU programs, more recent surveys of NFU programs indicate significant variability in practices. One conducted from 2009–2010 queried NFU program directors and/or NICU directors associated with the majority of pediatric residency programs in the United States with neonatology fellowship programs and residency programs without a fellowship program [5]. Responses revealed significant variability in eligibility criteria for clinic enrollment, age of clinic discharge, personnel in the clinic, training experiences, and scope of practice. The majority of programs were considered supplemental to primary care. Clinics were supported by multiple funding sources, including hospitals, state funding, and insurance billing. Most programs followed children until they were either 24 or 36 months. Over 90% did some type of neurodevelopmental assessment. The most common standardized tool was the Bayley Scales of Infant and Toddler Development III. The authors concluded that there is a need for standardized practices in follow-up, as well as formation of a national NFU program network that is representative of the patients who attend such clinics across the nation.

A similar survey in 2011–2012 extended to nonacademic NFU programs also reported variability in clinic structure, patient population, and funding [6]. Respondents to this survey reported that funding for both private and academic clinics came from multiple sources, including hospitals, the state, and foundation/grant funding. Academic programs tended to follow children for a longer period of time. Both types of programs reported similar barriers to provision of care, including a high no-show rate (median of 20%) and lack of funding for clinical activities. Interval of visits and how long these high-risk infants were followed varied widely. Some clinics followed infants for less than 1 year while others followed for greater than 5 years [5, 6]. Greater than 85% of programs utilized formalized referral guidelines to NFU. These guidelines were primarily based on birth weight and gestational age.

As a project of the Society of Developmental-Behavioral Pediatric NICU Follow-Up Workgroup of the Early Intervention Special Interest Group, 242 NFU directors were surveyed in 2014 [7]. Half of the programs participated in a research network (e.g., Vermont Oxford). Most (greater than 70%) conducted some kind of developmental screen or test. Many NFU programs were multidisciplinary with neonatologists and registered nurses representing the most available provider in clinic. Many clinics, however, lacked access to specialty physicians and allied healthcare providers even on a referral basis for many clinics. Like previous surveys, the Bayley Scales of Infant and Toddler Development was the most commonly used developmental assessment tool. Half of clinics screened for autism spectrum disorder using a validated tool. Forty-one programs reported a no-show rate of 26% or higher.

An important finding from these surveys is the high no-show rate to NFU [5, 6]. Adherence to NICU follow-up appointments has been historically problematic with low rates of sustained attendance [8–11]. The reasons are likely multifactorial and include child, sociodemographic, and maternal factors. Factors such as low birth weight and gestational age, significant medical morbidities, nonwhite race, and unmarried parents have been associated with nonadherence to HRIF appointments [10, 12–15]. One multi-site study reported greater distance from the HRIF program and single parenting as predictors of poor attendance to HRIF over several time points [16]. Another qualitative analysis suggested several barriers from the healthcare provider perspective, including lack of funding, scheduling difficulties, perceived low importance of HRIF, and fear of delivering bad news [17].

The effectiveness of NFU programs in terms of clinical outcomes has not been adequately studied using randomized clinical trials. The challenge lies in the variation of priorities and roles between programs with some having greater emphasis on research or education [1]. There is also variation in referral criteria with some programs focusing exclusively on very low birth weight patients while others include a wide range of neonatal medical problems [18]. Although uniform guidelines of care may not be feasible given the wide variation that is intrinsic in NFU, regionalized standards of care may be more realistic and meaningful. Care and outcome studies can be tailored to the needs of the local patient population. Deciding which children are to be followed is a function of several factors, such as research goals, funding, and availability of specialists. There is still value in creating standards of care; however, it may be more advantageous for these standards to be broad instead of specific, the minimum instead of the ideal (e.g., recommending developmental screening/testing instead of a specific battery or tool).

For the majority of children, the primary pediatrician or family physician remains central in the follow-up of high-risk infants. The medical home has the advantage of being embedded in the child and family's community where there is greater opportunity to collaborate and communicate with nonmedical service providers who are essential to the child's growth and development (e.g., early intervention programs). There is some evidence that suggests that the inclusion of comprehensive primary care within a NFU program yielded benefits including reduction in no-show rates, total healthcare costs, and the incidence of life-threatening illnesses [19].

The status of neonatal follow-up will likely remain in flux for years to come. There will certainly be more clarity in terms of standards of care as more outcome research is completed. Evidence that can show the short- and long-term benefits of NFU will lead to more stable streams of funding and reimbursement. Major opportunities exist to develop more uniform guidelines to meet the basic needs of these high-risk infants.

References

1. Sauve, R., & Lee, S. K. (2006). Neonatal follow-up programs and follow-up studies: Historical and current perspectives. *Paediatrics & Child Health, 11*, 267–270.
2. Committee on Fetus and Newborn. (2008). Hospital discharge of the high-risk. *Pediatrics, 122*, 1119–1126.
3. Vohr, B., Wright, L., Hack, M., Aylward, G., & Hirtz, D. (2004). Follow-up care of high-risk infants. *Pediatrics, 114*, 1377–1397.
4. Gong, A., Johnson, Y. R., Livingston, J., Matula, K., & Duncan, A. F. (2015). Newborn intensive care survivors: A review and a plan for collaboration in Texas. *Maternal Health Neonatology Perinatology, 1*, 24.
5. Kuppala, V., & Tabangin, M. (2012). Current state of high-risk infant follow-up care in the United States: Results of a national survey of academic follow-up programs. *Journal of Perinatology, 32*, 293–298.
6. Bockli, K., Andrews, B., Pellerite, M., & Meadow, W. (2014). Trends and challenges in United States neonatal intensive care units follow-up clinics. *Journal of Perinatology, 34*, 71–74.
7. Needleman, H., Contompasis, S., Steingass, K., Boyd, L., & Tang, B. (2015). *Creating systems of developmental follow up for NICU graduates: Lessons from the past, current challenges and collaborations for the future*. Las Vegas: Society of Developmental-Behavioral Pediatrics Annual Meeting.
8. Wolke, D., Sohne, B., Ohrt, B., & Riegel, K. (1995). Follow-up of preterm children: Important to document dropouts. *Lancet (London, England), 345*, 447.
9. Nehra, V., Pici, M., Visintainer, P., & Kase, J. S. (2009). Indicators of compliance for developmental follow-up of infants discharged from a regional NICU. *Journal of Perinatal Medicine, 37*, 677–681.
10. Ballantyne, M., Stevens, B., Guttmann, A., Willan, A. R., & Rosenbaum, P. (2012). Transition to neonatal follow-up programs: Is attendance a problem? *The Journal of Perinatal & Neonatal Nursing, 26*, 90–98.
11. Castro, L., Yolton, K., Haberman, B., Roberto, N., Hansen, N. I., Ambalavanan, N., et al. (2004). Bias in reported neurodevelopmental outcomes among extremely low birth weight survivors. *Pediatrics, 114*, 404–410.
12. Campbell, M. K., Halinda, E., Carlyle, M. J., Fox, A. M., Turner, L. A., & Chance, G. W. (1993). Factors predictive of follow-up clinic attendance and developmental outcome in a regional cohort of very low birth weight infants. *American Journal of Epidemiology, 138*, 704–713.
13. Slater, M. A., Naqvi, M., Andrew, L., & Haynes, K. (1987). Neurodevelopment of monitored versus nonmonitored very low birth weight infants: The importance of family influences. *Journal of Developmental and Behavioral Pediatrics, 8*, 278–285.
14. Tyson, J. E., Lasky, R. E., Rosenfeld, C. R., Dowling, S., & Gant, N. (1988). An analysis of potential biases in the loss of indigent infants to follow-up. *Early Human Development, 16*, 13–25.

15. Harmon, S. L., Conaway, M., Sinkin, R. A., & Blackman, J. A. (2013). Factors associated with neonatal intensive care follow-up appointment compliance. *Clinical Pediatrics (Philadelphia), 52*, 389–396.
16. Ballantyne, M., Stevens, B., Guttmann, A., Willan, A. R., & Rosenbaum, P. (2014). Maternal and infant predictors of attendance at neonatal follow-up programmes. *Child: Care, Health and Development, 40*(2), 250–258.
17. Ballantyne, M., Benzies, K., Rosenbaum, P., & Lodha, A. (2014). Mothers' and health care providers' perspectives of the barriers and facilitators to attendance at Canadian neonatal follow-up programs. *Child: Care, Health and Development, 41*, 722–733.
18. Tang, B. G., Feldman, H. M., Huffman, L. C., Kagawa, K. J., & Gould, J. B. (2012). Missed opportunities in the referral of high-risk infants to early intervention. *Pediatrics, 129*, 1027–1034.
19. Broyles, R. S., Tyson, J. E., Heyne, E. T., Heyne, R. J., Hickman, J. F., Swint, M., et al. (2000). Comprehensive follow-up care and life-threatening illnesses among high-risk infants: A randomized controlled trial. *JAMA, 284*, 2070–2076.

Part VI
Conclusion

Concluding Thoughts

Howard Needelman

Abstract Determination of the appropriate endpoint of evaluation of the neurodevelopmental outcome of the NICU graduate is not at all clear. The traditional use of modalities such as imaging is probably far too narrow to be used in measuring success and failure. Medical outcomes evaluations do not have nearly the usefulness of broader biopsychosocial evaluations. The use of the WHO ICF may better be used to describe any disabilities the child may have and those disabilities' relationship to impairments, activity limitations, and participation restrictions. Longer-term follow-up is also needed to more clearly understand the graduate and his family's satisfaction with the results of his NICU experience and the subsequent services he has received.

In *Guidelines for Perinatal Care* [1], it states that "when neurologic findings are suspect or developmental delays are suspected, neonates should be referred for more in-depth assessment, either to a neonatal follow-up program or to equivalent facilities or programs capable of providing detailed neuro-developmental assessments" (p. 281). Yet there is no consensus as to whom these programs should serve, how they should be structured, and what the appropriate tools for evaluation are. Deciding on the goal of the evaluation would seem to be the first step in determining the best choice among the various options available.

There is clearly a spectrum of outcomes that can be expected for an individual with suspected developmental delay and those patients with an NICU history, whether they be infants with congenital anomalies, premature infants, infants with a neonatal encephalopathy history, or essentially any other neonatal problem (as the issue leading to NICU admission will generally place the newborn at risk for such a delay). To maximize the care of these infants, it is reasonable to consider the follow-up clinic as a connector between the NICU environment and the health-care and

H. Needelman, M.D. (✉)
Munroe-Meyer Institute, University of Nebraska Medical Center, Omaha, NE, USA
e-mail: hneedelm@unmc.edu

H. Needelman, B. J. Jackson (eds.), *Follow-Up for NICU Graduates*,
https://doi.org/10.1007/978-3-319-73275-6_20

educational environment that exists in the community. The information gleaned in the follow-up clinic should be able to provide guidance for the care of the individual neonatal patient as he enters the community and also guide the NICU in the appropriate care of subsequent admissions.

As neonatal follow-up has matured as a field of interest, it has become increasingly clear that factors beyond the condition which cause the infant to be admitted to the NICU will have an important role in the child's developmental outcome. These other factors include socioeconomic status, maternal age and marital status, as well as a myriad of other conditions not directly related to neonatal intensive care. This biopsychosocial approach to follow-up has made it the shared province of various disciplines from maternal-fetal medicine to education to public health.

Because of this approach, it seems that follow-up can be appreciated better if examined not solely from a medical model of disability but rather from the combined approach of both a social and a medical model of disability. For some questions, relating to the utility of specific NICU interventions (e.g., the efficacy of head or body cooling), specific easily measured outcomes (e.g., mortality or abnormalities in imaging or EEG) can provide useful outcome data that is temporally related to the intervention. This same type of data, however, may not be appropriate for examining psychosocial outcomes. For example, in addition to the neonatal patient, the experience of having a child in the NICU can have lasting effects on the family. As many as 10% of mothers who have had a child in the NICU will have a positive screen for maternal depression when the child is 6–8 months of age [2]. Mothers describe symptoms of post-traumatic stress during the infant's NICU hospitalization [3]. Reports also describe paternal depression associated with a child's hospitalization in the NICU [3]. There is probably an increased incidence of divorce in these families. These factors will then also lead to detrimental effects on the child's development. Aiding evaluation of standardized medical outcomes such as mortality might best be described as a role of the World Health Organization (WHO) *International Statistical Classification of Diseases and Related Health Problems* (ICD-10) [4]. Psychosocial outcomes within the context of the infant and family in the community may better be examined using models like the WHO *International Classification of Functioning, Disability and Health* or ICF [3]. The ICF looks at both the individual's capacity, or ability in a standard environment, and performance, or functions in a usual environment. "In ICF, the term *functioning* refers to all body functions, activities and participation, while *disability* is similarly an umbrella term for impairments, activity limitations and participation restrictions" [4]. Figure 1 describes the combined relationship between ICD-10 and ICF as it relates to a health-/disease-related condition. The left side of the figure is appropriate to ICD-10, while the right side is more appropriate to ICF.

To effectively evaluate the status of the individual within the community involves the examination of the relationship between health conditions and contextual factors. In other words, how does the specific disease and intervention affect the specific individual given his unique characteristics and unique environment? No single aspect of the individual or his environment effectively describes a factor which can, on

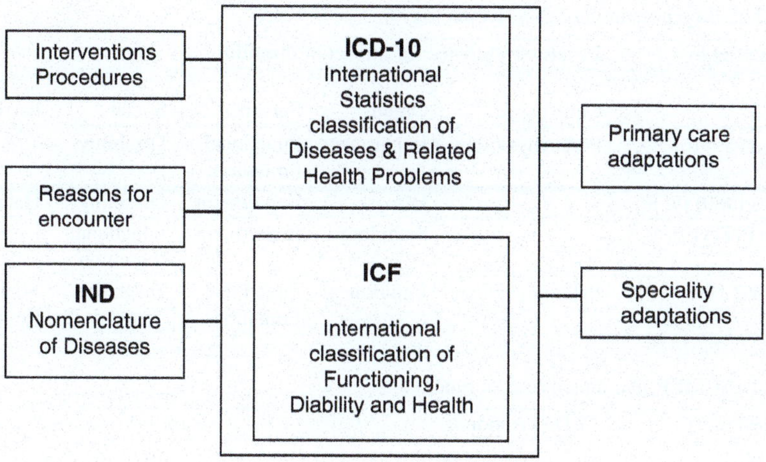

Fig. 1 The ICD-10 and the ICF

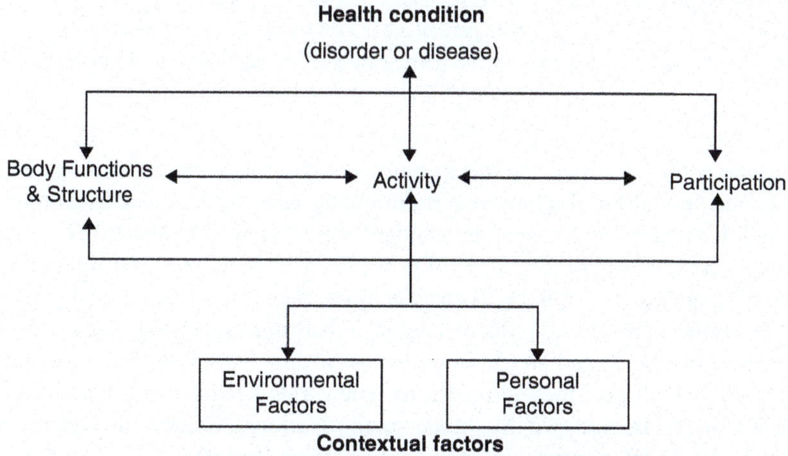

Fig. 2 Effect of health conditions and contextual factors on the individual

its own, describe the most useful intervention to optimize outcome for the individual. Thus, a biopsychosocial perspective seems an appropriate manner in which to address planning for the NICU graduate when seen in follow-up clinic. Figure 2 illustrates the relationship between health conditions and contextual factors and how they interact to effect the individual. Thus the effect of the condition, e.g. hypoxic-ischemic encephalopathy, will be on the individual's body function, his ability to perform activities, and his ability to participate in social activities. The graduate and his environment will ultimately determine the effect.

The key question seems to be at what level of disability has the intervention taken place.

Table 1 Differences in disability terminology

Terms used by	Levels of reference for impact of disability				
	Cells/tissue	Organ	Person	Social	External barriers
Nagi (1969) [6]	Pathophysiology	Impairment	Functional limitation	Disability	–
WHO (1980) [7, 8]	–	Disease	Impairment	Disability	Handicap
IOM (1991) [9]	–	Pathology	Impairment	Functional limitation	Disability
NCMRR 1993 [10]	Pathophysiology	Impairment	Functional limitation	Disability	Societal limitation

Table 2 NCMRR five dimensions of function

Dimension	Description
Pathophysiology	Interruption or interference of normal physiology and developmental processes or structures
Impairment	Loss or abnormality of body structure or body function
Functional limitation	Restricted participation in typical societal roles
Disability	Inability to participate in typical societal role functions
Societal limitation	Barriers to hill participation in society that result from attitudes, architectural barriers, and social policies

Perhaps a way of answering this question is asking what the appropriate measure of outcome should be. Studies have traditionally used such measures as mortality, ultrasound imaging, and scores on standardized tests such as the *Bayley Scales of Infant and Toddler Development* (BSID). While these outcomes may aid in the evaluation of results using an ICD-10 classification, they are not necessarily useful for the ICF. Butler et al. [5] note the difficulty in determining appropriate outcomes in disability research. Noted are different levels or dimensions of disability and variability in the labeling of these dimensions. The variability in terminology from Nagi to WHO to the Institute of Medicine to the National Center for Rehabilitation Research (NCMRR) is represented in Table 1 (from Butler).

A clear example of the utility of this dimensional view of disability is given by Butler in the evaluation of cerebral palsy. Table 2 describes the five dimensions of function as designated by the NCMRR, and Table 3 describes how these dimensions are relevant to the child with cerebral palsy.

It should be noted that while a single intervention may have effects in various dimensions, that is not always the case. For both outcomes research and the care of the individual, it is important however to determine which of the dimensions of disability the intervention is meant to address. Following that determination, however, it is often useful to evaluate the effect of the intervention in other dimensions.

For the individual and his family, the appropriate question is what is the desired outcome of the intervention. While the use of dexamethasone postnatally in premature infants was reported to cause impaired gray matter growth [11], the effect of worsening clinical outcomes is likely the major concern of families. Therefore, an

Table 3 NCMRR dimensions of function and cerebral palsy

Level	Examples
Pathophysiology	Cystic lesions and white matter loss as a result of periventricular leukomalacia of the premature infant's brain
Impairment	Spasticity, contractures, low endurance, perceptual dysfunction
Functional limitation	Awkward walking with fatigue, difficulty dressing, poor concentration and sustained listening, reading problems
Disability	Learning delays, education in restricted environment, limited sports activity, interference with dating and sexuality, not able to take communion at church, cannot participate in family activity by doing chores at home, unable to achieve independent living
Societal limitation	Exclusion from school/city team sports, denial of medical treatment or equipment by insurer, government action that blocks the building of independent living units for people with disabilities, failure of voters to support funding of wheelchair lifts for public buses

important indicator for the scientist, dexamethasone's pathophysiologic effect in the brain, probably only has clinical significance if there is an effect in other dimensions, such as the NCMRR dimension of disability. While Fischer et al. [12] have used scores on the BSID as an outcome for the effect of erythropoietin for neuroprotection, this is probably a measure of impairment or functional limitation and not disability or societal limitation. The parent or guardian of the graduate must be included in the discussion of the goal of intervention.

The difficulty in interpreting and comparing outcomes using different measures is illustrated in the Nebraska follow-up experience. The data has generally revealed that the greatest risk factors for a poor neurodevelopmental outcome are gestational age at birth, length of time on ventilation, and length of NICU stay. The major outcomes reported are generally scores on the BSID and the need for early intervention services. At 6–8 months corrected age, while these two outcome measurements trend in the same direction, the difference in what they suggest is striking. For example, while 50% of infants delivered at 24 weeks and 40% of those delivered at 25 or 26 weeks were involved in the early intervention system, scores on the BSID-III cognitive and language subscales were much more encouraging. For those 24 weeks at birth, mean BSID cognitive composite was 90 and language 86. For those at 25 weeks, the respective scores were 93 and 94, and for those at 26 weeks, they were 96 and 95. Therefore, the BSID data reveal an outcome measure easily within a single standard deviation of the mean, while the referral data place approximately 50% of these survivors in the 3% of infants requiring early intervention services. The disparity at this young age between the two outcome measures, Bayley scores and need for early intervention services, is illustrative of how measures of impairment, functional limitation, and disability can seem to give conflicting results.

Certainly, longer-term follow-up with more sensitive prognostic indicators than the BSID is needed to better evaluate especially disability and societal limitations. The Scandinavian literature has comprehensive long-term registries from which follow-up can be determined which give a hint of what follow-up results mean in adulthood.

In Norway, at ages 20–36 years, from approximately 900,000 newborns, there was a generally stepwise increase in cerebral palsy, "mental retardation"; schizophrenia; disorders of psychological development, behavior, and emotion; and other major disabilities as gestational age decreased [13]. From an examination of dimensions of disability, this increased effect as gestational age decreased was seen in "any disability affecting working capacity," supporting the notion that prematurity will increase societal limitations. Lindstrom et al. [14] in Sweden reported that, among more than 500,000 adults ages 23–29, there was an increasing incidence of disability at decreasing gestational age. Among outcomes examined were postsecondary education, employment, social welfare, and living with parents. Also considered were sickness pension, disability allowance, and disability assistance. The problem with these large long-term studies is, of course, the change in interventions in the NICU from the 1970s to the current time and the effect of socioeconomic status on outcome.

While these changes in care and variability in social status may have effects over time, it is less clear that parent and child preferences change. These preferences may be more wedded to cultural and religious factors and thereby influence what outcome needs to be addressed in an infant at risk. Saigal and her group have looked at longer-term outcomes in a North American population. In a 1999 study [15], neonatologists, neonatal nurses, parents, and adolescents, both those with a history of extremely low birth weight and controls, were presented four to five hypothetical health states and asked to rate them as to the quality of life they felt it provided. The scenarios varied from minimal disability, perhaps a learning disorder, to apparent cerebral palsy with associated sensory impairment. For those with a mild to moderate impairment, health professional and parents rated quality of life similarly. However, for those with severe impairment, parent's rating of quality of life was better than that rated by health professionals. The adolescents rated the milder disability as a lower quality of life than did the health professionals. However, the adolescents rated the quality of life with the more severe disability as better than did the parents or health professionals.

The data presented indicates the need to evaluate outcomes based on the needs and desires of the family as well as the apparent optimal interventions in the NICU as measured by standardized testing in the first years of life. It also makes clear the need to look at the long-term impact of neonatal conditions and interventions. To do this effectively, the follow-up program must partner with community programs to ameliorate the biopsychosocial sequelae of a NICU stay. Such programs must attempt to address all levels of disability. It is the role of the NICU follow-up clinic to help these at-risk children today and in the future.

References

1. American Academy of Pediatrics and American College of Obstetricians and Gynecologists. (2002). *Guidelines for perinatal care* (5th ed.). Elk Grove Village, IL: Authors.
2. Nabower, A., Pralle, A., Miller, K., Roberts, H., Anderson-Berry, A., & Needelman, H. (2016). *Maternal depression screening in NICU follow-up*. Philadelphia, PA: Poster presented at 2016 PAS annual meeting.

3. Lefkowitz, D. S., Baxt, C., & Evans, J. R. (2010). Prevalence and correlates of posttraumatic stress and postpartum depression in parents of infants in the neonatal intensive care unit (NIC). *Journal of Clinical Psychology in Medical Settings, 17*(3), 230–237.
4. World Health Organization. (2002). *Towards a common language for functioning, disability, and health*. Geneva: ICF.
5. Butler, C., Chambers, H., Goldstein, M., Harris, S., Leach, J., Campbell, S., Adams, R., & Darrah, J. (1999). Evaluating research in developmental disabilities: A conceptual framework for reviewing treatment outcomes. *Developmental Medicine and Child Neurology, 41*, 55–59.
6. Nagi, S. Z. (1969). *Disability and rehabilitation*. Columbus, OH: Ohio State University Press.
7. World Health Organization (1992–1994). *ICD 10*. Geneva, Switzerland: WHO.
8. World Health Organization. (1980). *International classification of impairments, disabilities and handicaps*. Geneva, Switzerland: WHO.
9. IOM. (1991). In A. M. Pope & A. R. Tarlov (Eds.), *Natiodisability in America: Toward a national agenda for prevention*. Washington, DC: National Academy Press.
10. NCMRR: National Institute of Health (1993). *Research plan for the National Center for Medical Rehabilitation Research*. NIH Publication no. 93-3509. Washington, DC: US Department of Health and Human Services.
11. Murphy, B. P., Inder, T. E., Huppi, P. S., Warfield, S., Zientara, G. P., Kikinis, R., Jolesz, F. A., & Volpe, J. J. (2001). Impaired cerebral cortical gray matter growth after treatment with dexamethasone for neonatal chronic lung disease. *Pediatrics, 107*(2), 217–221.
12. Fischer, H. S., Reibel, N. J., Buhrer, C., & Dame, C. (2017). Prophylactic early erythropoietin for neuroprotection in preterm infants: A meta-analysis. *Pediatrics, 139*(5), e20164317.
13. Moster, D., Terje Lie, R., & Markestad, T. (2008). Long-term medical and social consequences of preterm birth. *NEJM, 359*(3), 262–273.
14. Lindstrom, K., Winbladh, B., & Hjern, A. (2007). Preterm infants as young adults: A Swedish national cohort study. *Pediatrics, 120*(1), 70–77.
15. Saigal, S., Stoskopf, B. L., Feeny, D., Furlong, W., Burrows, E., Rosenbaum, P. L., & Hoult, L. (1999). Differences in preferences for neonatal outcomes among health care professionals, parents, and adolescents. *JAMA, 281*(21), 1991–1997.

Index

H. Needelman, B. J. Jackson (eds.), *Follow-Up for NICU Graduates*,
https://doi.org/10.1007/978-3-319-73275-6

The manufacturer's authorised representative in the EU is Springer Nature Customer Service Centre GmbH, Europaplatz 3, 69115 Heidelberg, Germany. If you have any concerns regarding our products, please contact ProductSafety@springernature.com

Printed and bound by CPI Group (UK) Ltd, Croydon, CR0 4YY
30/09/2026
02225587-0007